ENVIRONMENTAL ACCOUNTING IN THEORY AND PRACTICE

Economy & Environment

VOLUME 11

The titles published in this series are listed at the end of this volume.

Environmental Accounting in Theory and Practice

Edited by

Kimio Uno
Faculty of Policy Management
Keio University at SFC, Fujisawa, Japan

and

Peter Bartelmus
Environment and Energy Statistics Branch
United Nations Statistics Division, New York, NY, USA

KLUWER ACADEMIC PUBLISHERS
DORDRECHT / BOSTON / LONDON

Distributors

for the United States and Canada: Kluwer Academic Publishers,
PO Box 358, Accord Station, Hingham, MA 02018-0358, USA
for all other countries: Kluwer Academic Publishers Group,
Distribution Center, PO Box 322, 3300 AH Dordrecht, The Netherlands

A C.I.P. Catalogue record for this book is available from the Library of Congress

ISBN 0-7923-4559-2

Printed on acid-free paper

Kluwer Academic Publishers BV incorporates the publishing programmes of D. Reidel, Martinus Nijhoff, Dr W. Junk and MTP Press.

Printed in Great Britain

Contents

Preface

Policy failures in environment and development have been blamed on fragmented and eclectic policies and strategies. The 1992 United Nations Conference on Environment and Development, the 'Earth Summit' in Rio de Janeiro, called therefore for an integrated approach in planning and policy making to achieve long-term sustainable growth and development. The Conference also recognized in its action plan, the Agenda 21, that integrated policies need to be supported by integrated information, notably requiring the implementation of integrated environmental and economic accounting by its member States.

During the preparations for the Rio Summit, scientists and practitioners of national accounting met in a Special Conference on Environmental Accounting, organized by the International Association for Research in Income and Wealth (IARIW) in Baden, Austria. Their aim was to explore the need for and methodologies of adjusting national accounts for environmental reasons. National accountants had faced mounting criticism that conventional accounting neglected new scarcities in natural capital, as well as the social cost of environmental degradation. The result of their deliberations was a draft manual, later issued by the United Nations Statistics Division (UNSD) as a handbook of *Integrated Environmental and Economic Accounting.*

At the same time, the decade-long work on the revision of the conventional System of National Accounts (SNA) approached its conclusion. Just in time, the international working group, preparing the revised version of the SNA, recognized the significance of the new field of environmental accounting. It incorporated some of the findings of the Baden conference in its classifications and asset accounts and elaborated the links between the central framework and an environmental 'satellite' in the 1993 SNA. The result of this collaboration between environmental and national accountants is a System of integrated Environmental and Economic Accounts (SEEA) that is firmly anchored in the concepts and methods of the new SNA. This does not mean, however, that there is consensus on all the methodologies proposed in the SEEA: contrary to the SNA, the SEEA has not been adopted as an international recommendation but is considered to be work in progress.

A wide range of concepts, techniques, classifications and applications is still being discussed at national and international levels. Contrary to the first IARIW Conference, this discussion can now take into account considerable experience gained from pilot projects of SEEA applications and other, more limited case studies. This was indeed the main reason for convening a second Special IARIW Conference on 'Environmental Accounting in Theory and Practice', held in Tokyo, 5–8 March 1996. The Conference focused on implementation questions of the SEEA and the need for its modification for practical and theoretical reasons. As expected, a broad spectrum of approaches and views was presented. They reflect a distinct dichotomy of monetary versus physical accounting, and a pronounced need for gaining experience with the use of integrated accounts in modelling and policy analysis.

This book attempts to present a fair overview of this spectrum. It goes beyond a typical proceedings volume, cutting through the conference agenda for a clear presentation of the most pertinent or 'representative' (of a particular line of thought or methodology) parts of invited and contributed papers. Any misrepresentation is thus clearly the fault of the editors who, however, sought the approval of the authors on a first draft of the book. The intended result is a concise description of the state of the art in integrated environmental and economic accounting. At the same time, the book also points to emerging commonalities as a basis for future standardization of concepts and methods.

Peter Bartelmus
Carsten Stahmer
Kimio Uno

Part I

Introduction

1
Overview

PETER BARTELMUS

The main objective of the Special Conference of the International Association for Research in Income and Wealth (IARIW) on which this book is based was to take stock of concrete experiences in integrated environmental and economic accounting. Such an assessment was believed to cut through some of the mantras of theories and argumentation repeatedly seen and heard in publications and expert group meetings. As to be expected, this objective was only partially achieved. A number of commonalities did emerge from the different case studies and theoretical contributions. They included a consensus on using the national accounts system as the basic framework for environmental accounting, the use of certain valuations, notably net-pricing of natural resources and maintenance costing of environmental degradation, and the need to examine more systematically the analytical and policy uses of the accounting results.

Of course, dissenting voices were also heard. They reflected a fundamental dichotomy between those favouring an assessment of environmental concerns in physical measures and others advocating monetary valuation. The former considered valuation a matter of 'research' or 'modelling'; the latter believed that full integration of environment and economy can only be achieved by estimating the direct and social costs of environment–economy interaction. To some extent, this dichotomy represents different priorities expressed by developing and industrialized countries. The valuation of natural resources which are traded in markets is less controversial than the valuation of 'residuals' or 'externalities' that do not become part of a market exchange. It is not surprising, therefore, that many developing countries have made great efforts to measure the cost of losing their natural resource base. On the other hand, industrialized countries seem to be content with collecting physical indicators of environmental (pollution) impacts and expenditures for environmental protection.

All these questions, together with more technical problems of environmental accounting, form a rich basis for future research. An attempt was therefore made at the conference to obtain agreement on a coordinated research agenda – with little success as far as the setting of priorities and distribution of

---

The author is a staff member of the United Nations. The views expressed here are his own and not necessarily those of the United Nations.

K. Uno and P. Bartelmus (eds.), Environmental Accounting in Theory and Practice, 3–12,

research work are concerned. Nevertheless, the questions were put on the table, and the book is to serve as a menu for a wider audience. Further standardization of concepts and methods might be achieved where consensus is emerging, and research and experimentation should be encouraged where it is lacking.

This two-pronged approach is reflected in the work of the United Nations Statistics Division (UNSD) which developed a System of integrated Environmental and Economic Accounting (SEEA), closely linked to, but separate from, the globally adopted System of National Accounts (SNA). A number of pilot projects tested the SEEA and form the bulk of national experience with environmental accounting. Those studies revealed the need for a straightforward 'hands-on' description of the more practical recommendations of the generic SEEA handbook. Chapter 2 by Alessandra Alfieri and Peter Bartelmus presents an outline of the core chapter of an operational manual, building on commonly applied methodologies. The manual uses a step-by-step approach to lead the practitioners of national (environmental) accounting through the minimum requirements for the implementation of the basic versions of the SEEA. By contrast, the last part of the book, prepared by Kimio Uno, summarizes the main unresolved questions in a tentative research agenda. This agenda should be a good starting point for research and experimentation, leading eventually to the revision of the SEEA.

The main parts of the book are ordered so as to reflect the transition from concrete experiences to conceptual and analytical questions. Part II describes the approaches and results of environmental accounting in selected industrialized and developing countries. A number of equally interesting studies were presented for Colombia, Costa Rica, Ghana and Thailand but are not shown here, mainly because of space limits, but also to avoid repetition in describing similar approaches. Part III deals with an expanding area of research and application, notably in industrialized countries, i.e. the physical accounting of natural resources and space. Part IV addresses theoretical questions, especially the valuation of non-market environmental phenomena, and discusses possibilities of the analysis (through modelling) of the accounting results. Many questions raised in parts III and IV remain unresolved. The linkage of physical and monetary accounting is a particularly important concern as it might overcome the above-mentioned dichotomy between physical and monetary data systems. These and other questions of a conceptual, methodological and programmatic nature, raised in part IV, will need further exploration and are therefore included in the research agenda of part V. The following provides a brief summary of the key questions addressed in the individual contributions.

Country experiences

Chapters 3, 4 and 5 describe the results and problems implementing the SEEA and related approaches in countries at different stages of development. The description of the SEEA in Chapter 2 provides the methodological backdrop against which the scope, coverage and methods of the country studies can be assessed.

Katsuki Oda, Kiyoshi Arahara, Nobuyoshi Hirai and Hideki Kubo present trial compilations of the SEEA in Japan for 1985 and 1990. Their paper provides a detailed account of the concepts, methods and estimations applied. Given the relatively low significance of the use of domestic natural resources, a maintenance costing approach is applied for both environmental degradation and natural resource depletion. In addition, actual expenditures for environmental protection are compiled, including the cost of 'externalized' internal protection carried out within industrial establishments. The relatively low amount of environmental costs of about 2–3% of NDP (in 1990) is probably due to undercoverage of pollutants, the shift of the sustainability burden to foreign countries through importation of natural resources rather than domestic exploitation, and successful pollution control. It is planned to extend the coverage of environmental protection expenditures and pollutants and to improve methodologies through marginal rather than average costing and constant-price accounting. The use of accounting results for the analysis of Japan's dependence on foreign countries, regarding the sustainability of economic performance and growth, is also envisaged.

In contrast to the Japanese study which was carried out by the national accountants of the Economic Planning Agency, a pilot compilation for the Republic of Korea was conducted by a research institute, in cooperation with the United Nations. Seung-Woo Kim, Jan van Tongeren and Alessandra Alfieri discuss data availability, estimation procedures and data analysis. They call not only for the improvement and extension of data collection but also for the 'institutionalization' of environmental accounting within the official statistical services. As in the case of Japan, scant natural resource endowment, omission of possible depletion of fishing grounds and undercoverage of pollution account for the relatively low environmental cost compiled, amounting to 4.1% and 2.6% of NDP in 1986 and 1992 respectively (including environmental cost of private consumption: own calculation from project data). An attempt is also made to link environmental impacts to economic responses of protection expenditures by industries. However, no clear relation between the cost of environmental asset use and environmental expenses could be established.

A project of SEEA application in the Philippines, presented by Estrella V. Domingo, focuses on the natural resource base and its depletion. The project was carried out by the National Statistical Coordination Board (NSCB) which compiles the national accounts for the country. The results demonstrate the feasibility of compiling natural asset accounts in physical and monetary terms in a developing country without entering into primary data collection. In the words of the author: 'lack of data should . . . not prevent a country from pursuing environmental accounting'. The accounts for mineral resources, forests and fish indicate significant depletion since the late 1980s, with decreasing depletion rates in minerals and forests but increasing depletion rates for aquatic resources. Overall, environmentally-adjusted net domestic product (EDP) ranged between 96% and 99%, approximating NDP over time with a slightly higher growth rate than NDP. It is intended to expand the scope and coverage of the project to include environmental degradation from emissions of pollutants, as well as expenditures for environmental protection.

The next study describes a different approach to environmental accounting in the same country but conducted outside the official statistics system. Marian S. Delos Angeles and Henry M. Peskin present the results of a USAID-sponsored Environmental and Natural Resource Accounting Project (ENRAP) in the Philippines. The project applies a 'neoclassical' accounting approach proposed by Peskin. The objective is to go beyond national income accounts in costing the outputs of nature's environmental production system as well as its inputs of environmental damage. This expansion of the production boundary is the main difference from the SEEA application carried out by the national accountants. A comparison of the results of the two systems shows the significance of alternative concepts and definitions, with much lower environmental cost estimates in the ENRAP. Overall, the project estimates that the net effect (net benefit) from environmental services results in an increase of NDP for 1988 and a slight decrease for 1992. Total pollution control costs are also much greater than the expected benefits of environmental protection. The conclusion is that wholesale pollution control is not warranted and more detailed targets need to be established for particular environmental policies. Contrary to the NSCB study presented in Chapter 5, the authors conclude that environmental accounting should be addressed by an independent group (i.e. ENRAP) rather than the official statistical system.

The approach followed in the USA, described by J. Steven Landefeld and Stephanie L. Howell, differs from the previous studies by focusing only on mineral resources and by being 'modelled' on the SEEA. This modelling results in one major deviation from the SEEA by accounting for mineral resource discovery and environmental protection as capital formation. Such accounting largely offsets environmental depletion and degradation. Considering that SEEA accounting procedures would show an environmental (depletion) cost of between 15% and 78% of the net value added generated by the mining industry during 1977–1991 (according to our own calculation from the original data), this demonstrates again the influence of alternative concepts and accounting rules on environmentally adjusted economic aggregates. Other results of the asset accounts show the significance of subsoil assets whose (stock) value is estimated at two to four times the value of associated (produced) capital. The Bureau of Economic Analysis plans to implement two further phases in which renewable resources will be included and environmental asset depreciation will be estimated for air, water, land and forests.

The input–output framework of the national accounts is used by Deborah Vaughn Nestor and Carl A. Pasurka, Jr to measure the goods and services that constitute environmental protection (EP) activities in the USA. The framework is based on SEEA concepts, and thus produces data that are consistent and comparable with conventional national accounts aggregates. This is considered to be a major advantage over estimations of revenues and expenses outside the national accounts. The results show that the share of direct value added generated by EP ranged between 0.64% and 0.80% of GNP between and during 1977 and 1991; these shares more than double (to 1.53–2.12%) when indirect value added contributions to EP production are included. The disaggregation of the conventional input–output table was made by using survey results, engineering studies and other information from federal agencies. Current and future activities of the US Environmental Protection Agency will expand the approach to estimate emissions generated in production processes and explore the use of these data in a general equilibrium model.

The Dutch National Accounting Matrix including Environmental Accounts (NAMEA) shows environmental indicators of pollution and natural resource losses in physical units linked to the causing economic transactions of the national accounts. Steven Keuning and Mark de Haan describe the NAMEA and show how, for a more aggregate presentation and analysis, emissions of the different substances are converted into (Dutch) policy 'theme equivalents'. The result is an accounting format which is quite similar to the SEEA (as described in the physical data sheets of Chapter 2) but which excludes the asset accounts and stops short of monetary valuation of environmental costs. As a consequence, shares of direct and indirect contributions to themes (greenhouse effects, ozone layer depletion, acidification, eutrophication, waste) by different industries can be compared with the shares of these industries in GDP. All in all, these calculations indicate that a guilder of demand for food (including indirect contributions by agriculture) contributes most to the greenhouse effect while a guilder of demand for manufactured products generates most waste and ozone layer depletion. The further expansion of NAMEA to incorporate sociodemographic accounts in a System of Economic and Social Accounting Matrices and Extensions (SESAME) is currently being explored by Statistics Netherlands.

Physical and spatial accounting

Addressing both monetary and physical accounting, Alice Born's paper provides a pertinent transition to the following parts of the book, dealing with the concepts and methods of physical accounting on the one hand and monetary valuation on the other hand. Accounting for natural wealth in both monetary and physical terms is needed to obtain a comprehensive picture of the availability of natural resource stocks and material flows (in physical quantities) and of their economic significance for production and income generation (in monetary values). The monetary value of wealth and wealth per capita are key indicators of the sustainability of an economy; however, as time series of the quantities and present values of Canada's mineral resource stocks reveal, monetary wealth figures do not clearly indicate trends in the remaining reserves from which income is to be derived. 'From a national accountant's perspective, a natural resource is as much an economic as a physical concept'. Overall, the results of expanded wealth accounts (for produced and subsoil assets, and land) indicate that Canada's wealth and wealth per capita are being maintained and are increasing over time.

A physical accounting project in Namibia is based on the stock and use accounts of the SEEA. Glenn-Marie Lange describes work in progress on the compilation of accounts for water and water-based resources, marine fisheries, land and land-based resources, livestock, wildlife, and minerals. Given the dependency of the economy on natural resource exploitation, physical accounts are considered to provide insight into economic performance, assess the interaction of the economy with the environment and provide input into a model of the resource implications of national development strategies. Expansion into monetary accounting is planned to capture mineral resource rents, measure the value of water for different uses and to assess land values as an input into land use planning and land reform.

The above project of natural resource accounts in Namibia can be seen as the physical basis of integrated environmental and economic accounts. The following chapters describe approaches of material/energy and land (use/cover) accounting which go beyond the SEEA proposals in providing greater detail on material flows and spatial distribution of activities and impacts.

Walter Radermacher and Carsten Stahmer present three practical applications of a broad physical information system: material flow accounts, physical input–output tables and emission matrices linked to the input–output tables. A novel and 'unaccustomed' feature of physical accounts at the macro-economic level is the aggregation of material flows in weight units (tons). Such an approach cannot, of course, account for the ecological risks associated with particular material flows. On the other hand, some interesting results emerge from such accounting in Germany. They include a surprisingly high amount of 'unutilized' material deposits (slag, excavated soil) when compared with the useful (processed) material or ore extracted, a somewhat slower growth in materials 'bound' (accumulated) in the economy, and considerable increases in quantity throughputs through the economy (during 1960–1990). The next step in expanding material flow accounts is to introduce production and consumption processes in physical input–output tables that incorporate nature as a production factor. The first results of such a table are presented for Germany. By linking emission data to the extended input–output table direct and indirect (cumulative) emissions can be compiled for the different economic activities.

The physical accounts presented above as part of the official statistics of Germany are further explored by staff of a research institute in Wuppertal. Stefan Bringezu, Ralf Behrensmeier and Helmut Schütz assess what they call the 'ecological rucksack' (backpack) of imports and exports, i.e. the environmental impacts generated by domestic and foreign use of materials. These calculations are to extend the sustainability analysis of economic growth to the rest of the world. Both material intensity and environmental burden are also linked to industries and final demand categories. For Germany, the non-ferrous metal industry imports the highest total material input, i.e. the weight of raw materials and the associated 'rucksack' of material inputs that are not incorporated in products (wastes and emissions). The authors claim that physical accounting and analysis are a good substitute for economic valuation, indeed a superior approach to measuring the interaction between economy and environment.

It will take about 81 years to cover Germany with human settlements according to a business-as-usual scenario of land use extrapolation. Walter Radermacher uses this scenario and others to demonstrate the significance of accounting for land use and its pressures on the environment. A pilot project for a selected region is to assess the feasibility of presenting physical environment statistics in an accounting scheme that combines land use/cover accounts with geo-referenced environmental and economic data. Setting out from the core accounts, developed under a UN/ECE programme of physical accounting, supplementary (issue-orientated) accounts are being developed for Germany. Issues include land use by economic activities, partitioning of land and habitats by traffic 'lines', soil sealing and vegetation cover, and the degree of artificiality of landscapes. The main conclusion from the pilot project is that the linkage of land use accounts to economic activities obtains specific pressure indicators. However, the statistical linkage of pressure indicators to changes in the state of the environment requires further investigation.

Andrew Stott presents land use accounts for the UK. The accounts are based on the Countryside Survey, which included a land cover census, remote sensing data and a sample (field) survey. Accounts that measure rural land cover change provide the key paramenters for policies on agriculture, forestry, control of development, biodiversity and quality of life. However, these accounts do not measure changes in quality or value. To this end, 'biotope accounts' were formed which permit the classification of land by use intensity, according to an index of intensity of management. To assess the conservation value, in terms of botanical diversity, the mean number of species per plot has to be measured. The results show that a relatively small increase in use intensity can be accompanied by a considerable loss of botanical diversity within some biotopes. The paper also demonstrates that different scales of resolution yield different rates and trends in land cover change. National accounts may thus obscure ecological and environmental change which occur at the subnational level, but which can be captured in 'zonal landscape accounts'.

Concepts, methods and analysis

Valuation has emerged as the central and most divisive issue of environmental accounting. Peter Bartelmus assesses the discussion to date, setting out from a basic dichotomy between monetary environmental economics and accounting, and non-monetary or physical indicator development and use. Both environmental economists and environmentalists aim at integrating economic, social and environmental concerns in what they perceive as sustainable development. Integrated environmental and economic accounting achieves the highest degree of data integration by different categories of valuation: market valuation, maintenance costing, and contingent and related valuations. The abilities of these valuations to incorporate environmental concerns into the national accounting system are discussed from theoretical and practical (data availability) points of view. Operational concepts of sustainable economic growth are advanced, based on the key indicators of integrated accounting. Supplementary indicator sets are needed to capture non-economic aspects of sustainable development such as health, biodiversity, equity or political freedom. Given the current state of consensus in this new area of applied statistics, the author advocates a two-pronged approach of standardization of concepts and methods for capacity building at the national level while, at the same time, encouraging a pluralism of approaches and experimentation by the research community.

Anton Steurer, G. Gie, Christian Leipert, Carl Pasurka and D. Schäfer explore the treatment of one element of environmental accounting which is, in principle, already part of the conventional economic accounts. Environmental protection expenditures (EPE) are considered as a subset of 'defensive expenditures' that deserve greater visibility in national accounts. The authors argue that newly introduced EPE are fully contained in GDP at current prices and can thus be considered as a 'true share' of this aggregate. In constant prices, however, only final uses enter the GDP value since intermediate uses of EPE translate into price increases. This argumentation is distinguished from the question of actual increases or decreases of GDP resulting from EPE – an issue to be resolved through modelling.

The difficulties of linking 'costs caused' and 'costs borne' are discussed by Jyoti K. and Kirit S. Parikh in their overview of valuations in environmental projects. This discussion might be the much needed beginning of research into the linkages between cost–benefit analysis at the project or programme levels and national accounting at the sectoral and macroeconomic levels. The paper also shows the problems of measuring and valuing the benefit (of environmental functions) and damage (of function losses from pollution) valuations. There are considerable differences in the results between methods of contingent valuation and indirect market studies, though the differences 'do not look outrageous' to the authors. However, the range of these estimates (from one half to four times) might increase considerably at the national (accounting) level. Human capital assessment in monetary and physical terms is another issue addressed, whose incorporation into integrated accounts would provide for a broader analysis of the sustainability of production and income generation.

Michael Ward and Kirk Hamilton set out from a critique of the SEEA similar to the one offered by Bartelmus (Chapter 16). The SEEA is seen as a valid framework whose contents as to valuation need to be made more precise. Through differential analysis (rather than the difference equations used in Chapter 16) they come to similar assessments of the Hotelling rent valuation and the El Serafy method, advanced for the measurement of natural resource depletion. Both methods are practical techniques with the former catering to optimization of resource use and the latter to 'neutral' expectations of constant profits. However, the authors favour the income-oriented El Serafy method over the (net) production-oriented approach of rent valuation. This view contrasts with Bartelmus's assessment of the net-price method as more consistent with the national accounts concept of capital consumption. For the valuation of pollution impacts, the SEEA's 'cost caused' notion is recommended though its maintenance costing is rejected, again contrary to the findings of Chapter 16. Bartelmus, in fact, uses the Hamilton analysis to show how (marginal) maintenance costing can be justified for measuring environmental externalities. Genuine savings are advanced as an alternative to SEEA indicators of green GDP and capital accumulation, making similar deductions for environmental depletion and degradation costs.

'Soft' modelling, beyond direct observation, is already common practice in national accounting, according to André Vanoli. The costing of non-marketed outputs, hypothetical goods baskets in constant-price aggregates and the role of expectations in capital consumption measures are examples of modelling in the central national accounts. Much of the paper deals with modelling aspects inherent in contingent and maintenance valuations of different versions of the SEEA. The author presents a number of challenges to, as well as agreements with, some of the conclusions reached by Bartelmus (Chapter 16). Converging views are presented on the need to include soft or descriptive modelling in environmental accounting, a critique of demand-side (contingent) valuation and the non-viability of combining exchange values and welfare measures that include consumer surplus. On the other hand, the authors disagree on the deduction of maintenance cost (standing on weak theoretical grounds according to Vanoli), on the interpretation of maintenance costing as a 'moral' question (referring to desirable environmental action), and on the characterization of environmental accounting as 'largely a modelling construct'.

A critique of the SEEA is also provided by Henry M. Peskin. The author distinguishes between two main functions of accounting, namely score keeping or the recording of performance, and management or policy support. Both functions are addressed in the main comprehensive accounting systems that extend the national accounts, i.e. the SEEA and the Peskin framework of the Environmental and Natural Resource Accounting Project (ENRAP) (see Chapter 6 for an application of this framework). While 'having much in common', the author contends that the SEEA is consistent with the conventional national accounts, the SNA, but less so with (neoclassical) economic theory. This makes the ENRAP the 'more correct approach' whereas the

SEEA might be 'more practical to implement'. In practice, the process of implementation, i.e. 'accounting', is considered to be more important, owing to its data generation capability, than its results, i.e. the 'accounts'. As to policy uses and analysis, physical accounts may provide the appropriate inputs into regulatory policies and models of policy simulation and programming; cost and expenditure accounts, on the other hand, produce score keeping indicators such as a green GDP and are aimed at sectoral and macroeconomic impact analyses.

Bernd Meyer and Georg Ewerhart present a dynamic input–output model which assesses the implications of different CO_2 reductions for economic (GDP) growth and structural change. The model is considered superior to general equilibrium models as it assumes the long-term implementation of an environmental policy instrument, rather than a one-time adjustment. Under the model assumptions, notably full environmental cost incidence and cost compensation (from pollution permit revenues), the high CO_2 reduction scenario indicates that in West Germany a 30% (as compared to the 1990 emission figure) reduction can be achieved with a relatively small loss of about 3% of forecasted GDP. This would be done largely through structural change in consumption and production patterns. Further coverage of other pollutants and environmental policies is envisaged in future versions of the model.

Identifying research priority

The last part of the book builds on a historical perspective of environmental accounting and indicator development to advance a research agenda for the generation and use of environmental data. Kimio Uno describes the development of integrated environmental and economic accounting and of indicators of sustainable development, as requested by the Earth Summit in Rio de Janeiro in its Agenda 21. He then advances six priority items for research in these areas. The linkage of physical and monetary indicators is considered as an essential feature of a comprehensive information system of environment–economy interaction. International aspects of integrated accounting require further elaboration, given the significance of transboundary economic and environmental flows and impacts. The effects of structural change in production and consumption patterns need to be explored in intertemporal analysis of the results of environmental accounting and indicator compilation. Input–output tabulations, which are an extension of the conventional acounting framework, are a suitable means for measuring technological change and its effects on the environment, and the sustainability of economic growth. Social aspects of the quality of life, notably inequities in the distribution of income and environmental impacts, need to be explored through appropriate indicators. Finally, models of alternative scenarios of the sustainability of growth and development would have to take account of all of the research issues described above.

2
Implementation of environmental accounting: towards an operational manual

ALESSANDRA ALFIERI AND PETER BARTELMUS

Introduction

Accountability of economic activity for its impacts on the environment is at the heart of sustainable development. Accounting for both, socioeconomic performance and its environmental effects, is a first step towards integrating environmental concerns into mainstream economic planning and policies. This is because national accounts have provided the most widely used indicators for the assessment of economic performance, trends in economic growth, and the economic counterpart of social welfare. However, various flaws of conventional accounts may have sent the wrong signals to decision makers who, as a consequence, may have set society on a non-sustainable path of development. In assessing cost and capital, national accounts have neglected, on the one hand, new scarcities of natural resources which threaten the sustained productivity of the economy and, on the other hand, the degradation of environmental quality and its effects on human health and welfare. In addition, some expenditures for maintaining environmental quality are accounted as increases in national income and product; this is despite the fact that such outlays could be considered a maintenance cost to society, rather than social progress.

It has been argued that the traditional System of National Accounts (SNA) (Commission of the European Communities *et al.*, 1993) should be replaced by 'green' accounts. National accountants and mainstream economists disagree, pointing out that conventional accounts serve many short- and medium-term purposes of market observation and business cycle assessment. Moreover, these accounts are based on observable data, whereas environmental accounting would require numerous and controversial estimations and valuations.

In the absence of international consensus on how to incorporate environmental assets and the costs and benefits of their use into national accounts, the United Nations Statistics Division (UNSD) developed an SNA 'satellite' System of integrated Environmental and Economic Accounting (SEEA) rather than modifying the core SNA itself. This approach was confirmed by the United Nations Conference on Environment and Development in its Agenda

---

The authors are staff members of the United Nations. The views expressed here are their own and not necessarily those of the United Nations.

K. Uno and P. Bartelmus (eds.), Environmental Accounting in Theory and Practice, 13–32.

21. The Conference also recommended that systems of integrated environmental and economic accounting be established in all member States 'at the earliest date' (United Nations, 1993b, para. 8.42).

Reflecting the broad range of proposed methodologies for 'greening' the national accounts, UNSD issued an interim version of a handbook on *Integrated Environmental and Economic Accounting* (United Nations, 1993a) which incorporated most of those methodologies in different versions of the SEEA. The handbook was tested in a number of country projects[1] which demonstrated the feasibility of implementing the more practical (as far as data availability and policy use are concerned) modules of the SEEA. The country applications also revealed the need for a simpler guide to the implementation of the SEEA, especially in countries with limited statistical capacities and for data users that are not familiar with the complexities of national accounts.

UNSD, in cooperation with the United Nations Environment Programme (UNEP), proposed therefore to develop an operational manual which would provide easy access for both data users and producers to the concepts and methods of integrated accounting. To assist in this effort UNEP convened a working group, the so-called 'Nairobi Group', in October 1995. The Group met again in November 1996 to discuss a first draft of the manual. The core section of this draft, prepared by the UNSD, presents a step-by-step guide on how to implement the commonly applied modules of the SEEA. Those steps are presented below after a brief introduction of the overall structure and contents of the SEEA. The final section describes some of the experiences gained in formulating and organizing country programmes of integrated accounting.

The system of integrated environmental and economic accounting

The SEEA addresses the above-mentioned flaws of conventional national accounts by means of alternative versions or modules. This building-block approach allows SEEA users to choose among different approaches according to their priorities and statistical capabilities. The main objectives of the different SEEA versions are:

(a) Segregation and elaboration of all environment-related flows and stocks of traditional accounts. The purpose of this version is to present separately environmental protection expenditures. These expenditures have been considered as part of the costs necessary to compensate for the negative impacts of economic growth, in other words, as 'defensive expenditures' (Leipert, 1989).

(b) Linkage of physical asset accounts with monetary environmental accounts and balance sheets. Physical asset accounts cover the total stock or reserves of natural assets and changes therein, even if those assets are not (yet) affected by the economic system. Natural asset accounts thus provide the physical counterpart of the SEEA's monetary asset and flow accounts. They were pioneered by Norway (Alfsen *et al.*, 1987), and further developed by France as 'natural patrimony accounts' (Theys, 1989) and by German scientists as 'material/energy flow accounts' (Radermacher and Stahmer, 1996).

(c) Assessment of environmental costs and benefits. The SEEA expands and complements the SNA with regard to costing (1) the use (depletion) of natural resources in production and final demand, and (2) the changes in environmental quality, resulting from pollution and other impacts of production, consumption and natural events on the one hand, and environmental protection and enhancement on the other.

(d) Accounting for the maintenance of tangible wealth. The SEEA extends the concept of capital to cover not only man-made but also natural capital. Natural capital includes scarce renewable resources such as marine resources or tropical forests, non-renewable resources of land, soil and subsoil assets (mineral deposits), and cyclical resources of air and water. Capital formation is correspondingly changed into a broader concept of capital accumulation.

(e) Elaboration and measurement of indicators of environmentally adjusted product and income. Consideration of the costs of depletion of natural resources and changes in environmental quality allows the calculation of modified macroeconomic aggregates in different SEEA versions. Indicators thus compiled include, in particular, an environmentally adjusted net domestic product (EDP).

Integrated accounts can be used to assess two major aspects of economic policy:

(a) the sustainability of economic growth as conventionally measured by increases in NDP and its main determinant capital formation; and

(b) the structural distortion of the economy by environmentally unsound production and consumption patterns.

The former calls for macroeconomic policies that reorient economic growth toward a sustainable path. The latter aims at the internalization of environmental costs into the budgets of households and enterprises.

Conventional indicators of national income or product are typically used in the measurement and analysis of economic performance and growth. EDP or similar aggregates could therefore be introduced into such analyses. In particular, environmentally sustainable (allowing for depletion and degradation costs) economic growth could be defined as 'upward trend of EDP' (Bartelmus, 1994a, p. 70). Replacing conventional growth indicators, notably GDP or NDP, by EDP and expanding the scope of key variables such as capital in dynamic growth models could thus provide early warning signals about the trends and the limits of sustainable economic growth.

Given the inefficiencies of command and control measures in environmental protection and natural resource conservation, the application of market instruments of effluent charges, tradable pollution permits, etc. has generally been advocated. These instruments aim to internalize external (dis)economies into the budgets of households and enterprises in order to achieve an optimal allocation of scarce resources. Integrated accounting can help define those instruments and measure the appropriate levels of fiscal incentives (subsidies) or disincentives (charges etc.).

Implementation: a step-by-step approach

The proposed operational manual translates the complex methods of integrated environmental and economic accounting into a logical sequence of activities within the overall SEEA framework. This is achieved by means of a set of steps and related worksheets. The worksheets represent tabulations of the raw data that need to be compiled from different sources and processed in order to fit them into the ultimate accounts. Figure 1 shows the position of the different worksheets in the overall framework of the SEEA, described in generic terms above. Additions to and modifications of conventional economic accounts are indicated in shaded boxes. Given the operational nature of the manual, a detailed conceptual discussion, notably of alternative approaches, is generally avoided.[2] The following 10 steps encompass the full work programme required to establish the basic accounting framework, to modify it for environmental purposes, to compile physical and monetary stock and flow accounts and to present the results in final tabulations.

Step 1: Compilation of the conventional supply and use accounts

Worksheet (WS) 1 shows how the data systems for produced and non-produced (natural, non-financial) economic assets can be integrated into one framework of supply and use and asset accounts. Such integration is essential for environmental–economic analysis as it permits to extend and link conventional accounts and accounting identities, incorporating environmental assets and changes therein.

The SEEA tabulations will differ from the conventional accounts even in WS 1 with regard to its classifications. For environmental accounting, the International Standard Industrial Classification of All Economic Activities (ISIC) (United Nations, 1990) will show, at a higher level of detail, those industries that are particularly relevant for environmental analysis.

Step 2: Identification and compilation of environmental protection expenditures

Environmental protection (EP) expenditures are actual expenses incurred by industries, households, the government and non-governmental organizations (NGOs) to avoid environmental degradation or eliminate the effects after deg-

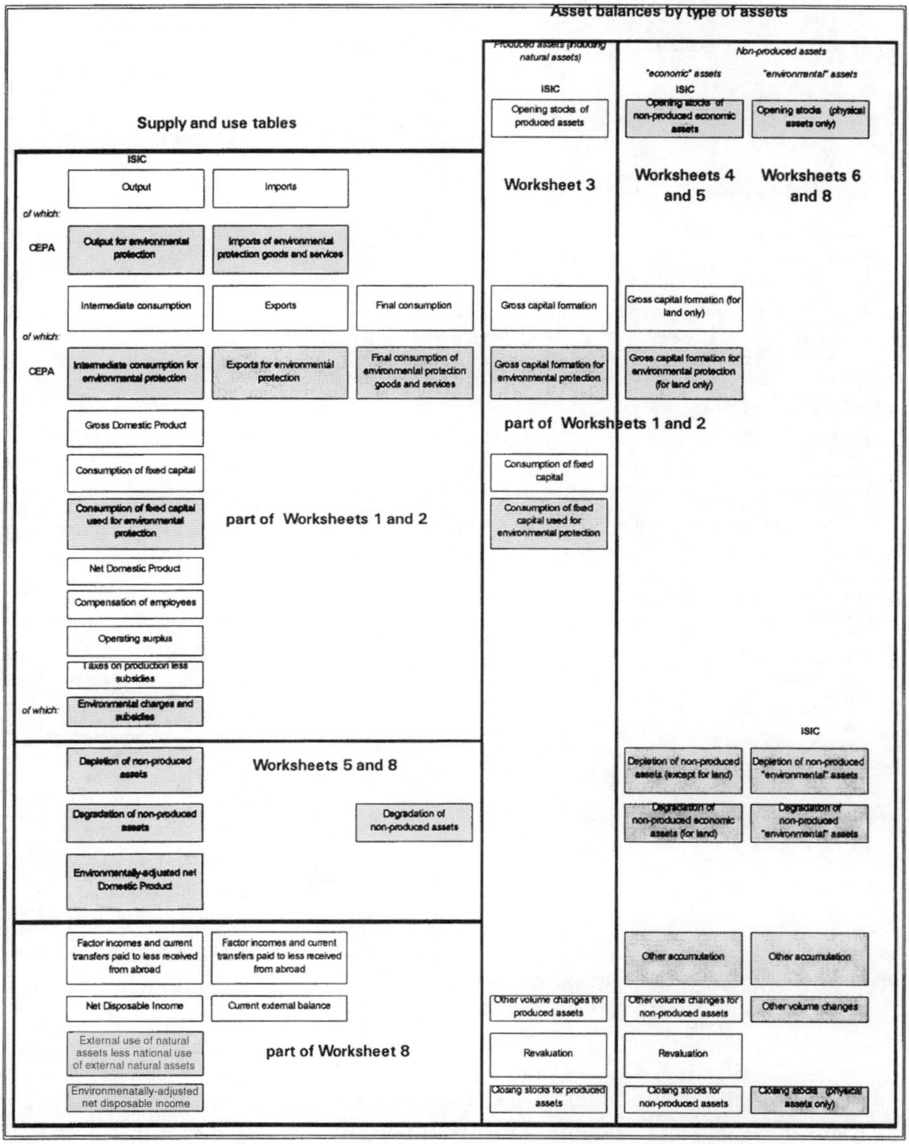

Figure 1. Framework for integrated environmental and economic accounting.

radation has taken place. EP products and services are included in the SNA, but are usually not identified separately in the conventional production and final use accounts. They are, therefore, displayed separately in WS 2 as a subset ('of which') of output, intermediate consumption, consumption of fixed

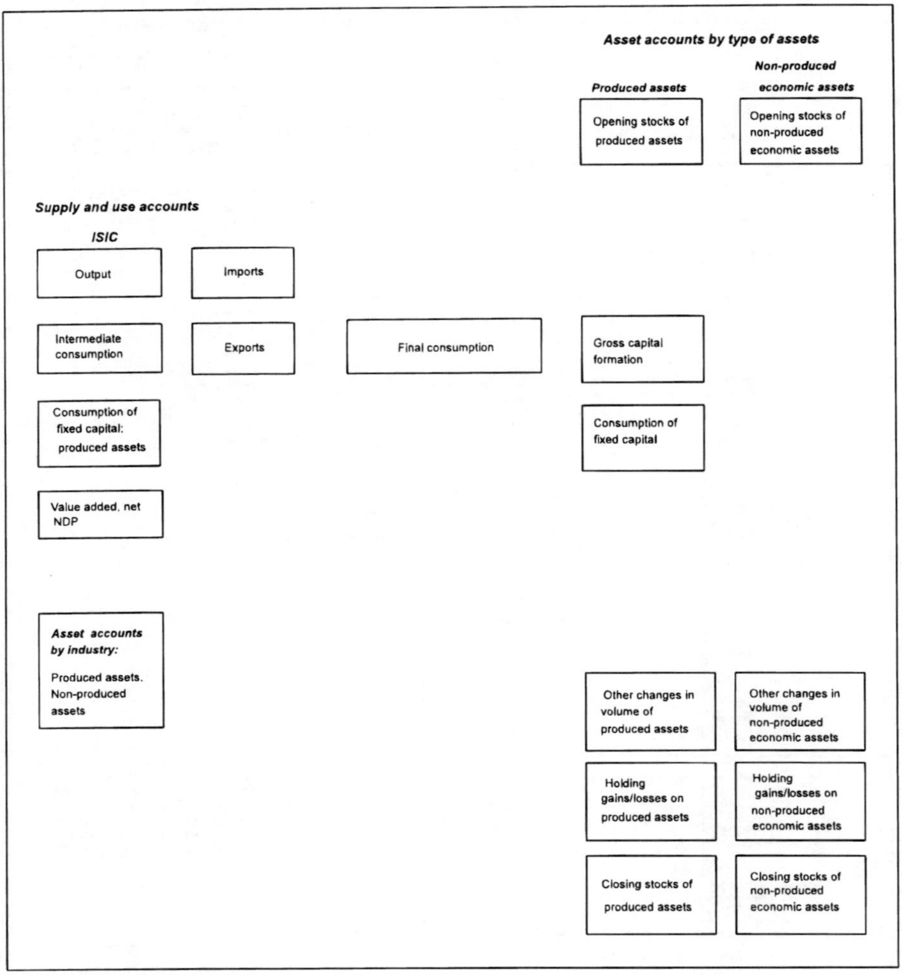

Worksheet 1. 1993 SNA, supply, use and asset accounts.

capital, capital formation, imports, exports and final consumption. A draft Classification of Environmental Protection Activities (CEPA) is proposed in the SEEA for the identification and measurement of EP expenditures. Though sometimes suggested (Daly, 1989; Leipert, 1989; Pearce *et al.*, 1989),[3] these so-called 'defensive expenditures' are not deducted from conventional accounts indicators.

WS 2 is the comprehensive framework for compiling expenditures on environmental protection services and facilities in the supply/use and asset accounts. In practice, it is hardly possible to collect or estimate the data for all blocks in WS 2, notably the intermediate consumption of environmental pro-

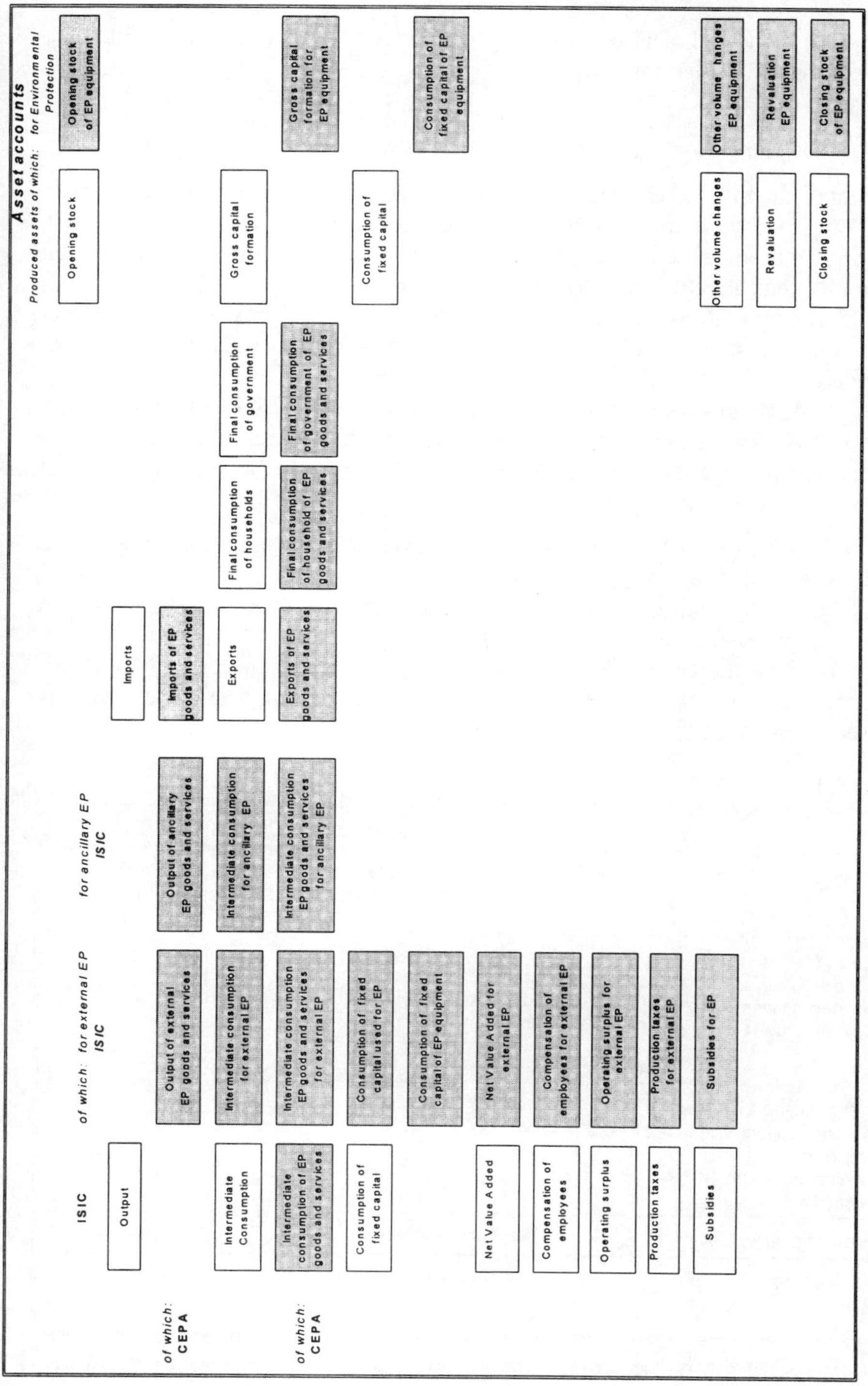

Worksheet 2. Environmental protection expenditures.

tection goods and services for the production of external and ancillary environmental protection outputs.

Step 3: Compilation of produced asset accounts

Another category of the more or less natural environment which is already part of the conventional flow and stock accounts is the produced natural asset. In order to assess comprehensively the level, distribution and changes in national wealth inclusion of the produced asset accounts (described in detail in the SNA), in addition to non-produced (economic and environmental) asset accounts (see steps 4 and 6 for the definition of these categories), is recommended.

WS 3 distinguishes between 'produced natural assets' and 'other produced assets'. SNA codes are shown in parentheses for the different transactions presented in the worksheet. The following activities have to be carried out to fill in the worksheet:

3.1 estimate asset/inventory stocks at the beginning of the accounting period (from wealth survey or research studies);
3.2 incorporate national accounts data on capital formation, changes in inventories and capital consumption;
3.3 introduce the economic appearance of produced assets, i.e. works of art, historical monuments etc. which previously had not been recognized as economic assets;

ISIC

	Produced natural assets	Other produced assets
Opening stock	Value of livestock for breeding, orchards, plantations, timber tracts	
Capital formation - Gross fixed capital formation (P.51) - Changes in inventories (P. 52,53) - Consumption of fixed capital (K.1)	- Acquisition less disposal of mature animals, trees etc., including acquisition of immature, animals, trees etc. produced on own account - Natural growth	
Other volume changes - Economic appearance of produced assets (K.4) - Catastrophic losses (K.7) - Other (K.8,9,12)	Reduction of value of fixed assets	
Revaluation (K.12)	Holding gains and losses	
Closing stock	Value of livestock for breeding, orchards, plantations, timber tracts	

Worksheet 3. Monetary asset accounts – tangible fixed assets, including produced natural assets.

3.4 assess other volume changes due to natural disasters or other destruction and uncompensated seizures by authorities (the latter are part of the 'other' row in WS 3);
3.5 estimate asset/inventory stocks at the end of the accounting period (in current prices);
3.6 calculate revaluation (holding gains and losses) as the difference between 3.1, modified by 3.2, 3.3 and 3.4, and 3.5.

Step 4: Compilation of physical natural resource accounts

The term 'natural resource' is used in the SEEA in a somewhat broader sense than SNA's definition of 'economic non-produced assets, natural assets'. In the SNA, they are those natural assets over which ownership rights are enforced and which provide economic benefits to their owners. The products of economic assets are generally valued in the market, either directly or indirectly (see below, step 5). In the SEEA, the ownership and control criteria are slackened to include also those resources that are currently exploitable or likely to be so, for economic purposes, even if no explicit ownership or control is currently exerted over these resources (e.g. in the case of fish in the oceans or commercially exploitable timber in tropical forests). Economic assets (in this broader, SEEA, sense) are thus distinguished from environmental assets; not so much because of any scarcity criteria, which apply to environmental assets as well, but because market values for economic assets are readily available and most of these assets are already defined and classified in the conventional accounts.

WS 4 records the stocks and changes therein in physical measures (km^2, tons) during the accounting period. Opening and closing stocks are measured as the quantity available (reserves, inventory) at the beginning and end of the accounting period. Changes in quantity are brought about by the direct use/exploitation of the asset, including extraction of minerals, logging, fish catch, water abstraction etc. Quantity change is a gross concept, distinguished from 'depletion' which represents exploitation of renewable resources beyond sustainable levels or yield. Natural resource depletion is the notion underlying environmental costing described below under step 5.

Changes in the quality of natural resources affect their productivity and value. Quality changes are thus necessary (physical) inputs in measuring environmental costs but are difficult to incorporate into physical asset accounts. They are therefore shown in shaded areas below the closing stocks of the worksheet.

Other accumulation and other volume changes in the SEEA are those quantitative changes in the asset accounts which continue to be accounted for outside the production and income accounts. They do not affect, therefore, value added and income generation (as 'cost'), but are important elements in the assessment of the availability of natural resources. Other accumulation is distinguished from other volume changes, with the former referring to changes

	Non-renewable resources		Renewable resources		
	Land/soil (km^2)	Sub-soil assets (tons)	Forests (economic functions) (m^3, tons)	Fishery resources (tons)	Water resources (m^3)
Opening stock	Area of land underlying buildings, land under cultivation, recreational land	Proven reserves	Volume of standing timber	Biomass	Volume
Changes in quantity	Land reclamation (asset increase)	Extraction of minerals (measured in ore or processed form)	- Logging (tons) - Clearing of forests (loss of timber)	Total catch	Water abstraction
Other accumulation	- Changes in land-use - Transfer of land from the environment to economic use	- Discoveries - reassessment of reserves due to changes in technology and relative prices	- Net natural growth - Transfers from the environment to economic use	Net natural growth	-Transfers (discovery) from the environment to economic use (m^3) - Replenishment
Other volume changes	- Changes in land use and land area due to natural, political or other non-economic causes - Transfer of land from economic use to the environment	Reduction in volume due to natural disasters or other non-economic factors	- Reduction in volume due to natural disasters or other non-economic factors (fires, floods, earthquakes) - Transfer of forest from economic use to the environment	Reduction in volume caused by natural disasters or other non-economic factors	Changes due to natural disasters (flooding etc.)
Closing stock	Area of land underlying buildings, land under cultivation, recreational land	Proven reserves	Volume of standing timber	Biomass	Volume
Changes in quality*	- Soil erosion or nutrient loss (tons) - Land/soil contamination including salinization and other changes in soil quality (km^2, ambient concentration)				Water quality index (index value)

* Quality measures are not part of the asset accounts, but are used in assessing the cost of productivity losses.

Worksheet 4. Physical asset accounts – non-produced 'economic' assets.

due to economic decisions or interest, and the latter referring to non-economic causes (political or natural events/disasters).

Step 5: Valuation of 'economic' natural resources

The SEEA applies different methods of monetary valuation to different categories of natural assets.[4] The manual proposes to use the so-called net-price and user-cost methods to determine the value of economic (in the accounting sense discussed above) assets and their use. WS 5 is thus the result of applying monetary unit values to the physical stocks and stock changes of a somewhat modified WS 4. The worksheet maintains the same column headings for the different categories of natural resources but modifies the row headings. Two further items 'depletion' and 'degradation' are introduced in particular. They are based on the (physical) items 'changes in quantity' and 'quality' of WS 4, but differ due to the introduction of sustainability criteria into the monetary accounts. It can be shown that these criteria are an extension of a sustainability (capital maintenance criterion) already inherent in conventional production and income accounting.[5] Thus, not all direct use of natural assets of resource extraction and waste/pollution disposal is to be valued here, but only that part that is not renewed or safely absorbed.

The following activities need to be carried out for the compilation of WS 5:

Application of the net price method

5.1 determine the market prices of different natural resource outputs:
  domestic or export price as applicable,
  price at the beginning and end of the accounting period and average during the period;

5.2 assess the total factor cost per unit of resource output:
  including normal return to capital,
  unit cost at the beginning and end, and average cost during the accounting period;

5.3 calculate the net price as the difference between 5.1 and 5.2;

5.4 apply market value or net price at the beginning of the accounting period to the opening stocks of non-produced economic assets;

5.5 apply average net price to volume changes of non-produced economic assets:
  exploitation/extraction (beyond sustainable use/yield), other accumulation, other volume changes;

5.6 apply market value or net price at the end of the accounting period to the closing stocks of non-produced economic assets;

5.7 calculate revaluation item as balance between opening and closing stocks and all other changes listed in WS 5;

5.8 enter environmental (depletion) costs into the SEEA (Figure 1).

	Land/soil	Sub-soil assets	Forests (economic functions)	Fishery resources	Water resources
Opening stock	See Worksheet 4	See Worksheet 4	See Worksheet 4	See Worksheet 4	(Use value of selected water bodies)
Acquisitions less disposals of non-produced non-financial assets	Acquisitions less disposals of land	Acquisitions less disposals of sub-soil assets	(n.a.)	(n.a.)	(n.a.)
Gross fixed capital formation	Expenditures on land improvement, including: - land reclamation - clearance of forest land - advantage of wetlands - prevention of flooding or erosion	n.a.	n.a.	n.a.	n.a.
Depletion	Capital consumption: decline in the value of land improvement	Value of extraction	Value of non-sustainable cut	Value of non-sustainable catch	(Value of non-sustainable abstraction)
Degradation	Change in (actual and imputed) market value due to contamination and erosion	(n.a)	(n.a)	(n.a)	(Change in market value due to change in quality)
Other accumulation	See Worksheet 4	See Worksheet 4	See Worksheet 4	See Worksheet 4	(See Worksheet 4)
Other volume changes	See Worksheet 4 (including effects of "natural" erosion)	See Worksheet 4	See Worksheet 4	See Worksheet 4	(See Worksheet 4)
Revaluation	Holding gains and losses	Holding gains and losses	Holding gains and losses	Holding gains and losses	Holding gains and losses
Closing stock	See Worksheet 4	See Worksheet 4	See Worksheet 4	See Worksheet 4	(Use value of selected water bodies)

n.a. = not applicable
(n.a.) = unlikely to be applicable

Worksheet 5. Monetary asset accounts – non-produced 'economic' assets.

Calculation of the user cost allowance

5.9 estimate the lifetime of the resource at current exploitation rates;
5.10 establish the discount rate as the opportunity cost of next-best investment;
5.11 compile the net return from the sales of the resource during the accounting period:
either directly,
or by multiplying the quantity sold with the net price of the resource;
5.12 calculate user cost allowance, applying the discount rate (5.10) and lifespan (5.9) established above to the current net return (5.11).

Step 6: Compilation of physical environmental asset accounts

Non-produced environmental assets are those for which neither ownership rights are enforced nor direct economic benefits are derived from their use. They include air, water, land in the wilderness, forests which are not commercially exploitable as well as those economic forests that exhibit significant environmental amenities, and wild fauna and flora; they might or might not be protected (from economic uses). There is obvious overlap, notably in the case of land and forest ecosystems, where a natural asset may exhibit both economic and environmental functions. As shown below, such overlap is sorted out, avoiding duplication of measurement, by appropriate valuation (maintenance costing) and combination of valuations.

WS 6 reflects the above-described expansion of economic assets to include environmental assets (and important environmental functions of economic assets) in the column headings. The row headings of WS 5 are maintained, though (environmental) quality changes become more important than quantity changes.[6] Most of the physical data presented in WS 6 are part of environment statistics systems, such as the Framework for the Development of Environment Statistics (FDES) (United Nations, 1984) which includes asset inventories as well as quantitative and qualitative changes and their effects on human welfare. Those statistical systems need to be consulted for further information on definitions, classifications and statistical data collection methods.

Step 7: Compilation of emissions by economic sector

The compilation of emissions, including discharge of wastes, from polluting sectors is required for applying the 'cost-caused' concept of maintenance costing of the SEEA (see step 8, below). WS 7 shows the tabulation of emissions, typically found in environment statistics. As in the case of renewable resources, only emissions that cannot be safely absorbed by environmental sinks should be recorded and costed in principle. One simplifying assumption (see below) for maintenance costing would be to consider all current emissions

	Land and terrestrial ecosystems (excluding forests) (km^2)	Forests and forest land in the wilderness (km^2)	Rare and endangered species of fauna and flora (no.)	Water and aquatic ecosystems (km^2)	Air
Opening stock	Area of land not covered in the "economic" asset accounts (WS 4)	Area	Population	Area, excluding aquifers and underground water	n.a.
Changes in quantity	n.a.	Clearing of environmental forests	- Number captured or killed - Number of species introduced	n.a.	n.a.
Other accumulation	Transfers from "environmental" to economic land, including drainage of wetlands	- Net natural growth - Transfers from environmental to economic forests	- Net natural growth - Status change (classification) of species (from environmental to economic)	n.a.	n.a.
Other volume changes	- Transfer of land from economic to environmental use - Changes in ecosystem boundaries (classification) - Changes in area due to natural, political or other non-economic causes	- Transfer of economic forest to environmental (protected) status - Area change due to natural disasters (fires, floods, earthquakes)	- Status change of species (from economic to environmental) - Changes in number due to natural disasters	- Changes in ecosystem boundaries (classification) - Area change due to natural disasters	n.a.
Closing stock	Area not covered in the "economic" asset accounts (WS 4)	Area	Population	Area, excluding aquifers and underground water	n.a.
Changes in quality*	- Soil erosion (tons) - Land contamination (loading and ambient concentration)	Forests affected by disease and acid precipitation (change in area or volume)	Species affected by disease (change in numbers)	Change in water quality (index)	Change in air quality (index)

* Quality measures are not part of the asset accounts, but are relevant for estimating environmental degradation cost, either for tracing the cost caused by economic activity or for applying damage valuations.

n.a. = not applicable

Worksheet 6. Physical asset accounts – non-produced 'environmental' assets.

	ISIC	Government	Households	Rest of the world
Air	(tons)	(tons)	(tons)	(tons)
CO				- from - to . . .
CO_2				
SO_2				
NO_x				
TSP				
etc.				
Water	(tons, m^3)	(tons, m^3)	(tons, m^3)	(tons, m^3)
BOD				- from - to . . .
COD				
selected pollutants				
Land	(tons, m^3)	(tons, m^3)	(tons, m^3)	(tons, m^3)
wastes				- from - to . . .
thereof: hazardous wastes				

Worksheet 7. Emission by sectors.

since the absorptive capacity of environmental media is already overloaded in most cases.

Step 8: Maintenance costing of environmental depletion and degradation

The SEEA proposes in principle three types of monetary valuation. Step 5 imputed market values to resources and their depletion. The present step applies a maintenance valuation to environmental costs caused by economic agents (government, households, enterprises). For measuring the monetary value of damages resulting from environmental degradation, i.e. the costs borne by economic agents, contingent and related (demand-side) valuations should be used. However, as shown in Chapter 16, cost-borne valuations are highly controversial and hardly applicable in national accounting.

	Land and terrestrial ecosystems (excluding forests) (km^2)	Forests and forest land in the wilderness (km^2)	Rare and endangered species of fauna and flora (no.)	Water and aquatic ecosystems (km^2)	Air
Depletion	n.a.	Value of forests destroyed	Value of species lost	(n.a.)	n.a.
Degradation	Value of land degradation (erosion and contamination by economic activity)	(n.a.) (only in cost-borne valuation)	(n.a.)	Value of water (quality) degradation	Value of air (quality) degradation
Other accumulation	Value of transfer to economic use	Value of transfer to economic use	(n.a.)	(n.a.)	n.a.
Other volume changes	Value of land degradation from natural and political events	Value of forests degraded or destroyed by pests, floods etc.	Value of loss and quality change of species (from natural events and disease)	Value of aquatic systems degraded by natural events (erosion, floods etc.)	Value of air quality change (e.g. from volcanic eruption)
Net external use of natural assets	n.a.	n.a.	n.a.	Value of the net transfer of water pollutants to the rest of the world	Value of the net transfer of air pollutants to the rest of the world

n.a. = not applicable
(n.a.) = unlikely to be applicable

Worksheet 8. Monetary asset accounts – non-produced 'environmental' assets (at maintenance cost).

WS 8 presents the monetary asset accounts for the environmental asset categories specified in WS 6. It estimates the costs of the most efficient (least-cost) practices and technologies applicable in maintaining the natural assets. The worksheet does not present the stocks of environmental assets (as in WS 6) in monetary terms. The reason is, of course, that maintenance valuation applies only to flow variables of stock changes that should have been avoided or mitigated. If a monetary valuation of environmental assets is desired other valuations such as existence or option values would have to be applied which usually resort to more controversial valuation techniques such as contingent valuation.

The following activities describe the estimation of the environmental (maintenance) costs and their introduction into the SEEA:

8.1 assess the minimum cost activities for avoidance/restoration of environmental degradation and depletion (WS 8, first two rows):
  avoid double counting in the case of conservation of resources with economic and environmental functions (forests),
  use cost/benefit analyses, environmental impact assessments, industrial surveys of environmental protection expenditure, and research on best available environmental technologies;

8.2 apply minimum unit costs to degradation and depletion impacts presented in WS 6 and 7;

8.3 enter environmental degradation costs into the SEEA (Figure 1).

Step 9: Aggregation and tabulation

Aggregation of the physical accounts is limited to particular natural resources and assets. Aggregation across assets for presentation in summary accounts requires a common numerator such as the market value or maintenance cost. The application of monetary valuation to physical stocks and stock changes permits such aggregation and the calculation of environmentally modified indicators such as environmentally adjusted value added or EDP. The results of the above-described compilations of monetary values of natural asset stocks and their depletion and degradation, i.e. the environmental cost of production, would be shown in a tabulation of Figure 1, filled in with data, notably from WS 5 and 8.

Step 10: Comparison of conventional and environmentally adjusted indicators

Further summary analysis could present the key aggregates compiled in step 9. For example, EDP I (in market values), EDP II (in maintenance cost or combined market values and maintenance cost) could be compared with NDP. Similarly, the components of NDP, capital formation and consumption can be compared with their environmentally adjusted counterparts.[7]

Programme formulation and institutionalization

At the outset of a national programme of integrated environmental and economic accounting there should be a clear perception of the overall objectives, the accounting framework, data availability and the mode of implementation. Such an approach would facilitate the effective coordination of data gathering by different agencies. Ideally, a national programme of environmental accounting should be long-term (10 or more years), since the statistics required take a long time to develop and the analysis of some environmental effects requires long time series. Elements of an implementing strategy could include pilot, benchmark and annual compilations, as well as special studies.

A pilot compilation of environmental accounts would start with the development of the accounting framework and supporting worksheets tailored to the particular concerns and objectives of the project. The initial pilot compilation would be based on existing statistics. Considerable data gaps can be expected at the start of the programme, requiring estimates that should be replaced by more reliable data in later compilations. However weak in terms of data, a pilot compilation serves important purposes. It familiarizes national staff with the concepts and methods of integrated accounting, assists in setting up coordination mechanisms of data collection and guides future data development. At the end of the pilot phase, data reliability, compilation methodology and coordination mechanisms should be assessed, and a course of action set for future work. Based on past experience, it is suggested that the pilot compilation be carried out as an interdisciplinary research programme in which the statistical office, and/or a particular research institute play key roles.

To date, practical experience has been limited to the first phase of environmental accounting, the pilot project. Given the costs of comprehensive benchmark compilations which can be assumed to be at least as much as a pilot project, annual compilations could be carried out in a reduced format. Reduced-format compilations would introduce aggregated environmental costs and capital indicators into summary economic accounts. Benchmark compilations would be similar in scope to pilot compilations, but would be conducted, not at the beginning, but in the course of the long-term programme, possibly every 5 or 10 years. Their purpose would be to update the economic–environmental database for time series and structural analyses.

The framework for environmental accounting can also serve in implementing special studies of particular sectors of the framework. One type of study could present an in-depth analysis of a particular natural asset (account) such as mineral resources, forests or water. A second type of special study could focus on the industries causing depletion and degradation of the environment. Those studies could deal either with one specific aspect of depletion or degradation across all industries, or with specific industries, assessing their contribution to different kinds of environmental impacts. The use of the SEEA framework would avoid the risk of non-compatibility with national accounts concepts and procedures which is a major drawback of ad hoc studies carried out outside the national statistical services.

Regional (sub-national) accounts could focus on an ecological zone of particular interest or value, or an administrative entity (province, state) in which the sustainability of development is at a particularly high risk. There is an advantage in compiling environmental data at the local/regional levels. However, this advantage might be offset by lack of information on consumption, production and capital formation in the region and on detailed transboundary flows, usually unavailable at subnational levels. The feasibility of such accounting at the regional (province) level is currently being explored by the UNSD in two countries (the Philippines and Indonesia).

Typically,[8] a first (pilot) project is set up in a national seminar. Given the broad, multidisciplinary nature of integrated accounting, the seminar should bring together data users and producers from a wide range of line ministries, research institutes and the statistical office. The seminar would serve several purposes:

(a) Identification of environmental and economic concerns and priorities to determine the scope and coverage of the project.
(b) Presentation of the SEEA, describing concepts and methods, data needs, resource requirements and the use of its results.
(c) Agreement on a collaborative work programme, including mechanisms of supervision and coordination.

As a follow-up, training seminars and workshops may familiarize staff, not only within the lead agency but also in the cooperating data producing institutions, with the technical concepts and data processing required for the SEEA implementation.

After the accounts have been compiled, a draft report is prepared. This report presents the accounts and describes the data gaps, problems encountered, solved or deferred. The report can be the basis for a second national seminar to discuss the results, present further analysis and interpretation of the results, and to make recommendations for follow-up programmes. Recommendations may include the allocation of a more permanent responsibility for recurrent environmental accounting to a particular agency. Considering that the SEEA has been designed as a satellite system of the SNA and that the SNA provides an integrative framework for the measurement of a large number of economic activities and transactions, the agency dealing with the compilation of the national accounts (such as the National Statistical Office or the Central Bank) is probably the best place for harbouring a regular programme of integrated environmental and economic accounting.

Notes

1. Case studies included Ghana (Powell, 1996), Mexico (van Tongeren *et al.*, 1991), Papua New Guinea (Bartelmus *et al.*, 1992), Thailand (Bartelmus and Tardos, 1992), the Republic of Korea

(see Chapter 4) and Japan (see Chapter 3). Other country projects are being carried out with UNSD assistance in Colombia, Indonesia and the Philippines (see Chapter 5).
2. The original methodological publications of the SNA (Commission of the European Communities *et al.*, 1993) and of the SEEA (United Nations, 1993a) should be consulted for details on the accounting concepts, definitions and classifications.
3. Such a deduction is questionable. The exclusion of defensive or any other undesirable activity from the economy would change the production boundary quite arbitrarily, since it is hardly possible to obtain consensus on what is desirable or regrettable in society.
4. Theoretical questions of valuation in environmental accounting are discussed in some detail in Part IV, notably Chapter 16.
5. In the SNA, capital maintenance is costed as the consequence of shifting previously produced capital into production where it is gradually 'consumed'. For non-produced (natural) capital such costing does not apply, and a more normative criterion of the long-term sustainability of production and income generation has to be introduced (see Chapter 16).
6. Figure 1 simplifies, therefore, the description of environmental cost by referring only to (natural resource) depletion, including productivity effects of degradation of natural resources, and to 'degradation' of environmental assets, including the quantitative depletion or destruction of these assets.
7. The country studies presented in Part II of the book provide examples of such summary assessments of the results of integrated accounting.
8. Based on the, admittedly limited, experience with about 10, mostly developing, countries.

Part II

Country experiences

3

Japan: the System of Integrated Environmental and Economic Accounting (SEEA) – trial estimates and remaining issues

KATSUKI ODA, KIYOSHI ARAHARA, NOBUYOSHI HIRAI AND HIDEKI KUBO

Introduction

The report of the World Commission on Environment and Development (1987) and Agenda 21 of the 1992 Earth Summit have heightened global interest in the concept of sustainable development and the development of integrated environmental and economic accounting.

In Japan, there is awareness of the need to develop integrated environmental and economic accounting that is both consistent with the System of National Accounts (SNA) (Commission of European Communities, 1993) and capable of assessing the burden of economic activities on the environment according to uniform criteria. For these reasons, the 1992 Economic Plan and the 1993 Country Action Plan of Agenda 21 state explicitly that the development of such an accounting system should be promoted. Related provisions and references are also found in the Basic Environment Plan of 1994.

The Economic Planning Agency (EPA),[1] which compiles the national economic accounts, considers the development of integrated environmental and economic accounting an important task and has been conducting research and development in this area since 1991. The EPA determined the basic structure of the integrated accounting system and published the first trial estimates for some items in 1995.[2] In this chapter, we will outline the trial estimates and summarize the remaining unresolved issues.

Outline of integrated environmental and economic accounting

The system of integrated environmental and economic accounting for Japan aims to describe interactions between the economy and the environment and

---

Economic Planning Agency (EPA), Government of Japan. The views expressed in this paper are those of the authors and do not necessarily reflect the views of the EPA. The authors thank Masakatsu Tamaru and Fumio Momose for their contribution to the trial estimation and Nobuhiko Kosuge for comments on an earlier draft, as well as Eva Hellsten, Statistics Sweden, Kimio Uno, Keio University and Peter Bartelmus, United Nations, for helpful comments on an earlier version.

K. Uno and P. Bartelmus (eds.), Environmental Accounting in Theory and Practice, 35–61.

to provide an analytical data system for policy use. In implementing the system, the EPA considered the special circumstances of Japan, data availability, and the feasibility of estimation, while basing the framework on the United Nations (1993a) handbook which is the international standard. As a result, out of the several versions suggested in the UN handbook, the system of integrated accounting was based on the following: (1) disaggregation of the SNA into environment-related items and others (version II); (2) the monetary valuation of the burden of economic activities on the environment (imputed environmental costs) according to the maintenance cost valuation method (version IV.2); and (3) the externalization of the internal environmental protection activities of industries (version V.6).

Integrated environmental and economic accounting for Japan was developed as a satellite account linked to the 1968 SNA (United Nations, 1968) and is presented in matrix form, composed of 43 rows and 40 columns[3] (see Table 1). The integrated environmental and economic accounts consist of flow and asset accounts. Flow data are recorded in rows 2–36 and asset data in columns 20–39. Flow and asset accounts are linked to each other by the changes of assets described in rows 2–24 and columns 20–39.

Environment-related disaggregation of the SNA

In order to protect the environment from the deleterious effects of economic activities (production and consumption) or to reduce their burden on the environment, goods and services related to environmental protection are produced by industries and the government and provided to other establishments, the government and households. Moreover, fixed capital is used for environmental protection purposes. The scale of environmental protection activities may depend on the overall scale of economic activities, available technologies, social concern with environmental issues and social regulations. The aim is to reduce the burden of economic activities on the environment as a contribution of the economy to an environmentally sound society.

Environment-related items are separated from the SNA in the following manner:

(a) Supply and use of goods and services related to environmental protection (output, imports, intermediate consumption, final consumption expenditures, capital formation, exports). Use of products (row 02) is divided into (03) goods and services related to environmental protection and (09) other goods and services. Goods and services related to environmental protection are further divided into (05) industries (external) and (07) the government. It is thus possible to record how much of the environment-related goods and services provided by industries (external) and the government is used in the production activities and capital formation of industries and the government, as well as in the final consumption expenditures of the government and households.

(b) Consumption of fixed assets related to environmental protection. The use of produced assets (row 15) is divided into (16) consumption of fixed assets related to environmental protection and (17) consumption of other fixed assets. The cost structure of the environment-related production activities of industries and the government and the net formation of environment-related fixed capital owned by industries and the government are thus recorded.

(c) Subsidies. Subsidies are divided into (row 31) subsidies related to environmental protection and (row 32) other subsidies. The purpose is to show the environmental protection measures of the government by means of subsidies.

(d) Cost structure of production activities related to environmental protection (intermediate inputs, consumption of fixed capital, indirect taxes and subsidies, compensation of employees, operating surplus). The production activities of industries (column 06) are divided into (07) the production activities related to environmental protection and (10) other production activities. The production activities related to environmental protection are further divided into (08) external activities and (09) internal activities. The government's production activities (column 11) are divided into (12) production activities related to environmental protection and (13) other production activities. The internal environmental protection activities of industries (column 09) are those environmental protection activities which are executed within establishments in conjunction with the production of goods and services. While the intermediate inputs for these purposes are included as part of the intermediate inputs for the production of goods and services in the SNA, the outputs of these activities are not recorded because they are not traded in the market. In order to record environmental protection activities on a wider basis, the cost structure associated with internal activities is estimated by treating them separately from other production activities.

(e) Stocks and accumulation of man-made assets related to environmental protection. Of produced assets (column 22), man-made assets are divided into (23) man-made assets related to environmental protection and (27) other man-made assets. Man-made assets related to environmental protection are further divided into (24) industries and (25) the government. The stocks and accumulation of environment-related man-made assets owned by industries and the government are thus recorded separately.

Externalization of internal environmental protection activities

In order to present the full amount of environmental protection, the services generated and consumed internally by industries need to be 'externalized'. To this end, a row (06) of services associated with externalized internal environmental protection activities is added to row (03) the use of goods and services

Table 1. Trial estimates of integrated environmental and economic accounting.

1990 (at current prices) (Billion Yen)

Serial No		Gross output (1)	Imports (Including Imp.duties) (2)	Trade and Transport Margins (3)	Disposition (Total Supply) (4)	Production activities (5)	Industries (6)	Industries: Environment-related (7)	Environment-related: External (8)	Environment-related: Internal (9)	Industries: Others (10)
(1)	Opening stocks										
(2)	Use of products	786,780.8	44,508.0	90,414.2	894,429.8	428,818.1	410,121.7	2,878.6	2,632.6	246.0	407,243.1
(3)	Environmental protection-related goods & services	9,790.0	0.0	295.9	10,085.9	5,112.5	4,431.5	0.0	0.0	0.0	4,431.5
(4)	Industries	7,213.9		259.9	7,509.8	4,653.0	4,026.5	0.0			4,026.7
(5)	External	6,639.7		295.9	6,935.6	4,078.8	3,452.3	0.0			3,452.3
(6)	Externalized internal	574.2			574.2	574.2	574.2	0.0			574.2
(7)	Government	2,576.1			2,576.1	459.5	405.0	0.0			405.0
(8)	Private non-profit institutions serving households				0.0	0.0	0.0	0.0			
(9)	Other goods & services	776,990.8	44,508.0	90,118.3	884,343.9	423,705.6	405,690.2	2,878.6	2,632.6	246.0	402,811.6
(10)	Industries	719,828.1	44,508.0	90,118.3	827,742.8	423,062.5	405,057.5	2,878.6	2,632.6	246.0	402,178.9
(11)	(Imports of lumber & other forest resources)		1,363.5								
(12)	(Imports of petroleum & other subsoil resources)		7,971.8								
(13)	Government	43,898.8			43,527.2	640.3	629.9	0.0			629.9
(14)	Private non-profit institutions serving households	13,263.9			13,073.9	2.8	2.8	0.0			2.8
(15)	Use of produced assets					62,819.9	59,298.3	527.9	349.0	178.9	58,770.4
(16)	Consumption of environmental protection-related fixed assets					1,061.4	527.9	527.9	349.0	178.9	
(17)	Consumption of other fixed assets					61,758.5	58,770.4	0.0			58,770.4
(18)	Use of non-produced natural assets(Imputed environmental costs)		0.0			6,395.1	6,393.3	0.0	0.0	0.0	6,393.3
(19)	Degradation of natural assets caused by residuals					4,816.4	4,814.6	0.0			4,814.6
(20)	Destruction of ecosystems					1,379.0	1,379.0	0.0			1,379.0
(21)	Depletion of resources					199.7	199.7	0.0			199.7
(22)	Effects on the global environment					0.0	0.0	0.0			
(23)	Other use of natural assets (amenity)										
(24)	Restoration of non-produced natural assets					0.0					
(25)	Shift of imputed environmental costs					2,053.6	2,057.7	0.0			2,057.7
(26)	(Environmental-related transfers)					0.0	0.0	0.0			0.0
(27)	Environmentally-adjusted net domestic product	355,290.2				377,108.1	339,585.0	3,807.4	3,658.1	149.3	335,777.6
(28)	Net domestic product	363,738.9				385,556.8	348,036.0	3,807.4	3,658.1	149.3	344,228.6
(29)	Net indirect taxes	30,567.9				30,063.3	29,954.9	145.3	152.5	(7.2)	29,809.6
(30)	Indirect taxes	35,211.7				34,707.2	34,598.8	162.8	162.8		34,436.0
(31)	(Less)Environment-related subsidies	(17.5)				(17.5)	(17.5)	(17.5)	(10.3)	(7.2)	
(32)	(Less)Other subsidies	(4,626.4)				(4,626.4)	(4,626.4)	0.0			(4,626.4)
(33)	Compensation of employees	233,373.7				233,373.7	195,961.3	2,883.4	2,726.9	156.5	193,077.9
(34)	Operating surplus	99,797.4				122,119.8	122,119.8	778.7	778.7		121,341.1
(35)	(Less)Imputed environmental costs(18+24+25)	8,448.7				8,448.7	8,451.0	0.0	0.0	0.0	8,451.0
(36)	Gross output	877,699.3				877,194.8	817,456.0	7,213.9	6,639.7	574.2	810,242.1
(37)	Adjustments relating to accumulation of natural assets										
(38)	Adjustments to imputed environmental costs										
(39)	Volume changes due to economic causes										
(40)	Other adjustments										
(41)	Volume changes due to non-economic causes										
(42)	Revaluation due to market price changes										
(43)	Closing stocks										

Table 1. Continued

1990 (at current prices)

(Billion Yen)

Serial. No		Production activities				Final consumption expenditures					Accumulation & stocks of non-financial assets
		Government			Non-profit institutions serving households		Government	Non-profit institutions serving households	Households		
			Environment related	Others						Consumer durable goods	
		(11)	(12)	(13)	(14)	(15)	(16)	(17)	(18)	(19)	(20)
(1)	Opening stocks									65,690.9	3,165,794.9
(2)	Use of products	13,786.9	1,090.9	12,696.0	4,909.5	279,316.7	38,806.6	3,135.1	237,375.0	14,842.8	139,054.5
(3)	Environmental protection-related goods & services	654.0	314.3	339.7	27.0	2,304.6	1,759.6	0.0	545.0	0.0	2,639.7
(4)	Industries	606.3	314.3	292.0	20.2	188.0			188.0		2,639.7
(5)	External	606.3	314.3	292.0	20.2	188.0			188.0		2,639.7
(6)	Externalized internal	0.0	0.0	0.0	0.0	0.0			0.0		0.0
(7)	Government	47.7		47.7	6.8	2,116.6	1,759.6		357.0		0.0
(8)	Private non-profit institutions serving households	0.0				0.0					0.0
(9)	Other goods & services	13,132.9	776.6	12,356.3	4,882.5	277,012.1	37,047.0	3,135.1	236,830.0	14,842.8	136,414.8
(10)	Industries	13,128.8	776.6	12,352.2	4,876.2	221,054.1			221,054.1	14,842.8	136,414.8
(11)	(Imports of lumber & other forest resources)										
(12)	(Imports of petroleum & other subsoil resources)										
(13)	Government	4.1		4.1	6.3	42,886.9	37,047.0		5,839.9		0.0
(14)	Private non-profit institutions serving households					13,071.1		3,135.1	9,936.0		0.0
(15)	Use of produced assets	2,508.0	533.5	1,974.5	1,013.6					0.0	(62,819.9)
(16)	Consumption of environmental protection-related fixed assets	533.5	533.5								(1,035.3)
(17)	Consumption of other fixed assets	1,974.5		1,974.5	1,013.6						(61,784.6)
(18)	Use of non-produced natural assets(Imputed environmental costs)	0.4	0.0	0.4	1.4	2,057.7	0.0	0.0	2,057.7	0.0	(8,452.7)
(19)	Degradation of natural assets caused by residuals	0.4		0.4	1.4	2,054.6			2,054.6		(6,870.9)
(20)	Destruction of ecosystems	0.0		0.0	0.0	0.0			0.0		(1,379.0)
(21)	Depletion of resources	0.0		0.0	0.0	3.1			3.1		(202.8)
(22)	Effects on the global environment	0.0				0.0					0.0
(23)	Other use of natural assets (amenity)										
(24)	Restoration of non-produced natural assets										
(25)	Shift of imputed environmental costs	(4.1)	(4.1)			(4.1)	(4.1)				4.1
(26)	(Environmental-related transfers)	0.0	(159.4)			2,053.6	(4.1)		2,057.7		
(27)	Environmentally-adjusted net domestic product	30,183.7	955.8	29,227.9	7,339.4	156.5			156.5		0.0
(28)	Net domestic product	30,180.0	951.7	29,228.3	7,340.8						
(29)	Net indirect taxes	43.6	13.4	30.2	64.8						
(30)	Indirect taxes	43.6	13.4	30.2	64.8						
(31)	(Less)Environment-related subsidies				0.0						
(32)	(Less)Other subsidies				0.0						
(33)	Compensation of employees	30,136.4	938.3	29,198.1	7,276.0						
(34)	Operating surplus										
(35)	(Less)Imputed environmental costs(18+24+25)	(3.7)	(4.1)	0.4	1.4						
(36)	Gross output	46,474.9	2,576.1	43,898.8	13,263.9						
(37)	Adjustments relating to accumulation of natural assets									0.0	9,207.7
(38)	Adjustments to imputed environmental costs										8,452.7
(39)	Volume changes due to economic causes										755.0
(40)	Other adjustments										248,987.4
(41)	Volume changes due to non-economic causes										0.0
(42)	Revaluation due to market price changes									(9,433.8)	248,987.4
(43)	Closing stocks									71,099.9	3,491,771.9

Table 1. Continued

1990 (at current prices) (Billion Yen)

Serial . No		Accumulation & stocks of non-financial assets: Produced assets (21)	Man-made assets (22)	Environment-related (23)	Industries (24)	Government (25)	Historical monuments (26)	Others (27)	Cultivated assets (28)	Cultivated forests (29)	Others (30)
(1)	Opening stocks	980,560.1	959,278.5	21,352.5	516.1	20,836.4	145.9	937,926.0	21,281.6	19,224.7	2,056.9
(2)	Use of products	134,571.2	134,339.5	2,639.7	520.5	2,119.2		131,699.8	231.7	0.0	231.7
(3)	Environmental protection-related goods & services	2,639.7	2,639.7	2,639.7	520.5	2,119.2		0.0	0.0	0.0	0.0
(4)	Industries	2,639.7	2,639.7	2,639.7	520.5	2,119.2				542.9	
(5)	External	2,639.7	2,639.7	2,639.7	520.5	2,119.2			0.0	542.9	
(6)	Externalized internal	0.0	0.0	0.0	0.0	0.0				0.0	
(7)	Government	0.0	0.0	0.0					0.0		
(8)	Private non-profit institutions serving households	0.0	0.0	0.0					0.0		
(9)	Other goods & services	131,931.5	131,699.8	0.0				131,699.8	231.7	0.0	231.7
(10)	Industries	131,931.5	131,699.8	0.0				131,699.8	231.7		231.7
(11)	(Imports of lumber & other forest resources)										
(12)	(Imports of petroleum & other subsoil resources)										
(13)	Government	0.0	0.0	0.0							
(14)	Private non-profit institutions serving households	0.0	0.0	0.0							
(15)	Use of produced assets	(62,819.9)	(62,562.0)	(1,061.4)	(527.9)	(533.5)		(61,500.6)	(257.9)	0.0	(257.9)
(16)	Consumption of environmental protection-related fixed assets	(1,061.4)	(1,061.4)	(1,061.4)	(527.9)	(533.5)					
(17)	Consumption of other fixed assets	(61,758.5)	(61,500.6)	0.0				(61,500.6)	(257.9)		(257.9)
(18)	Use of non-produced natural assets(Imputed environmental costs)	0.0	0.0	0.0	0.0	0.0		0.0	0.0	0.0	0.0
(19)	Degradation of natural assets caused by residuals										
(20)	Destruction of ecosystems	0.0							0.0	0.0	
(21)	Depletion of resources	0.0									
(22)	Effects on the global environment	0.0									
(23)	Other use of natural assets (amenity)										
(24)	Restoration of non-produced natural assets	0.0							0.0	0.0	
(25)	Shift of imputed environmental costs										
(26)	(Environmental-related transfers)	0.0	0.0	0.0	3.7	3.7					
(27)	Environmentally-adjusted net domestic product										
(28)	Net domestic product										
(29)	Net indirect taxes										
(30)	Indirect taxes										
(31)	(Less)Environment-related subsidies										
(32)	(Less)Other subsidies										
(33)	Compensation of employees										
(34)	Operating surplus										
(35)	(Less)Imputed environmental costs(18+24+25)										
(36)	Gross output										
(37)	Adjustments relating to accumulation of natural assets	925.0	0.0	0.0	0.0	0.0		0.0	925.0	194.7	730.3
(38)	Adjustments to imputed environmental costs	0.0							0.0	0.0	
(39)	Volume changes due to economic causes	925.0							925.0	194.7	730.3
(40)	Other adjustments	16,304.3	15,350.3	384.6	34.8	349.8	0.0	14,965.7	954.0	954.0	0.0
(41)	Volume changes due to non-economic causes	0.0	0.0	0.0			9.9		0.0		
(42)	Revaluation due to market price changes	16,304.3	15,350.3	384.6	34.8	349.8		14,965.7	954.0	954.0	0.0
(43)	Closing stocks	1,069,540.7	1,046,406.3	23,315.4	543.5	22,771.9	155.8	1,023,090.9	23,134.4	20,373.4	2,761.0

Table 1. Continued

1990 (at current prices)

(Billion Yen)

Serial . No		Accumulation & stocks of non-financial assets: Non-produced assets	Air	Water	Soil	Land use	Developed land	Agricultural & wooded land	Conservation regions	Subsoil resources	Exports
		(31)	(32)	(33)	(34)	(35)	(36)	(37)	(38)	(39)	(40)
(1)	Opening stocks	2,185,234.8				2,184,370.1	1,828,702.8	346,759.1	8,908.2	864.7	
(2)	Use of products	4,483.3	0.0	0.0	0.0	4,392.9	1,273.4	2,964.1	155.4	90.4	47,240.5
(3)	Environmental protection-related goods & services	0.0	0.0	0.0	0.0	0.0	0.0	0.0	0.0	0.0	29.1
(4)	Industries	0.0									29.1
(5)	External	0.0									29.1
(6)	Externalized internal	0.0									0.0
(7)	Government										
(8)	Private non-profit institutions serving households										
(9)	Other goods & services	4,483.3	0.0	0.0	0.0	4,392.9	1,273.4	2,964.1	155.4	90.4	47,211.4
(10)	Industries	4,483.3				4,392.9	1,273.4	2,964.1	155.4	90.4	47,211.4
(11)	(Imports of lumber & other forest resources)										
(12)	(Imports of petroleum & other subsoil resources)										
(13)	Government										
(14)	Private non-profit institutions serving households										
(15)	Use of produced assets										
(16)	Consumption of environmental protection-related fixed assets										
(17)	Consumption of other fixed assets										
(18)	Use of non-produced natural assets(Imputed environmental costs)	(8,452.7)	(6,680.2)	(190.7)	0.0	(1,379.0)	(975.0)	(343.3)	(60.7)	(202.8)	0.0
(19)	Degradation of natural assets caused by residuals	(6,870.9)	(6,680.2)	(190.7)	0.0						
(20)	Destruction of ecosystems	(1,379.0)				(1,379.0)	(975.0)	(343.3)	(60.7)		
(21)	Depletion of resources	(202.8)								(202.8)	
(22)	Effects on the global environment										
(23)	Other use of natural assets (amenity)										
(24)	Restoration of non-produced natural assets	4.1	0.0	0.0	4.1	0.0		0.0	0.0		
(25)	Shift of imputed environmental costs										
(26)	(Environmental-related transfers)										
(27)	Environmentally-adjusted net domestic product										
(28)	Net domestic product										
(29)	Net indirect taxes										
(30)	Indirect taxes										
(31)	(Less)Environment-related subsidies										
(32)	(Less)Other subsidies										
(33)	Compensation of employees										
(34)	Operating surplus										
(35)	(Less)Imputed environmental costs(18+24+25)										
(36)	Gross output										
(37)	Adjustments relating to accumulation of natural assets	8,282.7	6,680.2	190.7	0.0	1,379.0	25,613.4	(24,342.3)	107.9	32.8	
(38)	Adjustments to imputed environmental costs	8,452.7	6,680.2	190.7	0.0	1,379.0	975.0	343.3	60.7	202.8	
(39)	Volume changes due to economic causes	(170.0)				0.0	24,638.4	(24,685.6)	47.2	(170.0)	
(40)	Other adjustments	232,683.1	0.0	0.0	0.0	232,683.1	176,421.7	55,779.7	481.7	0.0	
(41)	Volume changes due to non-economic causes	0.0									
(42)	Revaluation due to market price changes	232,683.1				232,683.1	176,421.7	55,779.7	481.7	0.0	
(43)	Closing stocks	2,422,231.2				2,421,446.1	2,031,036.3	380,817.3	9,592.5	785.1	

related to environmental protection as a separate subcategory. The supply and use of these services are thus explicitly recorded.

Imputed environmental costs

Natural assets are used as a place of disposal for the residuals of economic activities (production and consumption) and as a source of resources and space necessary for economic activities. With the economic use of natural assets, however, natural assets change quantitatively (depletion) as well as qualitatively (degradation). The depletion of assets imposes a limit on the sustainability of economic activities that use those assets. The degradation of the natural environment also restricts the sustainability of economic activities by adversely affecting human health and other amenities through environmental pollution and the destruction of ecosystems.

In some cases, no payment is made for the economic use of these natural assets because they are not traded in the market. In other cases, even though the assets are traded in the market, valuation is made only from the point of view of economic use and differs from social valuation, which also incorporates the consideration of sustainability (external diseconomies).

In order to make a monetary valuation of these externalities of depletion and degradation of non-produced natural assets and to record them along with economic performance, the following items are separated from, added to, or integrated into the economic accounts:

(a) Use of non-produced natural assets. Along with (row 15) the use of produced assets, which is part of the gross value-added associated with production activities, the use of non-produced natural assets (row 18) is created for the monetary valuation of the burden of economic activities on the environment. This is further divided into the following five items within the system of integrated environmental and economic accounting: (row 19) degradation of natural assets caused by residuals: environmental pollution and other deterioration of the quality of the environment caused by the discharge of waste materials; (row 20) destruction of ecosystems: the deterioration of the quality of natural assets caused by land development and the felling of trees in excess of natural growth; (row 21) depletion of non-produced resources by extraction; (row 22) global environmental issues, such as global warming and the destruction of the ozone layer; (row 23) other uses of natural assets, such as the degradation of landscapes and other amenities.

(b) Disaggregation of cultivated assets and land use. In order to capture the destruction of ecosystems caused by the felling of trees in excess of natural growth, timber assets included in the SNA balance sheet are disaggregated so as to make explicit (column 29) cultivated forests. For the assessment of impacts on ecosystems, caused by land development, land is divided into (column 36) developed land, (column 37) agricultural and wooded land and (column 38) conservation regions, according to the type of land use.

(c) Expanded coverage of non-produced natural assets. In order to measure the burden of economic activities on the environment, particularly the degradation of natural assets caused by residuals, the concept of non-produced natural assets is expanded to include (column 32) air, (column 33) water and (column 34) soil.[4]

Environmentally adjusted net domestic product

Environmentally adjusted net domestic product (EDP, row 27) is the value added obtained by subtracting (row 35) the imputed environmental costs from (row 28) net domestic product (NDP). NDP is the value added obtained by subtracting consumption of fixed capital from gross domestic product (GDP). Consumption of fixed capital is the cost required for restoring the stock of fixed capital used in production activities to its opening level and is hence excluded in obtaining the net value added generated by economic activities. The imputed environmental costs are the costs of depletion and degradation of natural assets associated with economic activities and should thus be excluded from the net value added generated by economic activities, in full consideration of their impact on the environment.

Estimation procedures

Environment-related disaggregation of the SNA and the externalization of internal environmental protection activities

Supply and use of goods and services related to environmental protection

OUTPUT OF INDUSTRIES

Goods and services produced by external environmental protection activities (row 3, column 1) are estimated on the basis of the *Survey Concerning the Quantitative Analysis of Eco-businesses* (Environmental Agency, Japan, 1994). For 1985, the figures are estimated by the ratio of 1990 production to 1985 production in the principal industries that correspond to the below-mentioned four categories of eco-businesses. Then, the amount of fixed capital formation in construction and civil engineering related to the installation of environmental protection facilities is estimated.

The Environmental Agency covers the following goods and services:

(a) equipment that reduces the burden on the environment (e.g. pollution prevention equipment and energy-saving equipment);

(b) products that place little burden on the environment (e.g. recycled products made of used paper and heat insulating materials);
(c) services that contribute to environmental protection (e.g. processing of waste materials and environmental assessment);
(d) establishment of social infrastructure that places little burden on the environment (e.g. energy-saving buildings and regional air conditioning).

For services generated by internal environmental protection activities (row 6, column 1), the output of internal environmental protection activities (row 36, column 9) (to be explained later) is computed.

OUTPUT OF GOVERNMENT (ROW 7, COLUMN 1)

For the central government, output is obtained by excluding the capital formation budget from the environmental protection budget. For local governments, the output of sewage and waste disposal (under public management) is obtained from input–output tables and by the environmental protection budgets of single local government entities.

IMPORTS (ROWS 10–12, COLUMN 2)

A special entry is made for the value of imports of lumber and other forest resources, and petroleum and other subsoil resources that are closely related to the environmental problems in foreign countries.[5]

INTERMEDIATE CONSUMPTION

It is assumed that the environment-related production activities of industries (both external and internal) do not involve the intermediate consumption of environment-related goods and services (rows 3–8, columns 7–9: 0.0). As to the intermediate consumption by other production activities of industries (rows 3–8, column 10), the value of environment-related goods and services produced by external environmental protection activities is estimated by excluding total demand by all other activities from total supply. It should be noted that the output of internal environmental protection activities is assumed to be used up by the other production activities of industries as intermediate consumption. The value of environment-related goods and services produced by the government is computed by excluding the value of intermediate demand by public administration, schools and hospitals from the total value of intermediate demand for sewage and waste disposal (under public management), on the basis of the input–output tables.

The intermediate consumption of the government's production activities related to environmental protection (rows 3–8, column 12) is estimated by the cost of delegating sewage and waste disposal services to industries. The intermediate consumption of the government's other production activities (rows 3–8, column 13) is estimated by the provision of sewage, waste disposal (under public management), and waste disposal (industries) to the government on the basis of input–output tables.

FINAL CONSUMPTION EXPENDITURES (ROWS 3–8, COLUMNS 15–18)

For the central government, final consumption expenditures are estimated by excluding the capital formation budget and intermediate consumption from the environmental protection budget. For local governments, they are computed as the final consumption expenditures on sewage, waste disposal (under public management), and waste disposal (industries) from the input–output tables. For households, they are computed as the final consumption expenditures on sewage, waste disposal (under public management), and waste disposal (industries) from the input–output tables.

CAPITAL FORMATION

Of the goods and services related to environmental protection, fixed capital formation (rows 3–8, columns 22–25) is estimated from the survey of the Environmental Agency, Japan (1994). In order to distinguish between industries and the government, the ratio of private to public demand obtained from the research of Japan Association of Industrial Machinery Manufacturers (various years) is used.

Consumption of fixed capital related to environmental protection

The consumption of environment-related man-made assets is estimated by using the ratio of consumption of fixed capital to output in principal industries, obtained from the input–output tables, in the case of the external activities of industries (row 16, column 8). For the internal activities of industries (row 16, column 9), it is estimated by using the ratio of depreciation to the cumulative value of investment in environmental protection equipment during the previous 7 years in the electric power, water supply, and gas sector (8%). For the government (row 16, column 12), it is computed as the reserves for capital depreciation for sewage and waste disposal (under public management) from the input–output tables.

The cost structure of production activities related to environmental protection

EXTERNAL ENVIRONMENT-RELATED PRODUCTION ACTIVITIES OF INDUSTRIES

External environmental protection activities are assumed not to use goods and services produced by either the environment-related production activities of industries (both external and internal) or the production activities (related to environmental protection or otherwise) of the government as intermediate inputs (rows 3–8, 13 and 14, column 8: 0.0). The intermediate inputs from the other production activities of industries (row 10, column 8) are estimated from the input–output tables by using the ratio of intermediate inputs to output in the principal industries. The consumption of environment-related man-made assets (row 16, column 8) is estimated as explained above. Indirect taxes and subsidies, compensation of employees, and operating surplus (rows 29–34, column 8) are estimated from the input–output tables by using the ratios of the respective figures to output in the principal industries.

INTERNAL ENVIRONMENT-RELATED PRODUCTION ACTIVITIES OF INDUSTRIES

In this trial estimation, we estimated only the activities of industries that are implemented within establishments to prevent environmental pollution. Specifically, we used the value of investment in environmental protection equipment estimated in the *Survey of Investment in Environmental Pollution Prevention Equipment* (Ministry of International Trade and Industry, Japan, 1993). Estimates are made from the input–output tables using the figures for the electric power, water supply and gas sector.

Internal environmental protection is assumed not to use goods and services produced by either the external environment-related production activities of industries or the production activities of the government related to environmental protection or otherwise as intermediate inputs (rows 3–8, 13–14, column 9: 0.0).

The intermediate inputs from the other production activities of industries (row 10, column 9) are estimated by using the ratio of intermediate inputs to the cumulative value of investment in environmental protection equipment during the previous 7 years (11%). As stated earlier, we used the ratio of depreciation (8%) to estimate the consumption of man-made assets related to environmental protection, and the ratio of compensation of employees (7%) to estimate the compensation of employees. It should be noted that operating surplus is not computed (row 34, column 9: 0.0) and that the value of outputs is calculated as the sum of costs (rows 29–34, column 9).

The output of production activities as a whole (row 36, column 5), as well as the output of industries (row 36, column 6), are enlarged because of the externalization of the internal environmental protection activities of industries. However, the value added (both gross and net) is unchanged. The intermediate input of the other production activities of industries increases by the amount of the output of internal environmental protection activities (row 6, column 10) and decreases by the amount disaggregated as the intermediate input of the internal environmental protection activities (row 9, columns 9 and 10). The value added (gross and net) of the other production activities of industries declines by the amount disaggregated as the value added of the internal environmental protection activities (rows 15–17 and 29–33, columns 9 and 10), although the operating surplus remains unchanged (row 34, column 10).

PRODUCTION ACTIVITIES OF THE GOVERNMENT RELATED TO ENVIRONMENTAL PROTECTION (COLUMN 12)

For the central government, estimates are made on the basis of the environmental protection budget. For local governments, expenditures are estimated on the basis of the input–output tables for sewage and waste disposal (under public management) services.

Stocks and accumulation of environment-related man-made assets and other assets

MAN-MADE ASSETS (COLUMNS 23–25)

As stated earlier, fixed capital formation is estimated on the basis of the survey of the Environmental Agency (1994). At the same time, the opening and closing stocks of fixed capital are estimated by adjusting the figure for consumption of fixed capital associated with environment-related production activities of industries (both external and internal) and other elements.

CULTIVATED ASSETS (CULTIVATED FORESTS) (COLUMNS 28–30)

Cultivated assets (biological assets that can be artificially produced) are divided into cultivated forests and other assets in order to make explicit the burden on the environment from the felling of trees in excess of natural growth. Cultivated forests are estimated by multiplying the standing timber assets (the opening and closing stocks) in the SNA (asset accounts) by the ratio of the volume of cultivated forests to that of natural forests.

NON-PRODUCED NATURAL ASSETS (LAND) (COLUMNS 35–38)

In order to record the changes in use, land is divided into three categories: developed land, agricultural and wooded land, and conservation regions. Developed land refers to land underlying buildings in the SNA (asset accounts), such as residential land, commercial land, industrial land and so on. Agricultural and wooded land is estimated by adding the cultivated land to the timber tracts in the SNA and by excluding the cultivated forests and conservation regions. Conservation regions refer to natural parks and are estimated by using the ratio of their area to forests and wilderness.

Imputed environmental costs

The monetary valuation of the quantitative change (depletion) and qualitative change (degradation) associated with the economic use of non-produced natural assets permits the estimation of the environmental costs of economic activities. In the integrated environmental and economic accounts, environmental costs caused in economic activities are taken into account, but environmental costs borne by producers and consumers are not.[6]

Scope and coverage

DEGRADATION OF NATURAL ASSETS CAUSED BY RESIDUALS

Degradation includes the effects of environmental pollution on the quality of natural assets caused by economic activities (production and consumption). Data availability determined the coverage of environmental media and pollutants. National environmental standards specify SO_2, NO_2, CO, floating particles and photochemical oxidants as atmospheric pollutants. However, we estimated only SO_2 and NO_2, which are relatively easy to monitor (row 19, column 32). In addition to BOD and COD as overall measures of pollution, environmental standards specify hydrogen ion concentration, oxygen volume, intestinal bacilli, nitrogen and phosphorus as water pollutants. Only BOD and COD were estimated as readily available indicators (row 19, column 33).

DESTRUCTION OF ECOSYSTEMS

Depletion of cultivated forests by excess felling and the reduction of wooded land and conservation regions by land development cause a deterioration of the quality of natural assets. The imputed environmental costs of depleting timber resources, caused by the felling of trees in excess of natural growth (row 20, column 29) and land development (row 20, columns 35 to 38) were calculated as the costs for the destruction of ecosystems, instead of estimating directly the monetary value of the deterioration of the quality of natural assets.

DEPLETION OF NATURAL RESOURCES

The imputed environmental costs associated with the depletion of coal and other subsoil resources by extraction were estimated (row 21, column 39).

GLOBAL AND OTHER ENVIRONMENTAL EFFECTS

The system of integrated environmental and economic accounting includes rows for the effects on the global environment (e.g. global warming caused by the discharge of CO_2 and the destruction of the ozone layer by the discharge of freon gas). It also covers in principle other uses of natural assets such as the degradation of landscapes and other amenities. However, we did not estimate these effects and uses in this trial estimation.

Estimation methods

The maintenance cost valuation method, which is one of the methods used to estimate the imputed environmental costs, measures indirectly quantitative and qualitative changes in the environment associated with economic activities by estimating the required cost of maintaining the quantity and quality of the environment at a certain level. It is an estimation method consistent with the concept of sustainability. In view of data availability and the feasibility of estimation, the maintenance cost valuation method is used to estimate the imputed environmental costs by assuming certain activities that are necessary to maintain the sustainable quantitative and qualitative levels of the environment for the respective environmental media under investigation.

DEGRADATION OF NATURAL ASSETS CAUSED BY RESIDUALS

We estimated the required cost of removing the discharged residuals at the end-of-pipe in order to prevent the deterioration of the natural environment, while maintaining the current level of economic activities (production and consumption).

For air pollution, the sources of pollution are divided into stationary sources (e.g. factories) and mobile sources (e.g. automobiles). In the case of stationary sources, the cost of removal is set equal to the cost of operating air pollution prevention devices (the cost of depreciation, maintenance and operation). For SO_x, we estimated the volume of removal by desulphurization of crude petroleum as well as by exhaust desulphurization devices. For NO_x, we estimated the volume of removal by exhaust denitration devices. For each, the unit cost of removal is obtained by dividing the cost of removal by the volume of removal. The volume of discharge is estimated on the basis of existing reports (Yoshioka *et al.*, 1992; OECD, various years–1993). The imputed environmental cost is estimated by multiplying this by the unit cost of removal. For mobile sources, we assumed that the additional cost of placing exhaust devices in gasoline-operated vehicles was $US350 per vehicle g/km in 1981 (OECD, 1988). We used the same unit cost of removal for diesel and LPG-operated vehicles as for petrol-operated vehicles. The volume of discharge was estimated following Hayami (1992) and statistics of the Institute of Energy Economics, Japan (1993). The imputed environmental cost was estimated by multiplying total discharge by the unit cost of removal. It should be noted that the mobile sources are allocated to the final consumption expenditures of households in the case of family-owned vehicles and to the other production activities of industries in the case of other vehicles.

For water pollution, we estimated BOD and COD and used the larger of the two as a measure of the burden on the environment. Sources of pollution included household drainage, industrial drainage and livestock drainage. The method of estimating the imputed environmental cost is the same as that used for air pollution. Thus, we first estimated for each source the volume of polluting materials, the volume of removal and the volume of discharge. We then estimated the cost of removal as the cost of operating environmental protection devices. The unit cost of removal is obtained by dividing the cost of removal by the volume of removal. The imputed environmental cost is obtained by multiplying the unit cost with the volume of discharge (row 19, column 33).

For household drainage, we used the research results of Kunimatsu and Muraoka (1990) and of the Japan Association of Industrial Machinery Manufacturers (1993). We thus obtained for the entire population the volume of polluting materials and the ratio of removal by environmental protection devices. For industrial drainage, the estimates are made on the basis of the volume of polluting materials per shipment of products by type of industry (Ministry of Construction, Japan, 1994) and the rate of removal (Kunimatsu and Muraoka, 1990). For livestock drainage, we also used the volume of polluting materials and the rate of removal by natural purification found in existing research results (Kunimatsu and Muraoka, 1990). The cost of operating environmental protection devices is the combined cost of depreciation, maintenance and operation.

Destruction of ecosystems

The destruction of ecosystems refers to the depletion of timber resources associated with the felling of trees in excess of natural growth and with the reduction of wooded land caused by land development. Here, the imputed environmental cost (row 20, column 29) is set equal to the cost associated with the maintenance activities in the form of the reduction of, or abstention from, production activities.

In the case of the felling of cultivated forests, the output obtained from the felling of trees in excess of natural growth (excess felling) is considered as the value added forgone of the overall economy. This value is thus retained as the imputed environmental cost. It is therefore assumed that the volume of forests is to be maintained as found at the opening level by reducing the number of cultivated trees being felled to sustainable levels. We used research results of the Ministry of Agriculture, Forestry and Fisheries (1994, 1978 and 1993) to obtain the growth of cultivated forests and the amount of felling. The output of timber products is obtained from the input–output tables.

For land development, the estimation of environmental cost (row 20, columns 35 and 38) covered the conversion of agricultural and wooded land to developed land (residential, commercial and industrial land and so on), the change in the type of use within agricultural and wooded land, and the reduction of conservation regions. The output obtained from the associated development of land is considered as the value added forgone of the overall economy, hence as the imputed environmental cost. This assumes that the original use of land is to be maintained by abandoning the conversion of developed land, agricultural and wooded land, and conservation regions to other uses. We used the research results of the National Land Agency (1990 and 1993) to obtain the conversion of land use, and the input–output tables to estimate the output of land use.

DEPLETION OF NATURAL RESOURCES

We estimated the imputed environmental cost associated with the depletion of subsoil resources by extraction (row 21, column 39). The estimation covered coal, zinc and lime stone, applying the user cost method. The results of this method indicate that the cost estimates vary considerably from year to year because they depend on the annual change in volume of production (see below).

Main results

The system of integrated environmental and economic accounting was implemented on a trial basis for 1985 and 1990. The following shows the estimates for 1990 and compares the main results with 1985.

Environment-related disaggregation of the SNA and the externalization of internal environmental protection activities (results for 1990)

Supply and use of environment-related goods and services (Table 2)

Total output in 1990 was 877.7 trillion yen, including that of internal environmental protection activities of industries. The output of environment-related goods and services was 7.2 trillion yen for industries and 2.6 trillion yen for the government, for a total of 9.8 trillion yen, or 1.1% of total output. A total of 4.4 trillion yen of the produced environment-related goods and services was used in intermediate consumption of industries, and 0.7 trillion yen in the intermediate consumption of government. The total of 5.1 trillion yen of environment-related goods and services, used in intermediate consumption, amounted to 1.2% of total intermediate consumption (428.8 trillion yen). Final consumption expenditures for environment-related goods and services were 1.8 trillion yen for the government and 0.5 trillion yen for households, i.e. 0.8% of total final consumption expenditures (279.3 trillion yen).

Table 2. Economic activities and environmental protection (1990; trillion yen) (%).

Output	877.7
environment-related	9.8
(share)	(1.1)
Industries	7.2
External	6.6
Externalized internal	0.6
Government	2.6
Imports	44.5

Intermediate consumption	428.8
environment-related	5.1
(share)	(1.2)
Industries	4.4
Government	0.7
Final consumption expenditures	279.3
environment-related	2.3
(share)	(0.8)
Government	1.8
Households	0.5
Fixed capital formation	139.1
Man-made assets	134.3
environment-related	2.6
(share)	(2.0)
Industries	0.5
Government	2.1
Cultivated assets	0.2
Non-produced natural assets	4.5
Exports	47.2

Gross Domestic Product(GDP)	424.5
environment-related	5.8
(share)	(1.4)
Industries	4.3
Government	1.5
Consumption of fixed capital	62.8
environment-related	1.1
(share)	(1.7)
Industries	0.5
Government	0.5
Net Domestic Product(NDP)	363.7
environment-related	4.8
(share)	(1.3)
Industries	3.8
Government	1.0

Capital formation related to environmental protection was 0.5 trillion yen for industries and 2.1 trillion yen for the government, i.e. 2.0% of the total capital formation of man-made assets (134.3 trillion yen).

The cost structure of environment-related production activities (Table 2)

The total output of production activities of industries related to environmental protection was 6.6 trillion yen for external activities and 0.6 trillion yen for internal activities (total of 7.2 trillion yen). Intermediate inputs were 2.9 trillion yen, while gross product (gross value added) was 4.3 trillion yen. By subtracting consumption of fixed capital (0.5 trillion yen), net product (net value added) comes to 3.8 trillion yen. Compared with the aggregate figures for industries, environment-related goods and services accounted for 0.9% of total output (817.5 trillion yen), 0.7% of intermediate inputs (410.1 trillion yen), 1.1% of gross product (407.3 trillion yen), and 1.1% of net product (348.0 trillion yen).

The output of environment-related production activities of the government was 2.6 trillion yen, intermediate inputs were 1.1 trillion yen, gross product was 1.5 trillion yen, and net product was 1.0 trillion yen. Compared with the aggregate figures for government, environment-related goods and services accounted for 5.5% of total output (46.5 trillion yen), 7.9% of intermediate inputs (13.8 trillion yen), 4.5% of gross product (32.7 trillion yen) and 3.2% of net product (30.2 trillion yen). The gross product of environment-related production activities of industries and the government was 5.8 trillion yen, corresponding to 1.4% of GDP (424.5 trillion yen). Net product was 4.8 trillion yen, equal to 1.3% of NDP (363.7 trillion yen).

The stocks and accumulation of environment-related man-made assets (Table 3)

The opening stock of environment-related assets was 21.4 trillion yen, equal to 2.2% of all man-made assets (959.3 trillion yen). Although 2.6 trillion yen of capital formation was made in 1990, the closing stock was 23.3 trillion yen because of the consumption of fixed capital of 1.1 trillion yen, as well as the adjustment for revaluation due to market price changes during the year. This was 2.2% of all man-made assets (1046.4 trillion yen).

Table 3. Assets and asset accumulation (1990; trillion yen) (%).

	Opening stocks	Gross capital formation	(less) Consumption of fixed capital	(less) Imputed environmental costs	Adjustments relating to accumulation	Other adjustments	Closing stocks
Man-made assets	959.3	134.3	62.6	-	-	15.4	1,046.4
environment-related	21.4	2.6	1.1	-	-	0.4	23.3
(share)	(2.2)	(2.0)	(1.7)				(2.2)
Industries	0.5	0.5	0.5	-	-	0.0	0.5
Government	20.8	2.1	0.5	-	-	0.3	22.8
Cultivated assets	21.3	0.2	0.3	-	0.9	1.0	23.1
Cultivated forests	19.2	-	-	-	0.2	1.0	20.4
Non-produced natural assets	2,185.2	4.5	-	8.5	8.3	232.7	2,422.2
Air	-	-	-	6.7	6.7	-	-
Water	-	-	-	0.2	0.2	-	-
Soil	-	-	-	0.0	0.0	-	-
Land use	2,184.4	4.4	-	1.4	1.4	232.7	2,421.4
Developed land	1,828.7	1.3	-	1.0	25.6	176.4	2,031.0
Agricultural & wooded land	346.8	3.0	-	0.3	-24.3	55.8	380.8
Conservation regions	8.9	0.2	-	0.1	0.1	0.5	9.6
Subsoil resources	0.9	0.1	-	0.2	0.0	0.0	0.8

Imputed environmental costs and EDP (results for 1990)

Tables 4 and 5 present the estimates of imputed environmental cost and EDP. The total imputed environmental cost in 1990 was 8.4 trillion yen, and the ratio to GDP was 2.0% (or 2.3% in terms of NDP). Of this, 6.4 trillion yen (75.7%) was attributed to the production activities of industries and 2.1 trillion yen (24.4%) to the consumption activities of households.

By type of natural assets affected, 6.7 trillion yen (79.0%) relates to air pollution, 0.2 trillion yen (2.3%) to water pollution, 1.4 trillion yen (16.3%) to the

Table 4. **Environment-related external diseconomies (1990; trillion yen) (%).**

	[External diseconomies due to production activities]	[External diseconomies due to consumption activities]	[External diseconomies]
Imputed environmental costs	6.4	2.1	8.4
(ratio to GDP)			(2.0)
(ratio to NDP)			(2.3)
Degradation of natural assets caused by residuals	4.8	2.1	6.9
Air pollution	4.7	2.0	6.7
Water pollution	0.2	0.0	0.2
Destruction of ecosystems	1.4	-	1.4
Depletion of resources	0.2	0.0	0.2

Table 5. **GDP, NDP and EDP (1990) (trillion yen).**

Gross Domestic Product(GDP)	424.5
Net Domestic Product(NDP)	363.7
Consumption of Man-made and Natural assets:	
Consumption of fixed capital	62.8
Imputed environmental costs	8.5
Eco Domestic Product (EDP)	355.3

Note:Gross Domestic Product (GDP) and Net Domestic Product (NDP) do not exactly correspond to each other because of statistical discrepancies.

destruction of ecosystems and 0.2 trillion yen (2.4%) to the depletion of resources. Automobiles were the major source of air pollution (4.6 trillion yen was allocated to the production activities of industries, and 2.0 trillion yen to the consumption activities of households). Subtracting the imputed environmental cost (8.4 trillion yen) from NDP (363.7 trillion yen), gives an EDP of 355.3 trillion yen.

Comparison of results for 1985 and 1990

Environment-related disaggregation and externalization of environmental protection activities (Table 6)

The output of environment-related goods and services increased at an annual rate of 1.9% (annual average, unless otherwise indicated) during the 5-year period from 1985 to 1990, slower than the rate of growth of total output of 5.4%. As a result, the share of environment-related goods and services in total output declined from 1.3 to 1.1%. It should be noted that output increased by 1.4% for the external environment-related production activities of industries, decreased by 2.9% for the internal environment-related production activities of industries, and increased by 4.5% for the environment-related production activities of the government. As mentioned above, the output of internal environment-related production activities of industries was estimated by the sum of costs based on the cumulative value of investments in environmental

Table 6. Environmental protection (1985 and 1990; trillion yen) (%).

	1985	1990	Rate of growth per year
Output	675.0	877.7	5.4
environment-related	8.9	9.8	1.9
(share)	(1.3)	(1.1)	
Industries	6.8	7.2	1.0
External	6.2	6.6	1.4
Externalized internal	0.7	0.6	- 2.9
Government	2.1	2.6	4.5
Gross Domestic Product (GDP)	320.4	424.5	5.8
environment-related	5.1	5.8	2.6
(share)	(1.6)	(1.4)	
Industries	4.0	4.3	1.8
Government	1.1	1.5	5.4
Man-made assets (closing stocks)	752.8	1,046.4	6.8
environment-related (closing stocks)	14.4	23.3	10.1
(share)	(1.9)	(2.2)	

protection equipment during the previous 7 years. Thus, the fall in estimated output reflects the decline in the value of investment in environmental protection equipment.

While GDP rose by 5.8%, the gross product of goods and services related to environmental protection increased by only 2.6%. As a result, their share in GDP declined from 1.6% to 1.4%. The share of environment-related assets in the value of all produced assets rose from 1.9% to 2.2%.

Imputed environmental costs and growth of EDP (Table 7)

The imputed environmental cost increased by only 1.2%, from 8.0 trillion yen in 1985 to 8.4 trillion yen in 1990. Because the rate of increase was considerably slower than the rate of growth of GDP and NDP, the ratio to GDP declined from 2.5% to 2.0% (from 2.9% to 2.3% in terms of NDP). The cost attributable to the production activities of industries increased by 0.3%, while the cost attributable to the consumption activities of households increased by 4.3%. The imputed environmental costs, particularly those attributable to the production activities of industries, increased slowly because the costs associated with the depletion of resources declined considerably, from 1.3 trillion yen in 1985 to 0.2 trillion yen in 1990. If the cost associated with the depletion of resources is excluded, the total imputed environmental cost increased by 4.4%, with the cost attributable to the production activities of industries rising at

Table 7. Imputed environmental cost and EDP growth (1985 and 1990; trillion yen) (%).

	1985	1990	Rate of growth per year
Imputed environmental costs	8.0	8.4	1.2
(ratio to GDP)	(2.5)	(2.0)	
(ratio to NDP)	(2.9)	(2.3)	
By source Industries	6.3	6.4	0.3
Households	1.7	2.1	4.3
Imputed environmental costs (excluding cost of depletion of resources)	6.6	8.2	4.4
By source Industries	5.0	6.2	4.3
Households	1.6	2.1	4.9
By type Air pollution	5.0	6.7	5.8
Water pollution	0.3	0.2	- 5.9
Destruction of ecosystems	1.3	1.4	0.4
Depletion of resources	1.3	0.2	- 31.2
Environmentally adjusted domestic product (EDP)	268.7	355.3	5.7
(reference) Net Domestic Product (NDP)	276.6	363.7	5.6

4.3% and that attributable to the consumption activities of households at 4.9%.

The cost resulting from air pollution increased by 5.8%, water pollution cost declined by 5.9%, the cost associated with the destruction of ecosystems increased by 0.4% and the depletion cost of resources declined by 31.2%. The imputed environmental cost associated with the depletion of resources declined thus substantially (from 1.3 trillion yen in 1985 to 0.2 trillion yen in 1990). This reflects the fact that the extraction of coal declined by about one half over the same period, so that the expected exploitable period nearly doubled. This increase in the expected exploitable period caused the user cost to decline substantially. The cost of zinc in 1990 declined to about 60% of its 1985 level, while that of limestone increased by about seven times.

EDP increased at a rate of 5.7%, 0.1% more than the rate of growth of NDP.

Outlook

For Japan, this trial estimation was the first comprehensive attempt at implementing integrated environmental and economic accounts. It succeeded in establishing the basic accounts. However, improvements need to be made in the future regarding coverage, estimation methodologies and the structure of the accounts. The main remaining issues in this regard are discussed below.

Environment-related disaggregation of the SNA and the externalization of internal environmental protection activities

The following categories and issues are suggested for future discussion and development:

External environmental protection activities

In this trial estimation, output was estimated on the basis of the data provided by the Environmental Agency, Japan's *Survey Concerning the Quantitative Analysis of Eco-businesses* (1994). Sewage and waste disposal services and the production of pollution prevention devices have a relatively long history as an industry. Therefore, it is possible to obtain additional statistical information from input–output tables, budget documents and statistical reports of industry associations. However, it is difficult to define or specify such emerging industries as energy-saving systems and environmental purification services, rendering it impossible to obtain additional information. A clear definition of environment-related goods and services needs to be developed, and a more comprehensive survey conducted in order to implement classification by commodities and industries as suggested in the SNA/SEEA.

Internal environmental protection activities

Output and costs of internal activities were estimated by applying ratios of intermediate inputs and other costs to the cumulative value of investment in environmental protection equipment. Because it is difficult to obtain adequate information on recycling and internal activities other than pollution prevention activities, the long-term task for the future will be to conduct a survey on these items, based on a clearer definition of environmental protection activities.

Environmental protection activities of government

The survey of the environmental protection budgets of local governments covers only the prefectures and designated cities. For other municipalities, this trial estimation includes only the sewage and waste disposal services. There is a need to expand the coverage of statistical data.

Coverage of financial activities

The present accounts are designed to record only the production of goods and services and other non-financial activities. They do not record the environment-related loans of government financial institutions. In order to understand the environmental protection measures taken by the government on a wider basis, it is necessary to have a structure of accounts that covers not only the policy measures implemented through the production of environment-related goods and services and subsidies, but also their financial implications.

Valuation at constant prices

The output of internal environmental protection activities of industries was estimated on the basis of the cumulative value of investment in environmental protection equipment. For this reason, the estimated value of output declined from 1985 to 1990 because the amount of investment fell from the period 7 years prior to 1985 to the period 7 years prior to 1990. There is thus a problem with the estimation methodology. Moreover, because of the wider availability of environmental protection equipment, prices might have been falling. Expressing the environment-related disaggregation of the SNA, including internal environmental protection activities, at constant prices, is therefore an issue worth further exploration.

Imputed environmental costs

Expanding coverage

In the case of Japan, the share of mining in economic activity is so small that the estimation of the depletion cost of domestic subsoil resources has little significance. More significant are environmental issues of pollution and the destruction of ecosystems associated with land development. It is necessary, therefore, to expand the coverage of imputed environmental costs to include:

(a) air pollutants other than NO_x and SO_x (e.g. CO, floating particles);
(b) water pollutants other than BOD and COD;
(c) pollution issues other than air and water pollution (e.g. soil pollution, noise pollution, vibration, foul odour and land subsidence).

Global warming and the destruction of the ozone layer, which were not included in this trial estimation, are environmental problems of global proportion beyond the control of individual countries. Responding to them effectively is an extremely important contemporary challenge. The types of maintenance activities to be assumed in the estimation of the imputed environmental cost of global phenomena and the possibility of gathering data corresponding to assumed maintenance activities are issues for further investigation.

Landscapes and other amenities may be interpreted as pleasant environments that bring peace and comfort. However, their definition (in accounting terms) and coverage need to be further examined. It will also be necessary in the future to define and measure in physical and monetary terms, the environmental protection functions of forests, farms and other natural assets.

Improvement of estimation methodologies

We estimated the imputed environmental costs by the following methods, according to the type of natural assets:

(a) degradation of natural assets caused by residuals: residuals valued at cost of removal at the end-of-pipe while maintaining the level of economic activities;
(b) destruction of ecosystems: reduction of, or abstention from, production activities;
(c) depletion of resources: the user cost method.

In addition to these measures, the United Nations handbook shows the following measures for maintenance costing: the substitution of inputs with the same level of outputs from economic activities; the substitution of products from economic activities; and the restoration of deteriorated natural assets to their original state. The handbook suggests that the least-cost measure (best available technology) should be chosen from the different feasible measures. In the future, it will be necessary to consider other estimation methodologies based on other measures.

It should be noted that the imputed environmental cost associated with the discharge of residuals was estimated first, by obtaining the unit cost of removal by dividing the cost of removal by the amount of removal and then, by multiplying the unit cost by the volume of discharge. However, when additional residuals are actually removed, the marginal cost of removal is expected to increase. Thus, instead of using the valuation methodology based on average cost, as we did in this trial estimation, it will be necessary to examine possibilities of marginal costing.

Overseas environmental problems associated with Japan's imports

In this trial estimation, the imputed environmental cost associated with the depletion of resources declined substantially from 1985 to 1990 because of the greatly decreased extraction of coal. Because of the reduced amount of extraction, the expected exploitable period of domestic coal resources increased, hence lengthening the period of 'sustainability'. As a result, there was a reduction in the imputed environmental cost. However, the decreased extraction was not brought about only by the substitution of inputs or products associated with technical innovation or the improvement in energy efficiency. It was mainly a result of the weak international competitiveness of the domestic coal industry. As a consequence, the source of energy shifted from domestic production to coal and petroleum imports. From an environmental point of view, the imputed environmental cost was transferred from the domestic economy to foreign economies.

The imputed environmental cost is supposed to be connected with the economic activities that are the direct cause of the depletion of resources (in this case, production activities in the form of extraction of resources) and is not supposed to be connected with the economic activities that are the ultimate cause (in this case, consumption activities of goods and services that are produced with imported petroleum and coal). The imputed environmental cost of the depletion of resources associated with the extraction of imported petroleum and coal was therefore not imputed to the Japanese economy. However, one should be aware of the fact that the imputed environmental cost of the Japanese economy declined considerably because of the shift of supplies from domestic to foreign sources, when there was little change in the dependence of the Japanese economy on petroleum and coal. Environmental accounts should, therefore, be supplemented with further analyses of environmental problems in foreign countries, caused by Japan's imports.

Valuation at constant prices

The imputed environmental costs of water pollution declined from 258.1 billion yen in 1985 to 190.7 billion yen in 1990, a decline of 5.9%. These costs can be divided into the unit cost of removal and the volume of discharge as shown in Table 8. The imputed cost declined despite an increase in the volume of discharge, because the unit cost of removal declined considerably. The fall in the price of environmental protection equipment, owing to its wider availability, improvement in its capacity, and the economy of scale associated with the large fixed cost inherent in the sewage industry, is thus a major factor in the fall of the unit removal cost. In order to prevent these changes in the unit cost of removal from affecting the imputed environmental cost, we should consider expressing the imputed cost at constant prices.

Table 8. Environmental cost of water pollution (1985 and 1990)

	1985	1990	Average annual rate of growth (%)
Unit cost of removal (million yen per 1000 tons)	47.49	31.40	−7.9
Volume of discharge (in thousands of tons)	5435	6074	2.3
Imputed environmental cost (in billions of yen)	258.1	190.7	−5.9

Preparation of long-term time series

Because the main purpose of this trial estimation was to prepare the basic structure of integrated environmental and economic accounting for Japan, the estimates were made only for 1985 and 1990. It has turned out that there was little noticeable change in actual environmental protection expenditures or in the imputed environmental cost over the 5-year period. If it is possible to extend the estimates back into the 1960s, we might obtain results that show a considerable increase in environmental protection measures, along with a progressive worsening of environmental pollution. In order to evaluate the effects of environmental protection measures, it is therefore necessary to prepare long-term time series, along with the expansion of the scope of estimation and the improvement of the estimation methodologies.

Notes

1. The Department of National Accounts (DNA) of the Economic Research Institute of the Economic Planning Agency is in charge of compiling national economic accounts. The DNA is also given responsibility for the development of integrated environmental and economic accounting.
2. The DNA has been conducting research and development while consulting with a special group of scholars and experts on environmental accounting since 1991. In 1995, the DNA established the Council for Research on SNA (CRSNA), composed of scholars and experts, in order to examine the 1993 SNA. The CRSNA has seven committees, one of which is in charge of examining integrated environmental accounting. The DNA plans to assess the trial estimates in consultation with the committee for the recurrent compilation of environmental accounts.
3. The integrated environmental and economic accounts are compiled for the economy as a whole and do not at present include accounts broken down by industry or region. It is desirable to develop environmental accounts by region, since some environmental problems, such as water pollution and the excess felling of trees, have regional characteristics. We see the development of environmental accounts by region as a long-term task for the future, to be examined after the improvement of environmental accounts for the economy as a whole.
4. For air, water and soil, the accounts are restricted to data on degradation (i.e. flows) only.
5. With regard to cross-border issues, exports of environment-related goods are separately identified, but imports are not, due to lack of data. Cross-border environmental impacts, such as the import and export of residuals, are not taken into account when estimating imputed environmental costs due to the lack of data. The issues involved in the estimation of imputed environmental costs associated with Japan's import of resources are discussed in the 'Outlook' section.
6. See for a discussion of the different valuations proposed in the SEEA and the related concepts of cost caused and borne, Chapter 16.

4
Republic of Korea: SEEA pilot compilation

SEUNG-WOO KIM, JAN VAN TONGEREN AND
ALESSANDRA ALFIERI

Orientation and scope of environmental accounting in Korea

Following the success of an export-led growth policy during the 1970s and 1980s, the Korean environment began to face significant deterioration in the 1990s, due to a combination of rapid industrialization, population growth and urbanization. As in the past Korea had to pursue its economic development without a sufficient endowment of natural resources and technological accumulation, environmental concerns were not considered a priority in economic policy.

This chapter addresses the need to provide data to support policies which integrate governmental economic policies with environmental conservation. Based on international experiences and recommendations, the paper presents proposals for an environmental accounts framework which extends the present national (economic) accounts of Korea to reflect natural resource depletion and environmental degradation not valued in the market. The framework is based on the System of integrated Environmental and Economic Accounts (SEEA; United Nations, 1993a). It can be used to analyse the interactions between the environment and the economy (van Tongeren *et al.*, 1994). Using the proposed framework of accounts, a pilot compilation of environmental accounts for Korea was carried out for the period 1985–1992. The project was funded by UNDP, implemented by the Korea Environmental Technology Research Institute (KETRI), and technically supported by the United Nations Statistics Division (UNSD) and the United Nations Department of Development Support and Management Service (DDSMS).

The accounts deal with economic causes of air and water pollution and the environmental protection activities required in response to these concerns. Also dealt with is the depletion of the stock of natural forests, selected minerals and fish; the latter analysis is extended beyond the boundaries of Korea by analysing the effects of economic activities in Korea on depletable natural assets in other countries. Land use issues, showing how forest and other non-economic land is absorbed by agricultural, built-up and recreational land, as a consequence of economic development of Korea, are also discussed.

The views expressed here are those of the authors and not necessarily those of their respective organizations.

K. Uno and P. Bartelmus (eds.), Environmental Accounting in Theory and Practice, 63–76.

The framework includes selected national accounts data in monetary terms compiled by the Bank of Korea (1994), data in physical terms and imputed values of flows and stocks of natural assets, natural resource depletion and environmental degradation by industries. Data on environmental protection expenditures, implicitly included in the SNA aggregates, are presented separately as 'of which' elements.

This chapter presents a description of the accounting framework, the pilot compilation of environmental accounts based thereon, a preliminary analysis of the results of the compilation and suggestions for further work.

The accounting framework and its concepts

The accounting framework of the Korean environmental accounts, as for the SEEA, is the supply and use table of the 1993 System of National Accounts (SNA; Commission of the European Communities *et al.*, 1993). This table includes an integrated presentation of the supply and use and value added elements of the SNA. Supply includes output of industries and imports, and use covers intermediate consumption by industries, final consumption by households and government, exports and gross capital formation. Value added is presented in this framework as the difference between output and intermediate consumption of industries. To this central SNA framework, environmental accounts elements are added to obtain the SEEA framework presented in Figure 1 of Chapter 2. This chapter should be consulted for further details on the SEEA. Environmental accounts elements include data on the following:

(a) expenditures on environmental protection are made explicit as 'of which' items. They are reflected in intermediate consumption, capital formation and value added of industries, as well as in final consumption of households and the government;
(b) non-market uses of natural resources by industries and households, including depletion of minerals, as well as emissions, discharges or disposal of air and water pollutants and solid wastes by industries and households;
(c) output of industries and imports causing depletion of natural resources in the country and outside its borders are identified as an 'of which' category;
(d) asset accounts of non-produced natural assets including forests, minerals, fish, land, air and water.

The asset accounts for minerals, forests and fish include information on opening and closing stocks for each year and changes therein due to direct use, other economic decision or natural causes and changes in prices. Data on depletion, that is the value of the use of the resource above the sustainable level for renewable resources (forest and fish) and of the total use/extraction for non-renewable resources, are compiled. For water and air, only the cost of degradation is estimated. For land, the only imputed cost is that of degradation estimated as the cost of waste collection and treatment. All data are first compiled in physical terms and thereafter converted to monetary values, using unit values (see below). As there may be changes over time in the prices or nominal values of the assets, the asset accounts also include data on revaluation of the stock of natural assets.

The large data set covering the period 1985–92 consists of the central supply and use framework extended with environmental elements. It is supplemented with worksheets described in the next section. Two types of worksheets may be distinguished. The first contains economic data that are relevant to environmental analysis and are already included implicitly in the national accounts aggregates. These data refer to environmental protection expenditures, environmental charges and subsidies, and also to data on extraction products produced in Korea or imported. The second group of worksheets includes environmental data that are not covered by the national accounts, such as data on depletion, degradation and most data on the stocks and flows of asset accounts of non-produced assets. These data are generally specified in physical units and valued by applying unit net prices for depletion and unit 'best available technology' cost (current and levelized capital cost) for degradation. The latter method consists in estimating the imputed cost of treating the emissions, using the best available technology (e.g. construction and current expenses of waste and wastewater treatment plants, cost of scrubbers, filters, catalytic converters etc.).

Data, estimates and valuation

The 1985–1992 period allows the description of trends in economic development, environmental degradation and resource depletion. The Republic of Korea has well developed national accounts and economic statistics. However, data on environment and natural resources need to be further developed in order to improve the accounts and their analysis. The following describes the data used for the compilation of different elements of the SEEA framework.

Environmental protection expenses and environmental charges

As national accounts data, compiled annually by the Bank of Korea, do not provide enough information to identify environmental protection products and services in the economy, other data sources were used. *The Report on Mining and Manufacturing Survey* (National Statistical Office, Korea, 1987–1994) and *The Report on the Construction Work Survey* (National Statistical Office, Korea, 1986–1993b) were used to identify the supply and expenditure for environmental protection products and services for industries. For government and household expenditures, the *Government Revenue and Expenditures* (National Statistical Office, Korea, 1986–1993a) and *The National Survey of Family Income and Expenditures* (Ministry of Finance, Korea, 1985–1992) reports were the main data sources. Environmental protection expenditures are identified by industry (Korean Standard Industrial Classification, KSIC), degradation/depletion categories and environmental media affected.

Different data sources, in particular *The Report on the Construction Work Survey, Korea Environmental Yearbook* (Ministry of the Environment, Korea, 1989–1994) and *Survey on Environmental Protection Expenditure by Manufacturers* (only for 1991 and 1992) (Ministry of the Environment, Korea, 1992) were used to identify capital formation for environmental protection by industries. Data on government investments in environmental protection equipment were estimated, based on the amount of grants and revenue shares allocated by the central government to local governments (six major cities and nine provinces).

Environmental charges include mainly plastic waste disposal charges. Environmental subsidies cover tax reduction and exemptions for industries installing pollution abatement and control facilities. Data are obtained from the *Korea Environmental Yearbook.*

Data on environmental protection are included in the following worksheets:

(a) supply and use table with economic data, based on Bank of Korea (1994) national accounts;
(b) output and intermediate consumption of environmental protection activities for manufacturing, construction, and community, social and personal services are obtained from several sources, notably *The Report on Mining and Manufacturing Survey, The Report on the Construction Work Survey* and *Korea Environmental Yearbook*;
(c) capital formation on environmental protection equipment by industry (user) and environmental subject area (air, water, wastes, land, noise);
(d) government gross fixed capital formation and current expenditures on environmental protection by environmental subject area;
(e) internal environmental protection expenditures by industry and environmental subject area (installation and clean-up of scrubbers, maintenance of waste and wastewater treatment plants etc.);
(f) environmental protection expenditures of households by type (water purifier, waste treatment fee, sewage fee, mineral water, septic tank, air cleaner, three-way converter etc.);
(g) emission charges by industry;
(h) export and import of environmental protection products;
(i) tax reduction for installing pollution abatement and control facilities;
(j) exports and imports of extraction products by type (minerals, forestry and fishery products).

Asset accounts for produced assets

Produced asset accounts are compiled using the *National Wealth Survey of Korea* which is carried out every 10 years by the National Statistical Office

(1989). The last survey took place in 1987. For the other years, Phyo *et al.* (1993) estimated fixed capital stock using the Perpetual Inventory Method and the Polynomial Benchmark Year Method. However, their estimates did not include fixed assets for large animals and plants which were added, using data published in the *Statistical Yearbook of Agriculture, Forestry and Fisheries* (Ministry of Agriculture, Forestry, and Fisheries, Korea, various years).

Asset accounts for non-produced assets

The non-produced asset accounts include land, mineral resources, fish, forest, water and air. Due to lack of data on aquifers and other groundwater, depletion of water resources is not included. However, degradation of water in terms of emissions of wastewater is estimated. Asset accounts for forest and fish are compiled in physical and in monetary terms. No cost for the depletion of the assets is included. The extensive programme of reforestation and protection of forest by the government and efficient forest management are the reasons why sustainable yield has not been exceeded and therefore no depletion has taken place. For fish, data on biomass, essential for the calculation of maximum sustainable yield (MSY), are available only for one species, yellow corvina. Marine fish stocks are known to be decreasing as the catches per unit efforts (CPUE) have generally decreased since the 1970s. Due to data limitation, no cost of depletion is allocated to the fishing industry.

Stock data for the compilation of physical and monetary asset accounts for non-produced natural assets are obtained from various statistical yearbooks and national publications. However, few data are available on flows beyond direct use (harvest/extraction), i.e. natural growth of biota, catastrophic losses due to natural and economic causes, discoveries etc.

Data on asset accounts for non-produced assets, included in separate worksheets, refer to the following:

(a) asset accounts for land, by type of land area (agricultural land, forest land, built-up land, other), in physical and monetary terms and also including land prices by type of land area (paddy field, dry field, housing site, commercial site, forest land, manufacturing site, other);
(b) monetary asset accounts for forest resources by species (conifer forest, deciduous forest, mixed forest), in physical and monetary terms and including net price data;
(c) asset accounts for fish resources, in physical and monetary terms, and including net price data;
(d) asset account for mineral resources by type of mineral (coal, iron ore, limestone, tungsten ore, copper ore, kaolin, gold, silver), in physical and monetary terms, and including net price and user cost data with different discount rates.

Emissions

The accounts for environmental degradation cover emissions in physical and monetary terms into air, water and land by industry. Emissions into air include CO, NO_x, SO_2 and TSP, into water the amount of BOD discharged, and into land the amount of wastes discharged. The cost of emissions is calculated using the maintenance cost in the form of 'best available technology' for treating the emissions. Emissions into air are distinguished by emissions from stationary sources (e.g. heating, electricity generation etc.) and mobile sources (vehicles) to take into account different treatment costs. In the first case, the costs of flue gas desulphurization (FGD) and of the use of selective catalytic systems (SCR) are estimated. For mobile sources, the cost of the three-way catalytic converters for gasoline vehicles and the electric heaters for diesel vehicles is compiled. BOD emissions are valued, using the levelized annual cost of construction of an activated sludge treatment plant of 20,000–100,000 tons/day capacity with 15 years of life expectancy and a discount rate of 10%, and adding the current costs. The cost of discharging waste is estimated on the basis of landfill costs.

Data on emissions are obtained from *The Annual Survey Report on the Industrial Waste Water Discharge* (Ministry of the Environment, Korea, 1986–1992, 1994) for water, *The Report on Energy Census* (Korea Energy Economics Institute, 1987, 1990, 1992) for air and *Korea Environmental Yearbook* for waste discharge.

Data on emissions are recorded in the following worksheets:

(a) amounts of general wastes emitted by type (briquette waste, other) and treatment (untreated, landfill), treatment cost per unit (alternatively based on landfill and incineration cost) and total social cost;
(b) air pollutants emitted by type (CO_2, NO_x, SO_2, TCP, HC), by industry and (in the transport sector) by type of emitting unit (passenger cars, jeeps, bus, truck, special equipment vehicles), in physical and monetary terms, also including the avoidance cost per unit of pollutant;
(c) environmental 'social' cost of emitting domestic and industrial waste water, in physical terms and monetary terms.

Analysis

A structural analysis of the data is presented in four tables. The tables emphasize differences between traditional analyses, based on economic data, and similar analyses carried out on environmentally adjusted aggregates. As the data base covers a 7-year period, it is possible to show how these structural analyses evolve over time. For the purpose of this paper only two years are compared (1986 and 1992) although a much larger data base is available. No effort is made to carry out growth analyses, as past experience has shown little difference between the growth rates of economic aggregates and environmentally adjusted aggregates.

Table 1. Economic vs. economic–environmental analysis of main aggregates, 1992.

Economic analysis		Economic transactions related to natural assets as % of NDP						Economic-environmental analysis	
		Depletion			Degradation				
Concepts measured in economic analysis	(as % of NDP)	Forests	Minerals	Fish	Land	Air	Water	(as % of EDP)	Concepts measured in integrated economic-environmental analysis
NDP	100.00%							97.38%	**EDP (as % of NDP)**
of which:									
Environmental charges-subsidies	-0.05%				0.02%	0.00%	0.05%		
Value added of environmental protection industries	0.95%				0.32%	0.27%	0.36%		
Intermediate consumption/use of environmental protection products	1.74%				0.38%	0.66%	0.70%		
Use of natural assets (depletion & degradation) by industries	2.01%		0.04%		0.00%	1.96%	0.01%		
Final consumption of households	59.66%							61.88%	**Final consumption of households, adjusted for degradation impacts**
of which:									
Final consumption by households of environmental protection products	0.20%				0.04%	0.05%	0.11%		
Use of natural resources (degradation)	0.60%				0.01%	0.27%	0.33%		
Final consumption of government	12.01%							12.33%	**Final consumption of government**
of which:									
Final consumption by government of environmental protection products (*)	0.36%				0.29%	0.00%	0.05%		
Net fixed capital formation	29.44%							27.54%	**Net capital accumulation** (Net capital formation adjusted for natural capital consumption, i.e. depletion and degradation)
of which:									
Fixed capital formation on environmental protection equipment (**)	0.91%				0.23%	0.19%	0.49%		
Exports	31.93%							32.79%	**Exports**
less: Imports	33.04%							33.52%	**Imports**
of which:									
Imported products causing depletion in other countries	5.80%	0.44%	5.19%	0.17%					

(*) it includes cost of government expenditures on research and development and administration costs.
(**) It includes consumption of fixed capital on environmental protection equipment.

Economic vs. economic–environmental analysis of main aggregates

Tables 1 and 2 compare the conventional components of the net domestic product (NDP) (by expenditures) with those of environmentally adjusted net domestic products (EDP) taking into account costs of use of environmental assets. The percentages presented on the left-hand side refer to the breakdown of NDP, while those on the right-hand side reflect the breakdown of EDP. The differences between NDP and EDP analyses are explained by the economic transactions related to natural assets presented in the middle of the tables. These columns refer to the natural assets and environmental media that are the object of the study of this paper.

The following explains how the elements in the middle of the tables relate to the difference between the economic and environmental aggregates presented on both sides:

(a) NDP. Environmental charges less subsidies, value added of environmental protection industries, intermediate consumption/use of environmental protection products and uses of natural resources are identified as 'of which' elements of NDP. Environmental charges, value added of environmental protection activities and intermediate consumption/use of environmental protection products are identified, in the middle of the tables, according to the media they affect (land, air and water). The cost of uses of natural resources by industries refers to depletion of forests, minerals and fish, and

Table 2. Economic vs. economic–environmental analysis of main aggregates, 1986.

Economic analysis		Economic transactions related to natural assets as % of NDP						Economic-environmental analysis	
		Depletion			Degradation				
Concepts measured in economic analysis	(as % of NDP)	Forests	Minerals	Fish	Land	Air	Water	(as % of EDP)	Concepts measured in integrated economic-environmental analysis
NDP	100.00%							95.87%	**EDP (as % of NDP)**
of which:									
Environmental charges-subsidies	-0.20%				0.00%	0.00%	0.00%		
Value added of environmental protection industries	0.53%				0.18%	0.05%	0.30%		
Intermediate consumption/use of environmental protection products	0.94%				0.30%	0.18%	0.46%		
Use of natural resources (depletion & degradation) by industries	3.08%		0.02%		0.00%	3.04%	0.01%		
Final consumption of households	61.37%							65.12%	**Final consumption of households, adjusted for degradation impacts**
of which:									
Final consumption by households of environmental protection products	0.11%				0.00%	0.00%	0.11%		
Use of natural assets (degradation)	1.06%				0.01%	0.47%	0.58%		
Final consumption of government	11.12%							11.60%	**Final consumption of government**
of which:									
Final consumption by government of environmental protection products (*)	0.20%				0.15%	0.00%	0.02%		
Net fixed capital formation	20.92%							17.51%	**Net capital accumulation** (Net capital formation adjusted for natural capital consumption, i.e. depletion and degradation)
of which:									
Fixed capital formation on environmental protection equipment (**)	0.66%				0.10%	0.08%	0.48%		
Exports	41.86%							43.67%	**Exports**
less: Imports	35.28%							36.80%	**Imports**
of which:									
Imported products causing depletion in other countries	6.06%	0.64%	5.32%	0.11%					

(*) it includes cost of government expenditures on research and development and administration costs.
(**) It includes consumption of fixed capital on environmental protection equipment.

degradation of land, air and water; it is deducted from NDP in the derivation of EDP.

(b) Final consumption of households. Two 'of which' elements are identified for final consumption by households of environmental protection services (e.g. fees for the collection of household waste) and use of natural resources by final consumers (generation of wastes and emissions into air and water). The latter is added to final consumption of households in order to arrive at an environmentally adjusted equivalent. This addition deviates from the practice of the SEEA, which suggests shifting this use as an additional deduction from NDP. It has been assumed here, however, that households effectively consume more than they pay for and therefore the shift is not applied, but instead final consumption is increased (see also below).

(c) Final consumption of government. Government expenditures on environmental protection services are identified separately as 'of which' elements. They do not affect any of the environmentally adjusted aggregates.

(d) Net fixed capital formation. Only fixed capital formation on environmental protection equipment is identified as an 'of which' element. It refers to the gross concept as no data are available on consumption of fixed capital on environmental protection equipment. It includes equipment for the treatment or prevention of emissions into the different media. Net capital accumulation is the environmentally adjusted equivalent of net capital formation. It is obtained by deducting the cost of uses of natural assets by industries and by final consumers from net fixed capital formation.

(e) Exports and imports. Imports of products causing depletion in other countries e.g. timber, minerals and fish products, are identified separately and allocated, in the middle of the tables, to depletion of forests, minerals and fish. The purpose is to show the dependency of the Korean economy on natural resources of other countries.

The middle parts of the tables contain information on economic transactions related to natural assets. However, as explained above, not all transactions constitute adjustments in the conversion from NDP to EDP.

The following conclusions can be drawn from the data presented in Tables 1 and 2:

(a) The overall structural relations between economic and environmental analysis between 1986 and 1992 reveal an increase in the share of EDP to NDP from 1986 (95.87%) to 1992 (97.99%) (see (c)). The two expenditure components that change between NDP and EDP analysis are final consumption of households with and without environmental adjustments, and net capital formation as distinct from net capital accumulation; all other concepts remain unchanged in the tables between economic and environmental analysis. In 1986, consumption of households (environmentally adjusted) increased by approximately 2% (61.88–59.66%) when comparing its shares in NDP and EDP, while the share of net capital accumulation as compared with net capital formation decreased by approximately 2% (29.74–27.54%). In 1992 these percentages were respectively +4% (65.12–61.37%) and +3.4% (20.92–17.51%).

(b) The main environmental impacts (use of natural assets), identified in the middle of the tables, are those on air by industries and, to a lesser extent, by households as consumers, and on water (discharge of wastewater) and land (disposal of solid wastes) by households. The total impact of households is approximately one-third that of industries. These impacts, however, decreased as a percentage of NDP between 1986 and 1992. Thus, the value of emissions into air (at maintenance cost) decreased from 3.04% of NDP in 1986 to 1.96% in 1992 by industries and from 0.47% in 1986 to 0.27% in 1992 by households. The value of wastewater discharge by households decreased from 0.58% in 1986 to 0.33% in 1992.

(c) Environmental protection efforts increased between 1986 and 1992. Thus, intermediate consumption/use of environmental protection products as a percentage of NDP increased from 1986 to 1992 (0.94% in 1986 and 1.74% in 1992), closing the gap with the environmental impacts as percentage of NDP (3.08% in 1986 and 2.01% in 1992). Fixed capital formation on environmental protection equipment also increased from 0.66% in 1986 and 0.91% in 1992; and value added of environmental protection activities increased from 0.53% in 1986 to 0.94% in 1992. However, data deficiencies limit the analysis. Intermediate consumption/use of environmental protection by industries includes imputed 'output' of environmental protection generated by ancillary activities of industries. Intermediate consumption of 'external' environmental protection products is not available and was estimated as the difference between the supply of environmental protection products (output and imports) and the exports and final consumption of government and households. It was impossible to allocate it to the industries and it is presented in Tables 3 and 4 in the row of 'Non-allocated' together with 'internal' environmental protection expenditures which could not be allocated to any particular industry.

(d) Depletion of mineral resources is very low in both years. This is not surprising, because Korea is not a natural-resources rich country. Rather, it uses natural resources of other countries, which is reflected in import figures. In order to determine the influence on the resource depletion of other countries, imports as a percentage of NDP are presented at the bottom left-hand side of the tables. In 1986, the import percentages for these products were 0.64% for forest, 5.32% for minerals, and 0.11% for fish in other countries, and were somewhat lower in 1992, except for fish (i.e. 0.44%, 5.19% and 0.17%, respectively). In evaluating these percentages one should take into account that these figure are not based on net rent or user cost valuations, but are the total value of the products at which they are purchased abroad; they do not reflect, therefore, the cost of depletion. Furthermore, these values do not change the main aggregate of EDP, because imports were already deducted in the derivation of NDP.

Contribution of industries to net product in economic and economic–environmental analysis

Tables 3 and 4 present an analysis of data by industries. These tables compare economic analysis, based on SNA concepts, with environmental analysis based on environmentally adjusted data as defined in the SEEA. The left-hand side of the tables shows the percentage distribution of NDP by industries and in the right-hand side a similar distribution of EDP. The middle parts show the impacts of economic activities on natural resources and the economic responses thereto. Three main ISIC categories of economic activities are included: (1) agriculture, forestry and fishing; (2) manufacturing, electricity, gas and water, and construction and (3) services. Of the latter, only two categories of selected activities are included (i.e. manufacture of chemical products; electricity, gas and water; construction; transport and communication; public administration and defence). The data in the middle of the tables are expressed as percentages of value added of each of the corresponding industries.

To illustrate the analysis of the tables, the 'slice' corresponding to electricity, gas and water, may be taken as an example. The contribution of this

Table 3. **Contribution of industries to net product in economic and economic–environmental analysis, 1992.**

		Economic analysis % Distribution of NDP by industries	Economic transactions related to natural assets % of net value added					Economic - environmental analysis % Distribution of EDP by industries
			Current environmental protection expenditures (*)	Environmental charges - subsidies (**)	GFCF	Uses of natural assets	EVA	
Agriculture forestry and fishing	Total	7.69%	0.00%	0.01%	0.00%	0.35%	99.65%	7.86%
	Depletion							
	Land		0.00%	0.00%	0.00%	0.00%		
	Air		0.00%	0.00%	0.00%	0.35%		
	Water		0.00%	0.01%	0.00%	0.00%		
Mining	Total	0.37%	0.00%	0.01%	0.07%	11.78%	88.22%	0.33%
	Depletion					11.43%		
	Land		0.00%	0.00%	0.03%	0.00%		
	Air		0.00%	0.00%	0.00%	0.35%		
	Water		0.00%	0.01%	0.04%	0.00%		
Manufacturing, Electricity, gas and water, Construction *of which:*	Total	42.55%	0.00%	0.03%	0.29%	1.95%	98.05%	42.84%
	Depletion					0.00%		
	Land		0.00%	0.00%	0.07%	0.00%		
	Air		0.00%	0.00%	0.00%	1.95%		
	Water		0.00%	0.01%	0.22%	0.00%		
Manufacture of chemical products	Total	4.65%	0.00%	0.11%	0.35%	2.27%	97.73%	4.66%
	Depletion					0.00%		
	Land		0.00%	0.11%	0.03%	0.00%		
	Air		0.00%	0.00%	0.00%	2.27%		
	Water		0.00%	0.01%	0.32%	0.00%		
Electricity, gas and water	Total	1.83%	0.00%	0.00%	0.32%	16.79%	83.21%	1.56%
	Depletion					0.00%		
	Land		0.00%	0.00%	0.02%	0.00%		
	Air		0.00%	0.00%	0.00%	16.79%		
	Water		0.00%	0.00%	0.30%	0.00%		
Construction	Total	14.05%	0.00%	0.00%	0.32%	0.03%	99.97%	14.43%
	Depletion					0.00%		
	Land		0.00%	0.00%	0.16%	0.00%		
	Air		0.00%	0.00%	0.00%	0.03%		
	Water		0.00%	0.00%	0.16%	0.00%		
Services, total *of which:*	Total	50.18%	0.00%	0.00%	1.09%	0.23%	99.77%	51.40%
	Depletion					0.00%		
	Land		0.00%	0.00%	0.28%	0.00%		
	Air		0.00%	0.00%	0.00%	0.23%		
	Water		0.00%	0.00%	0.79%	0.00%		
Transport and communication	Total	6.13%	0.02%	0.00%	0.12%	1.60%	98.40%	6.19%
	Depletion					0.00%		
	Land		0.00%	0.00%	0.04%	0.00%		
	Air		0.02%	0.00%	0.02%	1.60%		
	Water		0.00%	0.00%	0.07%	0.00%		
Public administration and defense	Total	8.24%	0.02%	0.00%	5.71%	0.11%	99.89%	8.45%
	Depletion					0.00%		
	Land		0.00%	0.00%	1.32%	0.00%		
	Air		0.02%	0.00%	0.01%	0.11%		
	Water		0.00%	0.00%	4.29%	0.00%		
Non-allocated (% of the retail)	Total		99.04%	95.04%	26.85%	49.45%		
	Depletion							
	Land		100.00%	70.19%	25.58%	0.00%		
	Air		99.85%	0.00%	99.07%	50.13%		
	Water		100.00%	92.90%	0.03%	100.00%		
Total industries	Total	100.00%	1.74%	0.06%	0.91%	2.01%	97.38%	100.00%
	Depletion					0.04%		
	Land		0.38%	0.02%	0.23%	0.00%		
	Air		0.66%	0.00%	0.19%	1.96%		
	Water		0.70%	0.05%	0.49%	0.01%		

(*) Includes only internal environmental protection expenditures, external environmental protection exenditures are in Non allocated.

(**) Environmental subsidies appear only in Not-allocated.

Table 4. Contribution of industries to net product in economic and economic–environmental analysis, 1986.

		Economic analysis	Economic transactions related to natural assets					Economic - environmental analysis
		% Distribution of NDP by industries	% of net value added					% Distribution of EDP by industries
			Current environmental protection expenditures (*)	Environmental charges - subsidies (**)	GFCF	Uses of natural assets	EVA	
Agriculture forestry and fishing	Total	11.76%	0.01%	0.00%	0.01%	0.22%	99.78%	12.24%
	Depletion							
	Land		0.00%	0.00%	0.00%	0.00%		
	Air		0.00%	0.00%	0.00%	0.22%		
	Water		0.01%	0.00%	0.01%	0.00%		
Mining	Total	1.00%	0.02%	0.00%	0.02%	2.63%	97.37%	1.01%
	Depletion					2.47%		
	Land		0.00%	0.00%	0.00%	0.00%		
	Air		0.00%	0.00%	0.00%	0.15%		
	Water		0.02%	0.00%	0.02%	0.01%		
Manufacturing, Electricity, gas and water, Construction *of which:*	Total	40.25%	0.12%	0.01%	0.17%	2.77%	97.23%	40.82%
	Depletion					0.00%		
	Land		0.00%	0.00%	0.05%	0.00%		
	Air		0.00%	0.00%	0.00%	2.73%		
	Water		0.12%	0.01%	0.12%	0.03%		
Manufacture of chemical products	Total	5.12%	0.03%	0.09%	0.07%	1.92%	98.08%	5.23%
	Depletion					0.00%		
	Land		0.00%	0.08%	0.01%	0.00%		
	Air		0.00%	0.01%	0.00%	1.89%		
	Water		0.03%	0.00%	0.07%	0.03%		
Electricity, gas and water	Total	2.78%	0.01%	0.00%	0.01%	17.82%	82.18%	2.38%
	Depletion					0.00%		
	Land		0.00%	0.00%	0.00%	0.00%		
	Air		0.00%	0.00%	0.00%	17.82%		
	Water		0.01%	0.00%	0.01%	0.00%		
Construction	Total	7.24%	0.10%	0.00%	0.33%	0.12%	99.88%	7.54%
	Depletion					0.00%		
	Land		0.00%	0.00%	0.22%	0.00%		
	Air		0.00%	0.00%	0.00%	0.12%		
	Water		0.10%	0.00%	0.11%	0.00%		
Services, total *of which:*	Total	45.83%	0.04%	0.00%	1.01%	0.69%	99.31%	47.48%
	Depletion					0.00%		
	Land		0.00%	0.00%	0.07%	0.00%		
	Air		0.00%	0.00%	0.01%	0.69%		
	Water		0.04%	0.00%	0.93%	0.00%		
Transport and communication	Total	6.01%	0.02%	0.00%	0.05%	3.88%	96.12%	6.03%
	Depletion					0.00%		
	Land		0.00%	0.00%	0.01%	0.00%		
	Air		0.00%	0.00%	0.02%	3.88%		
	Water		0.02%	0.00%	0.02%	0.00%		
Public administration and defense	Total	7.20%	0.02%	0.00%	6.16%	0.38%	99.62%	7.48%
	Depletion					0.00%		
	Land		0.00%	0.00%	0.35%	0.00%		
	Air		0.00%	0.00%	0.06%	0.38%		
	Water		0.02%	0.00%	5.72%	0.00%		
Non-allocated (% of the total)	Total		92.79%	96.75%	19.60%	51.86%		
	Depletion							
	Land		100.00%	0.00%	50.96%	0.00%		
	Air		100.00%	0.00%	93.02%	52.51%		
	Water		85.30%	0.00%	1.22%	0.03%		
Total industries	Total	100.00%	0.94%	0.01%	0.66%	3.08%	95.87%	100.00%
	Depletion					0.02%		
	Land		0.30%	0.00%	0.10%	0.00%		
	Air		0.18%	0.00%	0.08%	3.04%		
	Water		0.46%	0.00%	0.48%	0.01%		

(*) Includes only internal environmental protection expenditures, external environmental protection expenditures are in Non allocated.

(**) Environmental subsidies appear only in Not-allocated.

industry to the NDP in 1992 was 1.83% while its contribution to EDP was 1.56%. The reduced contribution to EDP may be due to degradation impacts (emissions into air) caused by this industry, which amount in total to 16.79% of value added. On the basis of the available data and not taking into account the problem of non-allocation of cost to industries (see below), the economic responses by this industry to mitigate its environmental impacts are minor: environmental protection expenditures are negligible and environmental charges are not levied on this industry. Fixed capital formation on environmental protection equipment amounts to 0.32% of value added, of which 0.02% refers to environmental protection of land, and 0.30% to environmental protection of water. EVA (environmentally adjusted value added) as a percentage of net value added, presented in the middle of the tables, for this industry (83.21%) is below the average (97.33%) for all industries.

One of the difficulties in the analysis is the non-allocation to industries of all or large parts of some of the economic transactions and environmental impacts categories in the middle of the tables. Non-allocation, presented in the shaded row of the tables, is very serious for both 1986 and 1992. It is most serious in the case of environmental protection expenditures (1986, 92.79%; 1992, 99.94%) and environmental charges and subsidies (1986, 95.04%; 1992, 96.75%). It is somewhat lower, but still significant, for the use of natural resources (1986, 51.86%; 1992, 49.45%).

The conclusions drawn from the data presented in Tables 3 and 4, keeping in mind the large non-allocation of cost to industries, can be summarized as follows:

(a) As was already shown in Tables 1 and 2, there is an increase in the overall EVA/VA percentages for all industries between 1986 (95.87%) and 1992 (97.38%). The industries below these averages, in 1986, are electricity, gas and water and, in 1992, electricity, gas and water and mining. This might indicate that these industries have larger environmental impacts than the others.

(b) The average cost of natural asset use for all industries decreased between 1986 (3.08%) and 1992 (2.01%). This reflects similar decreases in the use of natural resources in manufacturing, electricity, gas and water, construction (1986, 2.77%; 1992, 1.95%) and in services (1986, 0.69%; 1992, 0.23%). However, at least for two industries the percentages were higher, i.e. mining (1986, 2.63%; 1992, 11.78%) and manufacture of chemical products (1986, 1.92%; 1992, 2.27%). This finding should be treated with some caution, as approximately half of the uses of natural resources was not allocated to industries.

(c) Total economic responses for all industries (current environmental protection expenditures, environmental charges, fixed capital formation on environmental protection equipment) increased between 1986 (1.61%) and 1992 (2.71%). This might explain the corresponding decrease in the cost of the use of natural assets (3.08% in 1986 and 2.01% in 1992). In 1992, the cost of the use of natural resources (2.01%) was even lower than the total of economic responses (2.71%).

(d) Economic responses do not seem to be in proportion to their environmental impacts. Thus, for manufacturing, electricity, gas and water, and construction, environmental impacts in 1986 and 1992 were 0.3% and 0.32% and the economic responses were 2.77% and 1.95% respectively. In the case of electricity, gas and water the cost of natural asset use in 1986 and 1992 was 17.82% and 16.79%, and the economic response was 0.02% and 0.32% respectively. With regard to services, the cost of natural asset use in 1986 and 1992 was 0.69% and 0.23%, while the totals of economic responses in this industry were 1.05% and 1.09% respectively. The same 'unrelatedness' between economic responses and the use of natural assets can be observed in the case of construction, public administration and transport and communication. In most instances, however, the economic responses have increased between 1986 and 1992.

Future work

The pilot compilation of the SEEA was carried out using existing statistics. It identified data deficiencies, especially for environmental protection expenditures, water resources and fish. Moreover, the coverage of industries was limited to those for which data were available. Environmental protection expenditures were estimated from different data sources for the output and intermediate consumption. This led to a discrepancy between supply and use of environmental protection goods and services. Moreover, the data collected by the Ministry of the Environment and the Korean Development Bank were not consistent with the classification used in the national accounts (KSIC).

With regard to fish, the study shows that, on one hand, imports of fish have increased over the years, while on the other hand, catches per unit effort (CPUE) have generally decreased since the 1970s. Fish stocks appear to decrease rapidly. A more in-depth study for estimating fish biomass for different species, and hence sustainable yield and depletion, is needed to assess the significance of fish stock depletion.

The natural extension of the pilot compilation, in addition to improvement in data collection, would be the institutionalization of the compilation. The study was carried out with the assistance of United Nations Statistics Division (UNSD), by KETRI, a research institute within the Ministry of Environment. Support from other institutions such as the National Statistical Office (NSO) and the Bank of Korea (BOK), responsible for the compilation of national accounts, was limited. In order to ensure a continuation in the compilation of integrated environmental and economic accounts a coordinating institution, possibly the BOK or NSO, should build up on the experience gained from the pilot compilation of the SEEA.

5
Philippines: Adaptation of the United Nations system of environmental accounting

ESTRELLA V DOMINGO

Introduction

Environmental accounting in the Philippines started in 1990 under the Environmental and Natural Resources Accounting Project (ENRAP). The USAID-funded project adopted the Peskin's framework (see Chapters 6 and 21). In 1995, the National Statistical Coordination Board (NSCB), the agency mandated to compile the national accounts, started a pilot project on the United Nations System of integrated Environmental and Economic Accounting (SEEA; United Nations, 1993a). The Philippine SEEA (PSEEA) project is a major component of the UNDP programme, Integrated Environmental Management for Sustainable Development (IEMSD),[1] which is implemented by the Department of Environment and Natural Resources (DENR). The NSCB is also currently undertaking a parallel study on the United Nations SEEA for ESCAP; the study is limited to the compilation of asset accounts for fishery, forest and mineral resources. The pilot study on the national PSEEA was programmed to be completed in 1996, after which a regional (sub-national) PSEEA will be piloted. The major activities for the development of the PSEEA are (1) the development of the PSEEA framework; (2) the compilation of the asset accounts for non-produced economic and non-economic (natural) assets in physical terms and with monetary valuation; (3) the estimation of emissions/discharges and other causes of degradation of environmental assets due to economic activities; (4) the identification and separation of environmental protection expenditures; and (5) the generation of environmentally adjusted net domestic product (EDP) and other aggregates.

At the time of the preparation of this report, the NSCB had compiled the asset accounts for three resources: fishery, forest and minerals. This chapter is therefore limited to the presentation of the Philippine experience in the compilation of these accounts. Guided by the SEEA framework, the asset accounts for fishery, forest and mineral resources have been compiled in physical and monetary terms in the following general format:

OPENING STOCK

- Changes due to economic activity (depletion/extraction)

K. Uno and P. Bartelmus (eds.), Environmental Accounting in Theory and Practice, 77–94.

- Other accumulation
- Other volume changes
- Revaluation (This term is included only in the monetary asset accounts)

CLOSING STOCK

The changes between opening and closing stocks may arise from economic decisions, i.e. changes due to economic activities, and other accumulation, or changes due to non-economic factors reflected in other volume changes. In the monetary accounts, revaluation is due to changes in prices of the asset between the beginning and the ending of the accounting period. More details of the accounting framework and intermediary tabulations applied are given in Chapter 2.

The Philippine experience demonstrates the feasibility of compiling natural resource accounts within the frameworks of the System of National Accounts (SNA; Commission of the European Communities *et al.*, 1993) and the SEEA. The Philippine study made use of existing data, which are mostly administrative, and results of special studies. No primary data were collected to meet the requirements of the initial compilation of the environmental accounts. Since considerable processed and unprocessed administrative data can be transformed into useful data for environmental accounting, lack of data should not prevent a country from pursuing environmental accounting. Rather, compiling the accounts will exert pressure on the statistical system to respond to needs of this new framework.

Fishery resource accounts, 1988–1992

Accounting framework

Fishery resources cover the cultivated fish stock and stock of other aquatic animals in fish ponds and farms and the non-cultivated fish stocks and other aquatic animals in the ocean and in inland and coastal waters. Fishery resources are cultivated in aquaculture and mariculture; the non-cultivated fishery resources are the object of marine and freshwater fisheries. Due to data limitations, this initial study was confined to marine fisheries only. For cultivated fishery resources, changes between opening and closing stock are the result of economic transactions, including additions to breeding stocks and inventories for consumption, and decrease in stocks due to harvesting the resource. Fish mortality because of natural death, disasters and ecological destruction of habitat constitute other volume changes in the aquaculture accounts. At present, estimates on cultivated fishery are still under study for later inclusion in the PSEEA.

For non-cultivated fishery resources, changes in stocks are due to depletion, other accumulation and other volume changes. Depletion is accounted for when fish caught exceeds the sustainable catch, that is when harvest exceeds long-term natural growth. Since fish is a renewable resource, the stock of fish can be increased if allowed to regenerate, but fish can only regenerate to a level limited by the carrying capacity of the ecosystem. Given the capacity of fish to regenerate and its natural life span, the use of this resource when replenished by natural growth is not considered depletion. Other accumulation is the conversion of fish stocks to economic control, i.e. their 'economic appearance' (in SNA terms).

Natural growth, natural mortality, net migration and mortality due to fishing and destruction of habitat within sustainability limits make up other volume changes in the stock level of the non-cultivated fishery resource.

Fish stock

The study recognizes the complexity of fishery resource accounting, taking into consideration the highly mobile nature of this resource. Fish stock data are not available at present. The Resource Ecological Assessment study conducted by the Bureau of Fisheries and Aquatic Resources (BFAR) in 1993, covering 12 major fishing grounds, provides results for only five area sites. This study estimates biomass and other parameters relevant in the construction of the asset account for selected areas only. However, stock assessment can only be meaningful if estimated by fishing ground. Given the incomplete results, to date, this study cannot be scientifically or statistically generalized for all fishing grounds. Fish stock will, therefore, be estimated only when the complete results of the fish stock assessment study become available.

Fish catch

Fish catch data are used to estimate depletion (see below). Data on fish catch (fish production) are currently provided by the Bureau of Agricultural Statistics (BAS)[2] from the Surveys of Commercial and Municipal Fish Landing Centers. Prior to 1980, these data were available from the BFAR, based on regional monitoring reports of licensed fishing vessels. For both series, catch data are classified into commercial and municipal fish production, and municipal fish production is further broken down into marine and inland fishery. For purposes of bio-economic modelling, longer time-series on fish catch are needed. These time-series are obtained by extrapolating the BAS data on fish catch, using trends of the BFAR data from 1980 backwards. Fish catch data are also used to estimate fishing effort. A correction factor of 20% is applied to reported fish production to account for undercoverage. Total (actual) catch for 1976–1993 is shown in Table 1.

Fishing effort

Data on fishing effort are required for computing sustainable catch. Current production surveys do not collect data on fishing effort[3] which is, therefore,

Table 1. Estimation of fishery resource depletion, 1976–1993.

Year	Effort (hp) (1)	Actual Catch (m.t.) (2)	Sustainable Catch (m.t.) (3)	Depletion (m.t.) (4)
1976	297,125	1,250,021	1,441,311	(191,290)
1977	315,059	1,358,525	1,477,571	(119,047)
1978	385,199	1,409,308	1,583,061	(173,753)
1979	525,161	1,363,548	1,658,353	(294,806)
1980	453,190	1,362,915	1,638,723	(275,809)
1981	551,659	1,445,709	1,657,263	(211,554)
1982	787,838	1,481,267	1,517,283	(36,016)
1983	718,181	1,548,365	1,576,930	(28,566)
1984	655,126	1,563,972	1,619,775	(55,803)
1985	649,477	1,556,542	1,622,976	(66,434)
1986	650,655	1,624,206	1,622,318	1,888
1987	653,145	1,688,926	1,620,911	68,015
1988	611,119	1,726,033	1,641,475	84,558
1989	561,349	1,823,409	1,655,890	167,518
1990	548,115	1,914,725	1,657,639	257,086
1991	455,734	2,008,007	1,640,049	367,957
1992	380,379	1,991,463	1,577,501	413,963
1993	405,231	1,978,350	1,603,754	374,596

Negative values (in parentheses) refer to years with zero depletion.

estimated indirectly using available parameters for commercial and municipal fishing.

For commercial fishing effort, corrected commercial fish catch in metric tons, by major fishing gear, is multiplied by the estimated parameters of horsepower per gross tonnage (hp/gt)[4] to arrive at horsepower equivalents (hp) for major fishing gear (purse seine, bagnet, ringnet, trawl, carrier vessels and muro-ami). To standardize the catch efficiency of the various fishing gear used, horsepower of other fishing gear is converted to purse seine equivalent (PS),[5] using the ratio of catch per unit effort (CPUE) by type of fishing gear to CPUE of purse seine.

For municipal fishing effort, the benchmark estimates of horsepower of municipal fishing vessels and total municipal fishermen by type are obtained from the 1980 Census of Fisheries. Both data categories are extrapolated, using ENRAP estimates on horsepower of municipal fishing vessels and number of fishermen, respectively. To translate the number of fishermen into fishing effort (hp equivalent), the following conversion factors are applied:

(a) hp of full-time fishermen = number of full-time fishermen × 0.18 hp × 6/24 h;
(b) hp of part-time fisherman = hp of full-time × 40%;
(c) hp of occasional fishermen = hp of full-time × 11%.[6]

The above constant factors were applied throughout the data series. The total horsepower equivalent for municipal fishermen equals (a) + (b) + (c).

To arrive at the total fishing effort in hp equivalent for municipal fishing, fishing efforts for municipal fishing vessels are added to the fishing efforts of fishermen. This is then translated into PS equivalent following the same methodology used to convert commercial fishing (hp) to PS equivalent. Total fishing effort in PS equivalent for marine fishery is the sum of estimated commercial fishing effort and municipal fishing effort, both in PS equivalent (see Table 1).

Sustainable yield (catch) and depletion

The level of sustainable catch is estimated using the following Fox model (FAO, 1989):

$$Y = E \times \exp(a + bE)$$

where: Y = catch or yield from the resource, E = fishing effort per unit time.

The sustainable catch level estimated by using the above equation is shown in Table 1 for 1976–1993. The regression coefficient for this equation was derived by establishing the relationship between the time series data on fish catch and fishing efforts, with an r^2 of 0.7960. The sustainable yield curve is presented in Figure 1.

Fish is a renewable biological resource. As an input to production it is only accorded a production cost when the harvest of fish exceeds the long-term

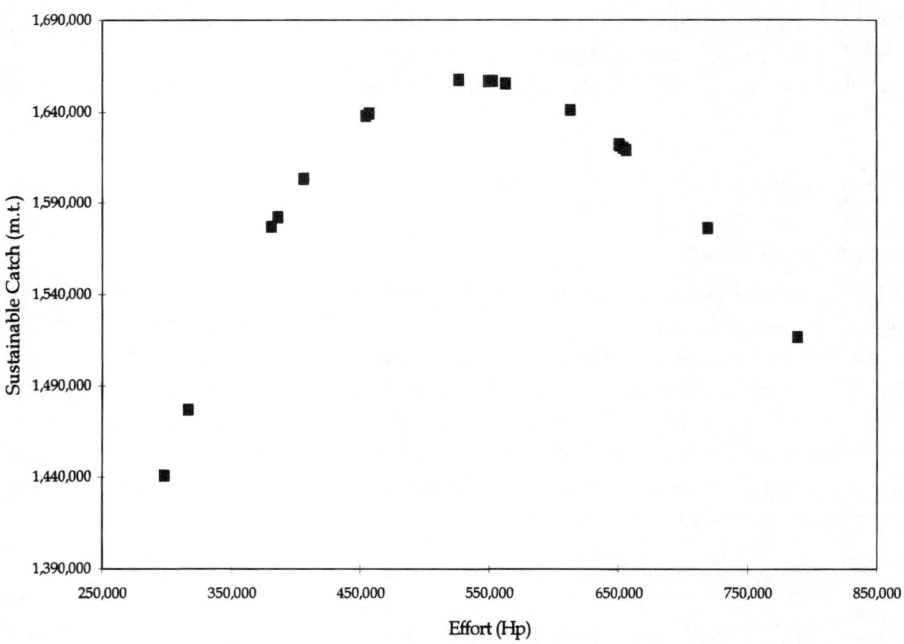

Figure 1. Sustainable yield–effort curve for marine fishery resources.

natural growth, that is when depletion occurs. In Table 1, depletion in physical terms (metric tons) is computed as the difference between actual catch and sustainable catch.

Depletion cost

In monetary terms, the value of depletion is estimated using the net price (net rent) method, defined as:

$$\text{net rent} = \text{net price} \times \text{physical volume of depletion}$$

$$\text{net price} = \text{per unit market price} - \text{marginal unit cost of fishing}$$

The Annual Survey of Establishments (ASE) of the National Statistics Office (NSO) provides the basis for arriving at the net price. From the gross output/total revenues (at producer's price) are subtracted the costs (at purchaser's price) consisting of intermediate inputs (mainly fuel), compensation of employees, indirect taxes, depreciation and 15% opportunity cost (as a normal return to fixed assets employed). Thus, in the absence of information on marginal costs, average costs have to be used.

The net price is derived by first estimating the net price at producer's price. Since ASE data do not provide information on the corresponding catch, the proportion of net price to gross output, which is 12.25%, is applied to the producer's price[7] of fish. The difference between purchaser's price (wholesale price) and producer's price, i.e. the trade and transport mark-up, is then added to the derived estimate of the net price at producer's price. For other years, the same procedure is followed, applying the same structure of net price to total output. Table 2 shows the total amount of depletion in physical and monetary terms.

Forest resource accounts, 1988–1994

Scope and coverage

Philippine forests are classified into different types: dipterocarp, pine, submarginal, mossy and mangrove. The dipterocarps make up roughly two-thirds of the total forest cover. In the present accounting study, the asset accounts for forest resources cover only the forest trees in forest land. Moreover, among the forest types, only dipterocarps and pine are included, primarily because of their economic and ecological significance, and partly because of data availability. Rattan, a non-timber forest product, is also included because of its significant economic contribution.

Physical accounts

The compilation of asset accounts for forest sets out from physical accounts to which monetary unit values are applied in a second stage.

Table 2. Fishery resource depletion, 1985–1993.

Year	Depletion (m.t.)	Net price[a,b] (pesos/m.t.)	Net rent[c] ('000 pesos)
1985	(66,434)	–	–
1986	1,888	4,801.44	9,065
1987	68,015	4,322.17	293,973
1988	84,558	2,244.73	189,810
1989	167,518	5,695.32	954,071
1990	257,086	6,742.99	1,733,529
1991	367,957	9,794.77	3,604,058
1992	413,963	8,397.71	3,476,340
1993	374,596	9,389.31	3,517,199

[a] Net price was derived by using the 1989 ASE. The series was generated by applying the change in producer's price to the derived net price of 1989.
[b] To translate the net price at purchaser's price, the trade and transport margin was added to the net price at producer's price.
[c] Net rent is the value of resource depletion.

Forest resource stock

The opening stock represents the stock of resources at the beginning of the accounting period. The base information on opening area and volume is available for 1988 from the Republic of the Philippines (RP) German Forest Inventory.[8] From the 1988 base information, the Forest Management Bureau (FMB) of the DENR extrapolates the forest resource stock for the succeeding years. These data are published annually in the Philippine Forestry Statistics (PFS). The PFS figures are adopted as the official opening stock (in hectares) for the accounting period. The closing stock for the accounting period is equal to the opening stock of the succeeding year.

Depletion

The cutting of forest trees in the Philippines has been extensive. Hence, it is assumed that current logging already exceeds the level of sustainable cut. Therefore, all logging, together with damages from logging, is considered depletion. Data on areas logged are taken from the 1990 Master Plan for Forestry Development (MPFD)[9] published by the DENR. For old-growth dipterocarp forest, logging was allowed only up to 1991. Beginning 1992, there was thus no reduction in the area due to logging, and depletion due to logging of old-growth dipterocarp is assumed to be zero. For second-growth forest, damage due to logging is 15% of the logged area in old-growth forest in the previous year. These area data are then multiplied by an assumed harvestable volume of 248.22 m^3 per hectare for old growth and 67.0 m^3 per hectare for secondary growth to arrive at the data on volume harvested.

Forest conversion

Forest conversion refers to clearing of forest to convert forest land to other uses, such as residential or industrial development. Data on forest conversion are taken from projections of the MPFD. The area converted from old growth to second growth is equal to the area logged in old-growth forest in the previous year less 15% for log landings, roads and skidding/yarding trails. Volume is derived by multiplying the area converted by the corresponding average harvestable volume per hectare.

Other volume changes

The data for other volume changes are taken from special studies. For old-growth forest, it is assumed that net stand growth is equal to zero, that is tree growth is equal to tree mortality. For a second-growth forest, the net stand growth is equal to opening volume multiplied by the annual growth rate taken from a study by Revilla (1976).

The 'other changes' under the volume changes denote the statistical discrepancy which is equal to the difference between the computed closing stock (taking into account the identified factors) and the opening stock in the succeeding year taken from official sources. The physical asset accounts for old- and second-growth dipterocarp are shown in Tables 3 and 4.

Monetary accounts

The net price method is adopted in valuing the stock of forest resources. Economic or monetary accounts are derived by multiplying the volume items of the physical accounts by the stumpage value of the forest. The stumpage value is the value of the standing timber and is equal to the market price of the product less the cost of harvesting and a margin for normal return to capital (assumed to be 30% of the logging cost).[10] The market price of timber is taken as a weighted price of the various species. The cost items include logging cost, pre-logging cost, post-logging cost, road construction cost, transportation cost and overhead costs.

The stumpage value represents the potential economic rent accruing to operators. The base information for stumpage value is taken from a local study covering only the year 1990. For years other than 1990, the corresponding stumpage value is derived by adjusting the 1990 value by the corresponding changes in weighted prices of logs, by species and also by changes in the corresponding costs of harvesting. Updated cost structures are taken from financial statements of logging companies, submitted to the Securities and

Table 3. Volume accounts of old-growth dipterocarp forests, 1988–1994 ('000 m^3)

	1988	1989	1990	1991	1992	1993	1994
Opening volume[a]	318,739	243,105	227,123	212,416	198,792	198,792	198,792
Changes due to economic activity	(24,152)	(12,411)	(8,688)	(8,688)	0	0	0
Reforestation	n.a.	n.a.	n.a.	n.a.	n.a.	n.a.	n.a.
Afforestation	n.a.	n.a.	n.a.	n.a.	n.a.	n.a.	n.a.
Depletion							
logged[b]	24,152	12,411	8,688	8,688	0	0	0
Other accumulation (±)							
Forest converted to non-forest use[c]	(3,475)	(2,482)	(2,433)	(2,408)	0	0	0
Other volume changes	(48,007)	(1,089)	(3,587)	(2,529)	0	0	0
Natural growth[d]							
Natural mortality[d]							
Other changes (s.d.)	(48,007)	(1,089)	(3,587)	(2,529)	0	0	0
Net change in volume	(75,634)	(15,982)	(14,707)	(13,624)	0	0	0
Closing volume	243,105	227,123	212,416	198,792	198,792	198,792	198,792

[a] Opening volume for a given year is based on the volume of old-growth forest in the preceding year (PFS).
[b] Volume logged equals logged area multiplied by 248.22 m^3/ha (mean harvestable volume of old-growth forest).
[c] Volume of forest converted equals converted forest area multiplied by 248.22 m^3/ha.
[d] Net-stand growth is zero in old-growth forest: tree growth equals tree mortality.

Table 4. Volume accounts of second-growth dipterocarp forests, 1988–1994 ('000 m^3)

	1988	1989	1990	1991	1992	1993	1994
Opening volume[a]	588,753	475,991	467,125	458,137	449,161	435,695	423,480
Changes due to economic activity	(1,068)	(978)	(503)	(352)	(352)	0	0
Reforestation	n.a.	n.a.	n.a.	n.a.	n.a.	n.a.	n.a.
Afforestation	n.a.	n.a.	n.a.	n.a.	n.a.	n.a.	n.a.
Depletion							
Damage from logging[b]	1,068	978	503	352	352	0	0
Other accumulation (±)	2,704	2,191	(1,179)	(1,960)	(1,893)	(3,812)	(3,745)
Old growth converted to second growth[c]	6,054	5,541	2,848	1,993	1,993	0	0
Forest converted to non-forest use[d]	(3,350)	(3,350)	(4,027)	(3,953)	(3,886)	(3,812)	(3,745)
Other volume changes	(114,397)	(10,079)	(7,306)	(6,665)	(11,222)	(8,403)	(7,308)
Net stand growth[e]	22,961	18,564	18,218	17,867	17,517	16,992	16,516
Other changes (s.d.)	(137,359)	(28,643)	(25,524)	(24,532)	(28,739)	(25,395)	(23,823)
Net change in volume	(112,762)	(8,866)	(8,988)	(8,976)	(13,466)	(12,215)	(11,053)
Closing volume	475,991	467,125	458,137	449,161	435,695	423,480	412,247

[a] The opening volume in a given year is based on the volume of old-growth forest in the preceding year.
[b] Area damaged from logging, multiplied by the harvestable volume of 67.0 m^3/ha.
[c] Area logged in old-growth forest in the previous year multiplied by density of 67.0 m^3/ha.
[d] Area converted to other uses, multiplied by the residual density of 67.0 m^3/ha.
[e] Net-stand growth equals opening volume, multiplied by the annual growth rate taken from Revilla (1976).

Exchange Commission of the Philippines. Values at constant prices are derived by dividing the stumpage value at current prices by the consumer price index (base year 1985).

The revaluation is equivalent to the differences in stock values due to changes in prices between the opening and closing periods. The monetary asset accounts for the total forest resource in current prices are presented in Table 5.

Mineral resource accounts, 1988–1992

Accounting framework

Mineral resources include metallic and non-metallic minerals. Metallic minerals found in the Philippines consist of copper, gold, chromite, nickel and other metals such as silver, lead, zinc, manganese, cobalt, iron and molybdenum. Non-metallic minerals found in the country are quarry resources and energy resources such as crude oil and natural gas.

The initial estimate of mineral resources for the asset accounts was limited to metallic minerals (gold, copper, chromite, iron, nickel and manganese). The account is compiled both in physical and monetary terms. However, valuation of assets in monetary terms is limited to gold, copper and chromite only. In physical terms, the account is presented in ore form and metal content.

The mineral resources asset accounts show the level of the stock of a particular mineral resource at a given point in time and record the transactions that cause the changes in its level. In principle, only proven reserves are considered in the construction of the asset accounts. Proven reserves are defined as 'the estimated quantities at a specific date, which analysis of geological engineering data demonstrate, with reasonable certainty, to be recoverable in the future from known reservoirs under the economic and operational conditions at the same date' (Commission of the European Communities *et al.*, 1993, p. 513). To operationalize this definition, positive and probable reserves are included, following the advice of mining engineers. Positive and probable reserves are those which have 81–90% and 71–80% confidence level of accuracy, respectively.

The end-of-year annual report of the Mines and Geo-Sciences Bureau (MGB) on the level of reserves is taken as the source for the closing stock for the year and the opening stock for the succeeding year. Since minerals are a non-renewable resource, the level of extraction during the period is treated as depletion. For other accumulation, relevant entries which are due to economic decisions are new discoveries and reclassification of reserves. Other transactions that are not due to economic decisions, but nevertheless affect the level of reserves, should be recorded under other volume changes; they include the closure of mines and abandonment of mine sites due to force-majeur conditions (i.e. natural calamities, war and strife, and situations created by other than economic considerations).

Table 5. Summary: economic accounts of forest resources at current prices, 1988–1994 ('000 pesos)

	1988	1989	1990	1991	1992	1993	1994
Opening stock	594,052,514	520,976,245	569,263,284	556,665,798	605,095,671	689,542,879	738,710,885
Changes due to economic activity	(24,529,229)	(14,851,981)	(10,404,307)	(11,766,271)	(541,817)	(299,577)	(270,905)
Reforestation	n.a.	n.a.	n.a.	n.a.	n.a.	n.a.	n.a.
Afforestation	n.a.	n.a.	n.a.	n.a.	n.a.	n.a.	n.a.
Depletion	24,529,229	14,851,981	10,404,307	11,766,271	541,817	299,577	270,905
Harvest/logging	23,877,361	14,173,344	10,040,522	11,404,248	159,548	171,187	132,761
Waste	128,810	128,937	76,356	134,792	119,651	128,390	138,144
Damage from logging	523,058	549,699	287,430	227,231	262,618	0	0
Other accumulation (±)	(2,363,002)	(1,840,322)	(3,507,767)	(4,432,764)	(1,477,811)	(3,186,031)	(3,251,821)
Old growth converted to second growth	2,963,997	3,114,963	1,628,770	1,287,641	1,488,170	0	0
Forest converted to non-forest use	(5,326,999)	(4,955,285)	(5,136,537)	(5,720,405)	(2,965,980)	(3,186,031)	(3,251,821)
Other volume changes	(102,835,307)	(6,658,794)	(8,405,080)	(7,723,201)	(8,540,012)	(7,043,010)	(6,392,373)
Natural stand growth	12,800,937	12,036,107	12,027,107	13,348,832	15,170,229	15,960,604	16,100,185
Stand mortality	(1,161,706)	(1,201,068)	(1,198,506)	(1,350,374)	(1,562,138)	(1,526,287)	(1,522,866)
Forest fires	(83,925)	(87,030)	(88,110)	(100,080)	(117,000)	(115,335)	(116,280)
Other changes (s.d.)	(114,390,613)	(17,406,803)	(19,145,571)	(19,621,579)	(22,031,103)	(21,361,992)	(20,853,412)
Net change in stock	(129,727,538)	(23,351,097)	(22,317,155)	(23,922,236)	(10,559,640)	(10,528,618)	(9,915,099)
Revaluation	56,651,269	71,638,136	9,719,668	72,352,108	95,006,849	59,696,123	27,053,110
Closing stock	520,976,245	569,263,284	556,665,798	605,095,671	689,542,879	738,710,385	755,848,395

The asset accounts in physical terms are translated into monetary values using both the net price and the user cost methods. Revaluation is imputed as a residual.

The physical accounts

The estimation of the physical asset accounts in ore form is straightforward. The volumes of reserves and extraction (depletion) for each mineral are taken from reports of mining firms, collected by the MGB as part of its regulatory function. Data on reserves are crucial in establishing the physical asset accounts and are readily available from the MGB. To estimate the reserves in metal content, a weighted average grade for proven reserves is applied to gold, copper and chromite. While it is recognized that ore bodies may contain more than one metal, only the primary metal is considered. Since only data on opening and closing stocks and extractions are available, other accumulation and other volume changes are derived as residuals.

The preliminary physical asset account for total mineral resources in ore form is presented in Table 6. Clearly, such information is of limited use, since metals of different value are simply added together in weight units. On the other hand, detailed accounts by type of resource are normally treated as confidential.

Table 6. Physical asset accounts of total minerals in ore form, 1988–1992[a] ('000 m.t.)

	1988	1989	1990	1991	1992
Opening stock	5,762,484	6,005,212	6,048,383	5,220,408	6,286,169
Extraction	71,903	72,127	61,574	51,268	46,390
Other accumulation	314,630	115,299	−766,402	1,117,030	−481,400
Other volume changes	–	–	–	–	–
Closing stock	6,005,212	3,024,192	5,220,408	6,286,169	5,758,379

[a] Includes only gold, copper, chromite, nickel and manganese.
Sources: Philippine Metallic Mineral Reserves, Mining Technology Division, MGB; Integrated Annual Report, Mineral Economic and Policy Division, MGB; NSCB Working Group estimates.

Asset accounts in monetary terms

To value the asset accounts in monetary terms, two methods of valuation are employed: the net price method (NPM) and the user cost allowance suggested by El Serafy (1989).

Following the NPM, the relevant unit resource net price is applied to the physical asset account measured in terms of metal content to arrive at their monetary values. Table 7 shows the results of this approach, including the environmental depletion cost in the 'extraction' row. As an alternative, the user costs were also calculated, applying discount rates of 5% and 10%.

Table 7. Monetary asset accounts of total minerals,[a] 1988–1992 (net prices; million pesos)

	1988	1989	1990	1991	1992
Opening stock	38,792	45,411	48,161	54,266	38,518
Extraction	2,247	2,133	1,426	862	477
Other accumulation	4,305	2,404	3,355	7,361	(8,031)
Other volume changes[b]	–	–	–	–	–
Closing stock	45,411	48,161	54,266	38,518	29,405

[a] Includes only gold, copper and chromite.
[b] There are no established data for other volume changes of minerals, so in the meantime these are left blank; consequently, revaluation cannot be estimated.
Sources: Philippine Metallic Mineral Reserves, Mining Technology Division, MGB; Integrated Annual Report, Mineral Economic and Policy Division, MGB; NSCB working group estimates.

Results

Given the preliminary estimates of depletion from the resource accounts, net domestic product (NDP) was partially adjusted to derive an environmentally adjusted net domestic product (EDP). The results presented in Table 8 indicate that EDP was at the least 97% in 1988 and gradually increased to 99% of NDP in 1992. This implies that depletion represented around 4% in 1988 and continuously declined to less than 1% in 1992. The annual growth rates of EDP were about 1% greater than NDP, except in 1991 when they were almost equal. Figure 2 shows GDP, NDP and EDP growth from 1988 to 1992.

The concept of extended net accumulation was also operationalized, using the basic structure of the SEEA as presented in the SNA (Commission of the European Communities *et al.*, 1993, p. 511). From this structure, the following accounting definitions can be derived:

$$A_{p.ec} = I = I_g - CFC \quad \text{for produced assets, and}$$

$$A_{np.ec} = I_{np.ec} - use_{np.ec} \quad \text{for non-produced assets}$$

$$\text{thus total net accumulation of economic assets} = A_{p.ec} + A_{np.ec}$$

where: $A_{p.ec}$ = net accumulation of produced assets, I_g = gross domestic capital formation, CFC = consumption of fixed capital or depreciation, $A_{np.ec}$ = net accumulation of non-produced assets, $I_{np.ec}$ = other accumulation of non-produced assets, $use_{np.ec}$ = depletion and degradation of non-produced assets.

Table 9 displays the definitory aggregates of the accounting identities. Data for produced assets were taken from the conventional national accounts while those for non-produced assets were taken from the preliminary estimates of the asset accounts for fishery, forest and mineral resources and are, therefore, limited in coverage. $I_{np.ec}$ in column 4 includes only data for forest and mineral

Table 8. GDP, NDP and EDP at current prices, 1988–1992 (million pesos)

Year (1)	GDP (2)	Depreciation (3)	NDP Level (4)	NDP Growth rate (5)	Depletion (6)	EDP Level (7)	EDP Growth rate (8)	EDP/NDP (9)
1988	799,182	67,162	732,020		26,966	705,049	–	96.32
1989	925,444	72,384	853,060	16.54	17,939	835,121	18.45	97.90
1990	1,077,237	82,456	994,781	16.61	13,564	981,217	17.49	98.64
1991	1,248,011	98,398	1,149,613	15.56	16,232	1,133,381	15.51	98.59
1992	1,351,559	109,082	1,242,477	8.08	4,495	1,237,982	9.23	99.64

Table 9. Net accumulation at current prices, 1988–1992 (million pesos)

Year	Produced assets: I_g (1)	Produced assets: CFC (2)	Produced assets: $A_{p.ec}$ (3) = (1) – (2)	Non-produced assets: $I_{np.ec}$ (4)	Non-produced assets: $Use_{np.ec}$ (5)	Non-produced assets: $A_{np.ec}$ (6) = (4) – (5)	Total net accumulation of economic assets $A_{p.ec} + A_{np.ec}$ (7) = (3) + (6)
1988	149,193	67,167	82,026	(531)	26,966	(27,497)	54,529
1989	199,900	72,384	127,516	(94)	17,939	(18,033)	109,483
1990	260,115	82,456	177,659	(3,973)	13,564	(17,537)	160,122
1991	252,327	98,398	153,929	4,971	16,232	(11,261)	142,668
1992	288,401	109,082	179,319	(12,544)	4,495	(17,039)	162,280

I_g = gross domestic capital formation, GDCF.
CFC = depreciation.
$A_{p.ec}$ = (I_g – CFC) net domestic capital formation, or net accumulation of produced assets.
$Use_{np.ec}$ = Depletion.
$I_{np.ec}$ = Other accumulation of non-produced assets.
$A_{np.ec}$ = Net accumulation of non-produced assets.

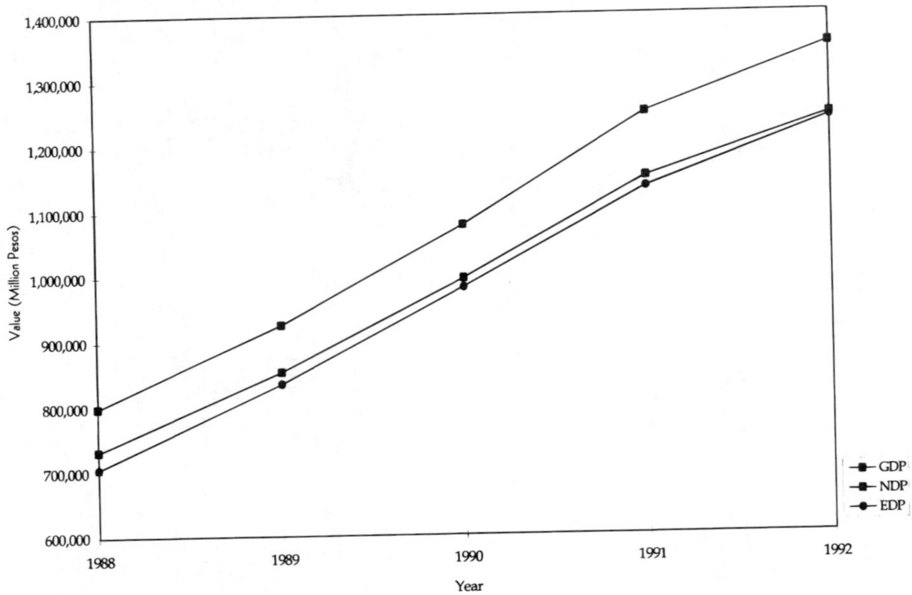

Figure 2. GDP, NDP and EDP (1988–1992).

resources and does not yet cover fishery, land and other resources. The negative entries in $I_{np.ec}$ are mostly due to conversion of forest to non-forest use, but do not reflect the value increase of assets into which they have been converted. For example, when a forest had been converted to agricultural land, the improvement in the land has been incorporated as I_g, but the value of the land per se has not yet been accounted for in $I_{np.ec}$, since there are no estimates yet for land as a resource. In spite of these data limitations and coverage, Table 9 demonstrates how the concept of extended net accumulation can be operationalized. Total net accumulation of produced and non-produced economic assets has been positive in the Philippines, indicating at least weak sustainability (of overall capital maintenance) of economic activity in the country.

Conclusions

The following are the main findings of the PSEEA project:

(a) it is feasible to compile the asset accounts using the SEEA framework;
(b) data for environmental accounting are inadequate;
(c) administrative data and special studies can be transformed into useful data for environmental accounting;
(d) involvement of planning, environmental and statistical agencies and resource institutions is critical in environmental accounting;
(e) training on the SNA, SEEA, environmental economics, resources and valuation methodologies is required;

(f) accounting work should be institutionalized within the agency that compiles the national accounts.

The operationalization of the PSEEA is being undertaken as an inter-agency effort with the NSCB assuming the lead role. The participants include senior and junior level government officials from agencies collecting and using environmental and natural resource data. Assisting the Inter-Agency Working Group are part-time foreign consultants from the United Nations Statistics Division (UNSD) and local consultants from the Philippine Institute for Development Studies (PIDS).[11] The UNSD provided assistance in the development and compilation of the PSEEA, while the PIDS provided assistance in valuing the three resources.

The environmental and data producing agencies provide the data and technical expertise. The inter-agency approach enables the NSCB to draw on the expertise of marine biologists and other natural resource specialists, economists, and statisticians. Training to build up knowledge and interest among the participants is an essential part of the programme. This approach is to ensure institutionalization of the environmental accounting process in the government bureaucracy.

Notes

1. The overall objective of the UNDP-IEMSD is to assist the government in systematically integrating environment and development concerns in policy making and in sectoral regional planning processes and to help formulate and test the means to make such integration possible. One of the major targets of the programme is to develop environment and natural resource accounts that can be institutionalized in the government bureaucracy.
2. The BAS is the agency that produces official data on fish production which is used by the NSCB to estimate the gross value added for fishery. The NSCB adopted the BAS production data while the ENRAP study (see Chapter 6) made use of the BFAR production data.
3. Fishing effort and fish catch were compiled by BFAR prior to 1980. This exercise was discontinued, however, by the BAS which took over data collection from BFAR.
4. Horsepower per tonnage by fishing gear was estimated by ENRAP based on incomplete data from BFAR.
5. Purse seine equivalent is used to standardize fishing efficiency for the different gears because purse seine accounts for a large proportion of marine fish landed.
6. The horsepower equivalent of a full-time fisherman is set at 0.18 per day (Southeast Asia average), based on a study by Karim (1985) on energy expenditure. The 6 hours is assumed to be the number of hours worked by a fisherman per day (full-time fishing). The 40% is the mean percentage of days of actual part-time fishing compared to full time fishing; 11% is the mean percentage of actual occasional fishing compared to full time fishing.
7. The producer's price for fish is obtained from the BAS Survey of Fish Landing Centers.
8. The RP-German Forest Inventory is a joint project of the two governments. The project paved the way for the preparation of new forest maps for the whole country, based on aerial photographs and satellite imagery, the inventory of standing volume, regeneration and minor forest products, and the compilation of forest statistics by region and province.
9. The MPFD is a nationwide aggregate blueprint for the development of the forestry sector across a 25-year horizon.
10. The margin for normal return to capital represents an allowance for normal profit and risk associated with the production activity and is approximated by the opportunity cost of capital or what the operator would earn from an alternative (next-best) investment.
11. The consultants from PIDS are part of the ENRAP.

6
Philippines: Environmental accounting as instrument of policy

MARIAN S DELOS ANGELES AND HENRY M PESKIN

In the Philippines, three phases of the Environmental and Natural Resources Accounting Project (ENRAP) have been implemented. The current Phase 4 includes more policy exercises, regional analysis and initial institutionalization. This chapter presents the findings of ENRAP in the following key areas: waste disposal services, pollution loads and control costs; pollution damage; natural resource depreciation; and non-market household production. Various applications to national policy and site-specific concerns have arisen from the ENRAP-generated database. Improvements in pollution management and shifts to better production technologies are called for with the anticipated high growth rates and trade liberalization. Assisted by local government units and non-government organizations, the government should widen its monitoring coverage to include households, agriculture and small- and medium-scale industries in pollution management strategies. Flexibility in policy making should allow the decentralized bureaucracies to respond to local environmental problems.

To address institutionalization requirements, the following data management activities implemented during ENRA and the corresponding problems encountered are discussed: estimation of natural resource depreciation, waste disposal services and their value; measurement of environmental damages; and direct nature services. Support is needed for improving data generation, storage and maintenance, adding more analytical capacity to various agencies and conducting more regional, site-specific accounting to support current decentralization efforts. Upgrading the Philippine statistical system will require following these key recommendations. The data requirements of ENRA are quite challenging, but only comprehensive data frameworks and models can track intersectoral linkages crucial to national planning.

The views, expressions and opinions contained in this report are the authors' and are not intended as statements of policy of either USAID or DENR.

K. Uno and P. Bartelmus (eds.), Environmental Accounting in Theory and Practice, 95–111.

Introduction

Three phases of the Philippine Environmental and Natural Resources Accounting Project (ENRAP) have been implemented, starting with forest resources and eventually covering most natural and environmental resources. The current Phase 4 includes policy exercises, regional analysis and initial institutionalization. Efforts are also underway to institutionalize the programme within the Philippine statistical system.

Since interactions between the economy and the environment are complex, understanding the system requires comprehensive models which need comprehensive data frameworks. Moreover, rational management of the environment requires information on both the benefits of protecting (or enhancing) the services of the natural environment and the opportunity costs of these protection policies. The Philippine Environmental and Natural Resource Accounts (ENRA) provide a comprehensive data framework that addresses these data needs by rectifying the following deficiencies in standard national accounting: lack of coverage of non-market household production, especially that which draws on natural resource inputs; non-measurement of the unpaid use of waste disposal services provided by air and water resources; lack of measurement of the benefits from direct use of environmental assets such as for recreation; failure to account for natural resource depreciation; and the neglect of damages arising from pollution. Two years, 1988 and 1992, were selected for the implementation of ENRA because of the availability of both economic and environmental information.

The ENRA cover two forms of non-market, household production: own-use agricultural production in mountain areas with 18–30% slope, and fuelwood gathering from forest lands. In 1988, the conventional accounts excluded farmers producing in these forest lands because of their small farm sizes, high mobility, and often illegal status. Unmonitored fuelwood production, through gathering of wood products from the forest, was estimated based on special surveys conducted by the Department of Energy. The monetary values of both forms of unmeasured household production are based on the value of labour expended in these forest-based activities.

The ENRA cover a majority of Philippine natural resources: (1) dipterocarp forests which compose 90% of Philippine forests; (2) small surface-dwelling fisheries which comprise 50% of fish production; (3) copper and gold which contribute 80% of mineral production value; and (4) upland soils that are located in areas of at least 8% slope, covering 70% of the total land area.

The value of waste disposal services provided by air and water resources was estimated by the cost of reducing uncontrolled residuals to harmless levels through least-cost, available technology. This cost, expressed in annualized terms, includes both capital cost, and maintenance and operating expenses. Damage to health and productivity are estimated through dose–response relationships adapted from the general literature and information generated by

Philippine studies on specific environmental problems.

Findings

The consolidated accounts

The consolidated ENRA supplement the traditional income accounts in a way that leaves the original tables intact. Modifications are made below the solid line in Table 1, where the 1988 and 1992 results are shown. The aggregate effects of all ENRA adjustments are very small. The net effect is an increase in net domestic product (NDP) for 1988, and a slight decrease for 1992. In fact, the difference between the original and the ENRA-modified NDPs is less than the 'statistical discrepancy'. The growth rate during the period 1988–1992, measured by the ENRA-adjusted NDP, is 6.7%, one point lower than the growth rate measured by the conventionally estimated NDP.

Net environmental benefit (disbenefit)

'Net environmental benefit' (NEB) is defined as the difference between the absolute value of the services generated by environmental assets (waste disposal and other positive services such as recreation) and any environmental damage due to the consumption of these services. It serves both as a balancing

Table 1. Modified Philippine national income and product accounts, 1988 and 1992 (in million pesos, 1988 prices).

EARNINGS APPROACH	1988		1992	
Compensation of employees		187,806		224,037
Indirect taxes		59,454		40,912
Depreciation (produced assets)		67,162		69,361
Net operating surplus		487,478		523,132
CHARGES AGAINST GROSS DOMESTIC PRODUCT (GDP)		801,900		857,443
Capital depreciation (-)		-67,162		-65,900
CHARGES AGAINST NET DOMESTIC PRODUCT (NDP)		734,738		791,543
Natural-resource Inputs to Unmarketed Household Production (+)		6,867		4,665
a. Upland agriculture	1,950			
b. Fuelwood	4,917		4,665	
Environmental Waste Disposal Services (-)		-26,584		-25,381
a. Air	-4,311		-5,102	
b. Water	-22,273		-20,279	
Net Environmental Benefit (Disbenefit)		25,455		24,866
Natural Resource Depreciation (-)		-2,508		-7,680
a. Forests	-936		-504	
b. Fisheries	-838		-6,560	
c. Minerals	-354		-26	
d. Soils	-380		-590	
CHARGES AGAINST MODIFIED NDP		737,968		788,013
Capital depreciation (+)		67,162		65,900
Natural resource depreciation (+)		2,508		7,680
CHARGES AGAINST MODIFIED GDP		807,638		861,593

PRODUCT APPROACH	1988		1992	
Personal consumption		558,765		653,271
Government consumption		72,183		79,904
Capital formation		149,193		183,384
Exports		226,910		277,685
Imports (-)		-215,292		-308,616
Statistical discrepancy		10,141		-28,186
GDP		801,900		857,443
Capital depreciation (-)		-67,162		-65,900
EXPENDITURE ON NDP		734,738		791,543
Unmarketed Household Production (+)		6,867		4,665
a. Upland agriculture	1,950			
b. Fuelwood	4,917		4,665	
Environmental Damages (-)		-2,709		-3,374
a. Air	-1,296		-1,943	
b. Water	-1,413		-1,431	
Direct Nature Services (+)		1,580		2,859
a. Diving (coral reefs)	1		3	
b. Visits to national forest parks	13		18	
c. Beach use	1,566		2,838	
Natural Resource Depreciation (-)		-2,508		-7,680
a. Forests	-936		-504	
b. Fisheries	-838		-6,560	
c. Minerals	-354		-26	
d. Soils	-380		-590	
MODIFIED NDP		737,968		788,013
Capital depreciation (+)		67,162		65,900
Natural resource depreciation (+)		2,508		7,680
MODIFIED GDP		807,638		861,593

entry as well as a measure of the net services of nature. These net services, summed over time and discounted, provide an indicator of the asset value of the environment. Changes in this asset value over the accounting period measure the economic depreciation of the environment.

Data limitations in the current measurement of NEB precluded its use in computing the depreciation of environmental assets. Instead, it is used to indicate areas for targeting pollution control. A positive value means that we are benefiting from the waste disposal services provided freely by air and water resources more than we are suffering from the damage that results from generating residuals. In the Philippines, the total value of waste disposal services exceeds the damage caused by waste disposal. Complete pollution control is not warranted, therefore: the opportunity costs of total control exceed the expected benefits. Less ambitious and carefully targeted control policies are called for.

Waste disposal services, pollution loads and control costs

As the Philippines is an archipelago, households, government and industries use the waste disposal services provided by water resources more than waste disposal services provided by air (Figure 1). As noted above, these services are estimated by the costs which would face users of these services if they were denied access to air or water. More specifically, pollution loads from all sectors are estimated, then estimates are made of the costs of reducing these loads to harmless levels (often about 90% of uncontrolled levels) associated with a locally available technology. Air pollutants have the following origins:

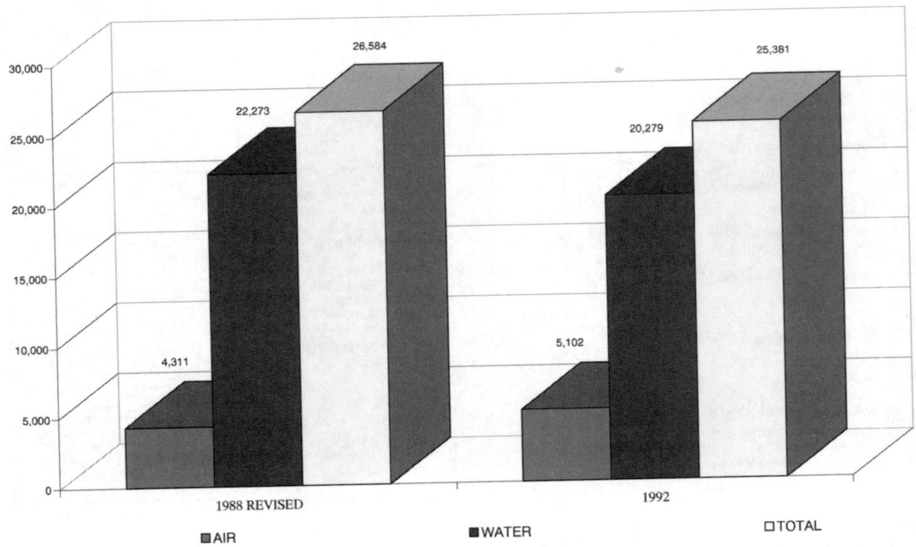

Figure 1. Waste disposal services from air and water, 1988 and 1992 (million pesos, 1988 prices).

fine particles or particulate matter 10 (PM_{10}) are generated mostly by industry, public transport and households; sulphur oxides (SO_x) by the power sector and industry; nitrogen oxides (NO_x) by households (private car use) and industry; and volatile organic compounds (VOC) and carbon monoxide (CO) by household activities (fuelwood burning and private car use). Households are the leading sources of organic water pollution, in terms of biochemical oxygen demand 5 (BOD_5). Surface runoff from agricultural and forest lands generates most of the suspended solids and nitrogen discharges.

The findings indicate that exclusive focus on industries as the target for pollution control would not improve air and water quality. Major reduction in water pollution requires targeting households and agriculture, perhaps through public investments for improving sewerage facilities and sanitary services and by encouraging agricultural management practices to limit soil erosion. In the long term, a broad-based growth would allow for good housing facilities and waste disposal as well as a reversal of lowland to upland migration rates. In the case of air pollution, changes in practices of both households and industries need to be encouraged.

These findings also reveal the need for a concerted effort at pollution management involving various sectors of society and several government agencies, which include not only the Department of Environment and Natural Resources (DENR) but also those in charge of sewerage services, agriculture and traffic control. Collaborative efforts between the Department of Agriculture and DENR would have to improve, considering that 33% of erosion results from agricultural land, and 67% from forest lands that have been degraded by upland agriculture, destructive logging and poorly constructed (logging) roads.

Pollution damage

Erosion from degraded uplands and pollution from mine tailings cause productivity losses to fisheries dependent on coral reefs, lower the lifespan of Magat, Pantabangan, Ambuklao and Binga dams, and reduce irrigated land area (Table 2a). Laguna Lake fishery decline results from pollution generated by households and industry. Damage in 1992 was lower than in 1988 because of less rainfall and lower pig, poultry and duck production. Health damage from air pollution consists of forgone earnings and medication expenses from worsened respiratory illnesses from PM_{10} and lead exposure; increased hypertension from lead; and premature deaths caused by both pollutants (Table 2b). Air pollution control is a priority for Metro Manila. In this metropolis, health damages from air pollution total P2557 million, exceeding the approximate cost of air pollution reduction by at least P1530 million. In contrast, the costs of complete water pollution control exceed the benefits. For example, reducing surface runoff costs P7.8 million and yields only P0.8 million of benefits. Control of household wastes costs P6.9 million and yields only P0.6 million of

Table 2a. Environmental damage from water pollution, 1988 and 1992.

	Value of Damages (million pesos,1988 prices)	
	1988	**1992**
Total	**1,413**	**1,430**
Health (increased incidence of water-borne diseases)	**596**	**615**
Foregone earnings from morbidity	43	131
Medication cost	91	138
Foregone earnings due to premature death	462	346
Offsite damages	**817**	**815**
Coral reefs (foregone fish production)	694	726
Reservoirs (reduced lifespan of dams)	58	58
Agricultural production (foregone rice production in reduced irrigated land area)	12	12
Inland fisheries (foregone fish production in Laguna de Bay)	53	19

Sources : Buenaventura *et al*.1996; Cortez *et al*. 1996

Table 2b. Health damage from air and water pollution, 1988 and 1992.

	Value of Damages (million pesos, 1988 prices)	
	1988	**1992**
Total Health Damages	**1,893**	**2,557**
From air pollution	**1,297**	**1,942**
Effect of PM and lead emissions, Metro Manila		
Foregone earnings from morbidity	48	38
Medication cost	658	1,194
Foregone earnings due to premature death	591	710
From water pollution	**596**	**615**
Increased incidence of water-borne diseases		
Foregone earnings from morbidity	43	131
Medication cost	91	138
Foregone earnings due to premature death	462	346

Source : Cortez *et al*. 1996

benefits. These results suggest that complete control is unwise. Rather, controls should be carefully targeted towards areas where the benefits attained are more commensurate with the costs.

Natural resource depreciation

The ENRAP III results indicate that renewable resource (forest and fishery) depreciation is higher than that of non-renewable resources (mineral and

Table 3. Depreciation of natural resources, 1988 and 1992 (in million pesos, 1988 prices).

	1988	**1992**
Fisheries	838	6,560
Forests	936	504
Minerals	354	26
Soils	380	590
Total	**2,508**	**7,680**

Sources: Francisco 1994 ; delos Angeles 1996; Logarta 1996; Padilla and Cortez 1996.

upland soils; Table 3). These depreciation estimates were generated as follows. Dipterocarp forest depreciation was estimated by changes in the asset values of old-growth and residual forests, assuming optimum cutting harvest cycles and associated growth and yield estimates. Mineral resource depreciation was calculated by application of user cost methods adapted from El Serafy (1989), using discount rates approximating the opportunity cost of private capital. The depreciation of small surface-dwelling fisheries was based on estimation of economic optimum catch levels and the value of rents generated. Lastly, the depreciation of the stock of upland soils was based on estimated lifespan and changes in agricultural land rent. Overall, the numbers generated through these methods are much lower than what would be derived through the application of replacement cost concepts, such as the net price method and the nutrient replacement values generated earlier by the World Resources Institute (WRI).

To attain sustainable levels of catch, fishing of small, surface-dwelling fish would have to be reduced by 50% through the allocation of fishing rights (Table 4). It is not necessarily optimal to conserve mineral resources. Mining should continue where economically viable, provided that the earnings are invested to enhance the economy's productivity and environmental controls are in place.

Overall, even with the addition of the depreciation of natural assets, capital formation in the Philippines remains positive. The decline in certain environmental assets is not, in itself, a cause of concern, considering that most of these assets have man-made and human capital substitutes. To the extent that certain natural assets are considered unique and non-substitutable, they should, of course, be protected at levels to assure their adequacy for future generations.

Table 4a. Overfishing of small surface-dwelling fish, 1948–1988.

	Maximum Economic Yield (MEY)	Actual Open-Access	
		Values	Percent Deviation from MEY
Fishing effort (horsepower)	261,600	537,900	106
Catch (metric tons)	569,000	457,000	(20)
Revenues (million pesos)	7,414	5,958	(20)
Rents (million pesos)	7,128	0	(100)

Source: Padilla and de Guzman 1994

Table 4b. Implications of fishery decline, 1988–1998.

a.	Foregone earnings	2,069	Million pesos (1988 prices)
b.	Displaced workers	120,630	Fishers

Note: Values are deviations from projected 1988 levels based on current growth targets.

Source: Padilla and Cortez 1996

Non-market household production

Unmeasured household production, including the subsistence use of the uplands for food consumption and the forests for fuelwood, is an important addition to national production for both years. These activities are also considerable sources of forest land degradation. As Figure 2 reveals, inclusion of household activities has a considerable effect on the aggregate economic measures. Including these items in the official income accounts would have increased outputs for the agriculture and forestry sectors by 2% and 30%, respectively. There is also a need to improve the monitoring of informal forest-based activities. Official statistics reflect only 14.2 million m^3 of the 40.4 million m^3 estimated depletion of forested land.

Policy implications from ENRAP-generated studies

Various applications to national policy and site-specific concerns have arisen from the data base generated by ENRAP.

Policy adjustment

Exploring growth targets

Simulation of a high-growth scenario suggests greatly increased pollution, assuming the technologies in place as of the late 1980s. Thus, high-growth

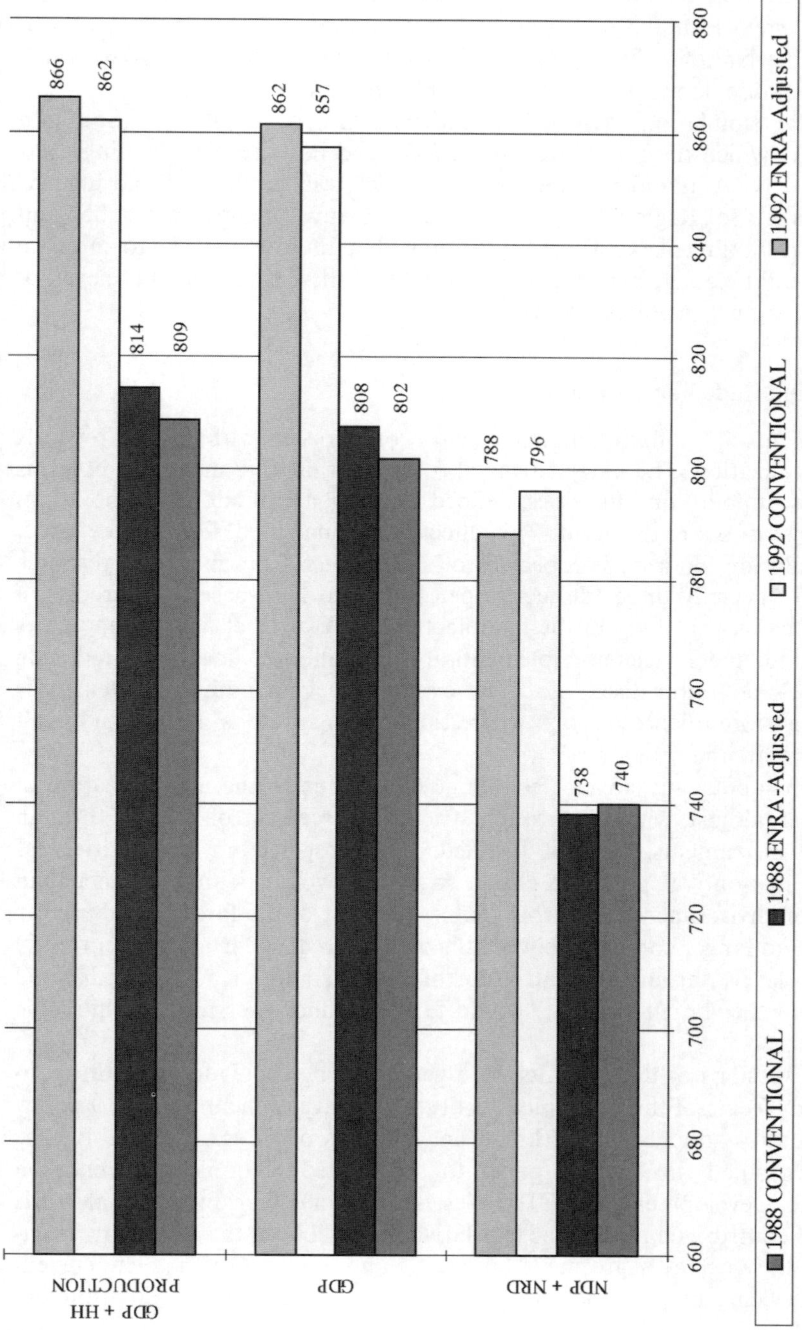

Figure 2. Alternative Philippine NDP and GDP: effects of household production and natural resource depreciation on national measures (1988 and 1992) (in billion pesos, 1988 prices). NRD = national resource depreciation; HH = household.

targets may require improvements in pollution management and shifts to better production technologies. These findings were recently used as inputs into the current formulation of the Philippine Agenda 21 Action Plan coordinated by the National Economic and Development Authority (NEDA) which chairs the Philippine Council for Sustainable Development.

A similar simulation exercise was implemented for Region 11 (Southern Mindanao), which the NEDA regional office used in its recent planning exercises. An ENRA-modified input-output model was used for exploring the growth targets for Region 11, where considerable investments, both public and private, are taking place. The results show that there is a need to improve environmental management practices specially at these early stages of development, to minimize problems in the near future.

Anticipating trade liberalization

Improvements in pollution management would occur with moves towards trade liberalization. The likely effects of trade reforms were measured in terms of changes in pollution intensities, defined by the share of additional pollution control cost of sectoral output. Two effects were indicated. On the one hand, higher pollution intensity is expected to be generated from increased resource extraction, specially under de facto open access to land-based resources. On the other hand, shifts within the manufacturing sector to cleaner technologies are expected under stricter implementation of pollution laws, thus reducing this sector's pollution discharges. The overall effect on pollution intensity is less with a more open economy, where the exchange rate is flexible and tariff reform is more uniform.

Concerns about any adverse economic effects due to the internalization of environmental cost, whether through stricter enforcement of laws or through economic instruments, are not justified. For most sectors, competitiveness would be lost only at pollution charge rates that would be much higher than the cost of environmental controls (Table 5). Most of the forest-based sectors which are no longer competitive would have to be reduced from their currently inefficient levels through liberalization of imported logs and finished wood products in the short-term. This would further reduce pressure on Philippine forests.

Earlier versions of these results were generated by the Trade and Environment Study Team of the Philippine Institute for Development Studies, a policy research entity attached to NEDA, using findings of ENRAP Phase II. The team contributed the country paper to the United Nations Conference on Trade and Development (UNCTAD)-assisted studies worldwide (Intal *et al.*, 1995) and contributed also to the legislative discussion on the Philippine ratification of the General Agreement on Tariffs and Trade (GATT). The current results are being used in the ongoing ASEAN-wide Trade and Environment Study.

Table 5. Likely effects on competitive advantage under full cost internalization.

PSIC [a] No.	Sectors with Significant Foreign Trade	Share of Pollution Control Cost to Output (%)	Hypothetical Environmental Tax Measured as Control Cost Share to Output, at Which Competitiveness is Lost (%)		
11	Agricultural crops	2.3	coconut/copra	:	10.0
14	Fishery	0.1	aquaculture	:	50.0
15	Forestry	28.9	forestry	:	10.0
21	Metallic ore mining [b]	8.0	copper	:	n.e. [c]
32	Textile, etc.	0.4	textile	:	0.5
33	Wood manufacturing	1.3	lumber	:	already uncompetitive
34	Paper and printing	0.7	paper and paper products	:	already uncompetitive

a Philippine Standard Industry Classification

b Gold and other precious metals, 9%; copper ore mining, 8%; other metallic mining, 34%.

c/ Not estimated

Source: Medalla and Intal 1996

Pollution control

The gasoline lead reduction target

An example of deriving price reform implications from the accounting results was applied on the gasoline lead reduction targets. During the transition between now and the full use of unleaded gasoline targeted for 1998, the use of low-lead gasoline would still result in health damages for Metro Manilans. Users could be charged pollution fees equivalent to the damage. These fees could then be used to finance the subsidy granted to unleaded gasoline producers, currently partly financed through general taxation. The resulting price increase should cause more shifts to unleaded gasoline use, resulting in less health damage, provided effort is also exerted to reduce the emission of aromatics.

Including small- and medium-scale industries in pollution management strategies

The government needs to widen its monitoring coverage with the assistance of other parties, such as local government units (LGUs) and non-government organizations (NGOs). Current monitoring of water pollution discharges focuses on those establishments that generate at least 30 m^3 of wastewater daily. However, for some sectors, such as pig production, most of the polluters are small-scale and are thus unmonitored.

Support of regional analyses

Detailed sectoral work conducted by ENRAP reveals the differences in distribution of pollution loads at the regional level. Thus, flexibility in policy making should include mechanisms that would allow for the recently decentralized bureaucracies to respond to local environmental problems expediently.

Achieving rice self-sufficiency

The ENRAP explored the environmental implications of achieving self-sufficiency in rice. The results show that the two objectives of self-sufficiency and the minimization of environmental damage are not necessarily conflicting because farmers who raise upland rice tend to construct water conservation structures that also enhance soil conservation. In addition, proper preparation of the planting area allows for slow leaching of nutrients into the lower soil levels, minimizing the discharge of such nutrients through surface runoff.

Managing air pollution and groundwater depletion problems in Metro Cebu

Air pollution dispersion modelling was implemented for Metro Cebu, the country's second largest city. The results indicate those areas where the concentration of SO_x and PM_{10} pollutants exceeds ambient standards. Damage to health from air pollutants in Cebu was estimated to be about 7% of the Metro Manila estimate.

The groundwater resources of Cebu island have long exhibited signs of the effects of depletion, including saline water intrusion in specific areas and drying up of wells. The ENRAP estimate of groundwater depreciation yields a monetary measure of what the approximate price should be for continued extraction. The current private cost of extracting water from the island's aquifers reflects only the obstruction costs. Implementing a charge for groundwater use would require improvements in the monitoring of actual use rates.

Estimation and data issues

To explore institutionalization requirements, the various data management activities implemented during the ENRAP are discussed.

Estimating natural resource depreciation

The initial step in formulating depreciation tables is building the physical accounts. This process allows us to trace the factors that contribute to changes in resource stocks. The information required for physical accounting includes:

(a) regular reports on the status of natural assets;
(b) the flow of goods and services from these assets per unit of time;
(c) reduction of stocks due to both legal and extralegal activities; and
(d) stock additions through natural growth and investments in renewal or rehabilitation.

Of the four items, only item (b) is consistently available from Philippine official records which mostly track the legal activities. The following problems were experienced with the other three:

(a) lack of regularity, completeness and comparability of resource inventories;
(b) insufficient monitoring of extralegal draw-down, particularly of banned use (the government usually ceases monitoring when activities have been banned[1]); and
(c) dearth of statistically derived information on stock additions – most information is from case studies which differ markedly from each other.

Estimates of resource stocks, which rely on specialized techniques such as ground assessment of forest stands, fishery stock assessment and remote sensing techniques, are needed periodically, not just on an annual basis. There is a general tendency, however, among resource specialists to aim for complete inventories, and not to rely on sampling techniques. Lack of sampling makes periodic data gathering unnecessarily expensive. Comparability of information generated over time needs to be explicitly planned for. There may be a need for different sampling frames that would allow for representativeness at the national and sub-national levels. Here, collaborative work among the research community, various offices of DENR, the National Statistical Coordination Board (NSCB), and the National Statistics Office (NSO) should be strengthened.

ENRAP made use of available information, usually case studies for item (c), and special information sources for items (c) and (d). An important consideration in using specialized information sets is the statistical validity of the data. Incomplete discussion of methodology, data annotation and availability of guidance from the original data generators pose problems. Retrieval of these specialized studies, particularly when obtained from unpublished sources, has been difficult. Since such literature has limited circulation, it has likely not benefited from serious review and may, therefore, have grave limitations.

Stocks and flows were subsequently transformed into value terms through specific methods that depend on characteristics unique to the resource. The key variables to valuation are:

(a) output and output prices;
(b) input and input costs;
(c) profit and risk margins.

Item (a) is readily available from official statistics, except for undermeasurement of household use and non-measurement of smuggled output (e.g. illegally cut logs, fish selling at sea, small-scale, mined gold bought by foreign traders instead of by the Central Bank, etc.). Item (b) has similar problems. In addition, often only the value of inputs is reported while the corresponding physical quantities are not easily derived. Item (c), normally computed as a residual, may be grossly understated for monopolistic or oligopolistic operations. Important sources for these data are the records of the Securities and Exchange Commission, to which access is limited.

Much of the above information required for measuring natural resource depreciation is also needed for natural resource management, whether at the micro- or at the macro-level. The challenge to generate such information is greater now that a policy of granting more access to numerous, smaller-scale users and to LGUs is underway.

At the same time, the existing management information sets that DENR requires from licence holders (such as annual operations plans) and from its monitoring and enforcement of regulations could be built upon. These systems need to be computerized, however, so that data can be processed in a timely manner and are stored securely. Such computerization will require strengthening DENR statistical offices.

Estimating waste disposal services

Monitoring pollution discharges is conducted by DENR in the process of enforcing pollution control laws. To cope with limited resources, DENR has adopted a deliberate bias towards monitoring the large polluters, rather than the small polluters. This bias is probably not warranted. Even in the case of point sources of toxic and hazardous pollutants, much is generated by machine shops, paint shops, metal finishers, and other small enterprises.

Faced with limited monitoring information, ENRAP applied pollution coefficients, drawn from various sources, to national estimates of output rates or input use (for process pollution) and energy use (for pollution associated with the specific fuel types). Because of the need for such engineering and economic data, it is apparent that a large part of ENRA data requirements does not fall within the narrow definition of 'environmental statistics'. Unfortunately, there is a lack of pollution coefficients that reflect local production and engineering practices, economics (factor and output prices), and pollution control and abatement policies.

The current effort of building the ENRAP Environmental–Economic Database (EED) strives for more localized estimates of pollution loads. Thus, ENRAP is building a geographically based database that integrates all relevant data and information gathered and generated by the project thus far. Data from two regional studies, one in Region 11 and the other in Region 7 (with a focus on Metro Cebu), designed to study the importance of different mixes of economic activities and their locations on environmental problems, were integrated into the main database.

In addition to secondary data sources, ENRAP gathered primary sectoral data at the establishment level through telephone inquiries and/or surveys (limited to less than 200 establishments). Establishment lists for key sectors provided by industry associations (Philippine Sugar Commission, Mines and

Geosciences Bureau, Fertilizer and Pesticides Authority, Forest Management Bureau, Department of Energy, etc.) were also checked against those of the census lists and government agencies with regulatory and monitoring functions over sectors. After validation (for example, with regard to sector classification based on primary economic activity), more accurate and complete lists were generated.[2] Basic data include production volume/capacity, employment, fuel consumption, pollution control devices in place, abatement cost, effluent and emission concentrations, etc. When there is sufficient confidence in the information, pollution load coefficients and output levels were refined. An important component of ENRA is the treatment of the household sector as a source of pollution. Household-based consumption activities generate pollution, including fuelwood burning and private car transport.

The method used by ENRAP to estimate the value of waste disposal services follows economic principles: the costs of control are used as a proxy for the establishment's willingness to pay for waste disposal services. The incremental costs of pollution control include both the prospective capital costs (expressed in annualized terms) and maintenance and operating costs.

Measuring environmental damage

Environmental damage was estimated through the application of dose–response relationships which measure the effects of pollution on health, ecosystems and economic productivity.[3] In some cases, such as the health impacts of water pollution and the effects of siltation on coral reef fisheries, expert opinion was sought on either the extent or the attribution of damage. A major problem is still the lack of reliable time series on even the narrow set of environmental statistics such as ambient air or water quality. The second is lack of information on human exposure to deteriorating environmental quality. Proper design of the corresponding information system will require superimposing economic with environmental statistics with such exposure information that reflects local conditions and ecosystems.

An important start may be provided through the adoption by the official statistical system of a coding scheme for all respondents of standard surveys or censuses that would tag the observation units on an ecosystem address. The same coding scheme would have to be applied to a Geographic Information System (GIS) and other natural resource data. Given the Global Positioning System (GPS) technology now available, the method should be feasible to implement in the near future.

Direct nature services

Information on these services is not directly available from the formal statistical agencies. Data are generated through highly specialized surveys such as those conducted by economists in connection with contingent valuation studies. Few such studies have been attempted in the Philippines. In the future, this component of ENRA is likely to be undertaken by specialized groups with assistance from the official statistical system.

Concluding remarks: moving the Philippine statistical system towards ENRA

While the interactions between the economic and environmental systems are complicated, they are nonetheless important to understand. Accounting for these interactions is necessary to make decision making more environmentally responsible and to make environmental targets more economically feasible. In the long-term, official agencies will have to fill in the various roles initiated by the ENRAP programme. The statistical system will likewise need to examine seriously expanding its coverage to include the basic information needed for the conduct of ENRA. This will necessarily mean handling larger data sets and more resources for generating and maintaining the key statistics. On the other hand, it could also require agencies to implement special projects. Improved coordination (including standardized definition of terms) among the various components of the Philippine statistical system would minimize duplication of effort among official and unofficial groups in the gathering of data.

Regardless of the components of an ENRA statistical system, support is needed for improving the data generation, storage and maintenance; adding more analytical capacity to various agencies and groups; and conducting more regional, site-specific accounting to support current decentralization efforts. Thus, the eventual evolution of the statistical system will require following the key recommendations previously mentioned. While these recommendations will require higher budgetary allocations, they will allow for economies of scale and scope in the longer term.

The major problem that should be tackled first is the lack of a GIS-based data set which relates physical environmental loads, identifiable by location of source, to impacts, and impacts to economic values; identifies loads and impacts on the pertinent receivers (airsheds, water bodies, land); and regularly reveals the state of natural resources. Pollution-related data are available, but only for some years, in the case of the National Capital Region (NCR), Laguna Lake and Pasig River. Forest cover statistics on the other hand are outdated, with 1992–1993 as the latest GIS data for certain regions, and 1988–1989 at the national level.

The data requirements of ENRA are quite challenging. However, only comprehensive data sets allow for tracking intersectoral linkages which is important to national planning. Moreover, the analysis derived from such an information system benefits from economies of scope. Future potential analysis may also be tailored to site-specific issues that are crucial to LGUs.

Notes

1. A recent exception is the attempt by DENR to examine the effectivity of the logging ban and the controls on the transport of logs.

2. In general, associations have explicit and implicit membership criteria, while government lists are also limited to those establishments actually submitting reports. Thus, both sources have limited coverage.
3. This relationship shows how (exposure to) environmental quality and socioeconomic characteristics affect health or the production of an ecosystem-based commodity.

7
USA: Integrated economic and environmental accounting: lessons from the IEESA

J STEVEN LANDEFELD AND STEPHANIE L HOWELL

In April 1994, the United States Bureau of Economic Analysis (BEA) introduced a framework of Integrated Economic and Environmental Satellite Accounts (IEESAs) designed to cover the interactions of the economy and the environment. Modelled on the United Nations handbook on *Integrated Environmental and Economic Accounting*, they are constructed as satellite accounts to supplement, rather than replace, the existing accounts. This paper considers BEA's experiences in the development of the IEESAs, and suggests some lessons that may be useful for countries considering similar plans.[1] Over the years, the national economic accounts have benefited from discussion and critique of concepts, source data, and estimating methods. The same is to be expected for the IEESAs.

Background

Measures of economic activity such as gross domestic product (GDP) and national wealth shape perceptions and policies in profound ways. Better understanding of the critical interaction between the economy and the natural environment requires better measures of that relationship. However, the construction of accounts that illustrate these interactions involves a diversity of problems, including the controversial nature of the issue itself, theoretical approaches and data limitations. Countries attempting to move forward with environmental–economic accounts should carefully consider each of these issues in developing their research and implementation plans.

Determining the objective of the accounts is an obvious starting place, but one deserving of special mention, as the atmosphere surrounding environmental discussions lends urgency to this first step. The state of the environment has been a source of international concern for several decades, but approaches to its evaluation and maintenance or restoration still diverge widely, and the intensity of the emotions on all sides of the debate creates an atmosphere where any participation may imply advocacy of some form.

Economic vs welfare accounts

Since the founding of the US national accounts there has been an ongoing debate regarding the treatment of natural resources and the environment, as

K. Uno and P. Bartelmus (eds.), Environmental Accounting in Theory and Practice, 113–129.

well as the treatment of a whole set of broader welfare-based measures of economic and social progress. One school, exemplified by Kuznets (1946), favoured development of a much broader set of welfare-orientated accounts that would focus on sustainability and address the externalities and social costs associated with economic development.[2] Another, exemplified by Jaszi (1971), insisted that the national accounts must be objective and descriptive and thus based on observable market transactions. Jaszi felt that, conceptually, the accounts should be extended to treat the economic discovery, depletion, and stocks of natural resources symmetrically with plant and equipment and other economic resources. The absence of observable market transactions and the wide uncertainty and subjectivity associated with such estimates led him to conclude, however, that they should not be included in the accounts.[3]

In the 1960s and early 1970s another more environmentally focused move to broaden the accounts arose out of concern about environmental degradation and fears that the world was running out of resources and approaching the 'limits to growth'.[4] Externalities associated with economic growth also prompted renewed interest in broader social accounting. Work by Nordhaus and Tobin (1973), among others, on adjusting traditional economic accounts for changes in leisure time, disamenities of urbanization, exhaustion of natural resources, population growth and other aspects of welfare produced indicators of economic well-being. However, the seemingly limitless scope, the range of uncertainty and the degree of subjectivity involved in such measures of non-market activities limited the usefulness of, and interest in, these social indicators. It was felt that inclusion of such measures would sharply diminish the usefulness of traditional economic accounts for analysing market activities. Attention subsequently focused on more readily identifiable and directly relevant market issues, such as the extent to which expenditures that relate to the protection and restoration of the environment (and other so-called defensive expenditures) are identifiable in the economic accounts.

The United Nations system of environmental and economic accounting

The development of the United Nations system of environmental and economic accounting (SEEA) and the use of supplemental, or satellite, accounts went a long way towards resolving the long-standing impasse between those who advocated broader sets of accounts and those concerned with maintaining the usefulness of the existing economic accounts. The supplemental accounts allowed conceptual and empirical research to move forward with estimates that can be linked to the existing accounts, but without diminishing their usefulness.

The SEEA, as described in the United Nations (1993a) handbook, is a flexible, expandable satellite system. It draws on the materials balance approach to present the full range of interactions between the economy and the environment. The SEEA builds on, and is designed to be used with, the System of

National Accounts (SNA) (Commission of the European Communities *et al.*, 1993). Like the 1993 SNA, the SEEA is primarily concerned with the implications of the environment for production, income, consumption and wealth.

The SEEA has four stages, each successively providing a more comprehensive accounting for the interaction between the economy and the environment. The four-stage presentation recognizes the need to develop concepts, to inventory and augment source data, and to adapt the implementation to differing analytical needs. The starting point is the 1993 SNA, which disaggregates, or provides additional detail on, environmentally related economic activities and assets. The second stage begins with the physical counterpart of the first stage. It maps, in physical terms, the interaction between the environment and the economy, providing the physical quantities to which prices are applied to derive the economic values included in the economic accounts. These physical accounts also provide a bridge to natural resource accounting and to materials and energy balances accounting. It then links the physical quantities to monetary values. The first two stages of the SEEA record the effects of the economy on nonproduced or environmental assets, either as other changes in the volume of assets or as changes in the distribution of income among the factors of production; these changes do not explicitly affect gross domestic product, final demand or net domestic product.

The third stage provides far more comprehensive and explicit measures of the interaction between the economy and the environment. It does so, first, by the use of alternative valuation techniques – that is, alternatives to the use of values tied to the market – and second, by the more explicit introduction of environmental effects on the measures of national production, investment, income, and wealth.

The fourth stage consists of further extensions of the SEEA. These extensions are provided for the purpose of 'opening a window on further analytical applications', and they will require further research. They include household production and the use of recreational and other unpriced environmental services in household production.

BEA's integrated economic and environmental satellite accounts

In constructing its IEESAs, BEA built on several key lessons from the social accounting experience of the 1970s and on the framework of the SEEA. First, such accounts should be focused on a specific set of issues. Second, given the kind of uses to which the estimates would be put, the early stage of conceptual development and the statistical uncertainties (even if the estimates are limited to the environment's effects on market activities), such estimates should be developed in a supplemental, or satellite, framework. Third, such accounts should not focus on sustainability or some normative objective, but should cover those interactions that can be tied to market activities and valued using market values or proxies thereof. Fourth, in keeping with the focus of the existing accounts, the supplemental accounts should be constructed in such a manner as to be consistent with the existing accounts and thus allow analysis of the effects of the interactions between the environment and the economy on production, income, consumption and wealth.

The existing economic accounts do not provide normative data, and neither do the integrated economic and environmental accounts of the BEA. The IEESAs either report market values or proxies for market values. If a problem with property rights leads to the undervaluation and overexploitation of a resource, a set of integrated economic accounts will not reveal the right price or the correct level of stocks. They will, however, provide the data, for example, about changes in the value of the stocks and the share of income to be attributed to the resource, needed for objective analysis of the problem.

Scope

In accordance with the first criterion, BEA limited the IEESAs to those interactions that directly affect the economy and are thus relevant to the objective of economic accounts. From this standpoint, the environment can be thought of as consisting of a range of natural resource and environmental assets that provide an identifiable and significant flow of goods and services to the economy. The economy's uses of these productive natural assets and the goods and services they provide can be grouped into two general classes. When use of the natural asset permanently or temporarily reduces its quantity, this is viewed as involving a flow of a good or service, and the quantitative reduction in the asset is called depletion. When use of the natural asset reduces its quality, the qualitative reduction in the asset is called degradation. However, the use of natural assets describes only part of the interaction between the economy and the environment. There are also feedback effects, such as the reduction in the future yield of crops, timber, fisheries etc. from current pollution or overharvesting. Materials balance and energy accounting highlight both the use of the natural assets and the feedback effects from the use; thus, they capture the full interaction between the economy and the environment. In the case of environmental assets, the feedback is more complicated, with effects that often fall on other industries and consumers.

Integrated economic and environmental accounting aims to provide a picture of these interactions, both uses and feedbacks, between the economy and the environment. However, while this picture has numerous elements and is complex, by definition it does not cover many of the transformations and interactions within the environment itself, for example, the disposal of waste products from wild fish and mammals or the conversion of natural carbon dioxide into oxygen by plant matter on land and in the oceans.

Compliance with the first criterion resulted in accounts that were objective, rather than normative. They describe activities which bear upon the market in the monetary terms of the market, without implying any conclusions about whether the reflected situation is 'right'. Put simply, the IEESAs attempt to answer the analytical questions that are raised by the interactions of the economy and the environment, such as:

(a) The nation's wealth includes natural resources, such as oil and gas reserves and timber, that are used in production. At what rate are these resources being used?
(b) The income of producers in the mineral industries includes a return to the drilling rigs, mining equipment, and other structures and equipment engaged in them and a return to the mineral. What share is attributable to the mineral?
(c) Economic activity adds to the proved stock of natural resources by exploration and technological innovation. How much of the use of natural resources in production has been offset by these additions?
(d) Households, governments, and business all make expenditures to maintain or restore the environment. What share of their spending is for the environment?
(e) The economy disposes of wastes into the air and water, and the resulting degradation of the environment imposes costs, such as lower timber yields and fish harvests and higher cleaning costs. What are these costs? Which sectors bear them?

Structural features

In accord with the second criterion, the IEESAs have two main structural features. First, natural and environmental resources are treated like productive assets and only the economically productive aspects of the resources are considered. These resources, along with structures and equipment, are treated as part of the nation's wealth, and the flow of goods and services from them is identified and their contribution to production measured. Second, the accounts provide substantial detail on expenditures and assets that are relevant to understanding and analysing the interaction. Fully implemented IEESAs would permit identification of the economic contribution of natural and environmental resources by industry, by type of income and by product. Ultimately, accounts by region would add an important analytical dimension.

Productive assets

BEA's decision to treat natural and environmental resources like productive assets in the IEESAs is based on their similarity to man-made capital: for labour and materials are devoted to producing fixed assets, and they then yield a flow of services over time. For inventories, stocks are held pending further processing, sale, delivery, or intermediate use. An example of a fixed natural resource is trees; a natural resource inventory is livestock raised for slaughter.

Table 1. IEESA asset account, 1987 (billions of dollars). This table can serve as an inventory of the estimates currently available for IEESA. In decreasing order of quality, the estimates that have been filled in are as follows: For made assets, estimates of fixed reproducible tangible stock and inventories, from BEA's national income and product accounts or based on them, and pollution abatement stock, from BEA estimates (rows 1–21); for subsoil assets, the high and lows of the range based on alternative valuation methods, from the companion article (rows 36–41); and best-available, or rough-order-of-magnitude, estimates for some other developed natural assets (selected rows 23–35 and 42–47) and some environmental assets (selected rows 48–55) prepared by BEA based on a wide range of source data described in this article. The 'n.a.'--not available--entries represents a research agenda.

			Change				
	Row No.	Opening Stocks (1)	Total, net (3+4+5) (2)	Depreciation Depletion, Degradation (3)	Capital Formation (4)	Revaluation and Other Changes (5)	Closing Stocks (1+2) (6)
PRODUCED ASSETS							
Made assets	1	11565.9	667.4	-607.9	905.8	369.4	12233.3
Fixed assets	2	10535.2	608.2	-607.9	875.8	340.2	11143.4
Residential structures and equipment, private and government	3	4001.6	318.1	-109.8	230.5	197.4	4319.7
Fixed nonresidential structures and equipment, private and government	4	6533.6	290.1	-498.1	645.3	142.9	6823.7
Natural resource related:	5	503.7	23.1	-19.2	30.3	12.0	526.8
Environmental management	6	241.3	8.4	-7.0	10.6	4.7	249.6
Conservation and development	7	152.7	3.6	-4.4	5.3	2.7	156.4
Water supply facilities	8	88.5	4.8	-2.5	5.3	2.0	93.3
Pollution abatement and control	9	262.4	14.7	-12.2	19.7	7.3	277.1
Sanitary services	10	172.9	12.8	-5.6	13.7	4.8	185.8
Air pollution abatement and control	11	45.3	0.6	-4.1	3.5	1.3	45.9
Water pollution abatement and control	12	44.2	1.3	-2.5	2.6	1.2	45.5
Other	13	6029.9	267.0	-478.9	615.0	130.9	6296.9
Inventories/1/	14	1030.7	59.3	...	30.1	29.2	1090.0
Government	15	184.9	6.8	...	2.9	3.8	191.7
Nonfarm	16	797.3	62.4	...	32.7	29.7	859.7
Farm (harvested crops, and livestock other than cattle and calves)	17	48.5	-9.9	...	-5.5	-4.4	38.6
Corn	18	10.2	0.3	...	-1.1	1.4	10.5
Soybeans	19	5.0	-0.1	...	-1.0	0.9	4.9
All wheat	20	2.6	0.0	...	-0.2	0.2	2.6
Other	21	30.7	-10.1	...	-3.2	-6.9	20.6
Developed natural assets	22	n.a.	n.a.	n.a.	n.a.	n.a.	n.a.
Cultivated biological resources	23	n.a.	n.a.	n.a.	n.a.	n.a.	n.a.
Cultivated fixed natural growth assets	24	n.a.	n.a.	n.a.	n.a.	n.a.	n.a.
Livestock for breeding, dairy, draught, etc.	25	n.a.	n.a.	n.a.	n.a.	n.a.	n.a.
Cattle	26	12.9	2.0	n.a.	-0.3	2.3	14.9
Fish stock	27	n.a.	n.a.	n.a.	n.a.	n.a.	n.a.
Vineyards, orchards	28	2.0	0.2	n.a.	0.0	0.2	2.2
Trees on timberland	29	288.8	47.0	-6.9	9.0	44.9	335.7
Work-in-progress on natural growth products	30	n.a.	n.a.	...	n.a.	n.a.	n.a.
Livestock raised for slaughter	31	n.a.	n.a.	...	n.a.	n.a.	n.a.
Cattle	32	24.1	7.5	...	0.0	7.5	31.6
Fish stock	33	n.a.	n.a.	...	n.a.	n.a.	n.a.
Calves	34	5.0	0.9	...	-0.5	1.4	5.9
Crops and other produced plants, not yet harvested	35	1.8	0.3	...	0.1	0.2	2.1
Proved subsoil assets/2/	36	270.0 - 1066.9	57.8 - -116.6	-16.7 - -61.6	16.6 - 64.6	58 - -119.6	299.4 - 950.3
Oil (including natural gas liquids)	37	58.2 - 325.9	-22.5 - -84.7	-5.1 - -30.6	5.8 - 34.2	-23.1 - -88.3	35.7 - 241.2
Gas (including natural gas liquids)	38	42.7 - 259.3	6.6 - -57.2	-5.6 - -20.3	4.1 - 14.9	8.1 - -51.8	49.4 - 202.2
Coal	39	140.7 - 207.7	2.2 - -3.4	-5.4 - -7.6	4.4 - 6.3	3.2 - -2.1	143.0 - 204.2
Metals	40	* - 215.3	67.2 - 29.5	-0.2 - -2.2	2.2 - 9.2	65.2 - 22.5	38.5 - 244.8
Other minerals	41	28.4 - 58.7	4.3 - -0.8	-0.4 - -0.9	0.1 - 0.0	4.6 - 0.1	32.8 - 57.9
Developed land	42	n.a.	n.a.	n.a.	n.a.	n.a.	n.a.
Land underlying structures (private)	43	4053.3	253.0	n.a.	n.a.	n.a.	4306.3
Agricultural land (excluding vineyards, orchards)	44	441.3	42.4	n.a.	-2.8	45.2	483.7
Soil	45	n.a.	n.a.	-0.5	n.a.	n.a.	n.a.
Recreational land and water (public)	46	n.a.	n.a.	-0.9	0.9	n.a.	n.a.
Forests and other wooded land	47	285.8	28.8	n.a.	-0.6	29.4	314.6
NONPRODUCED/ENVIRONMENTAL ASSETS							
Uncultivated biological resources	48	n.a.	n.a.	n.a.	n.a.	n.a.	n.a.
Wild fish	49	n.a.	n.a.	n.a.	n.a.	n.a.	n.a.
Timber and other plants of uncultivated forests	50	n.a.	n.a.	n.a.	n.a.	n.a.	n.a.
Other uncultivated biological resources	51	n.a.	n.a.	n.a.	n.a.	n.a.	n.a.
Unproved subsoil assets	52	n.a.	n.a.	n.a.	n.a.	n.a.	n.a.
Undeveloped land	53	n.a.	n.a.	-19.9	19.9	n.a.	n.a.
Water (economic effects of changes in the stock)	54	...	n.a.	-38.7	38.7	n.a.	...
Air (economic effects of changes in the stock)	55	...	n.a.	-27.1	27.1	n.a.	...

n.a. not available.
* The calculated value of the entry was negative.
1. The estimate for inventories differs from the NIPA estimate by the amount of government inventories added and cattle and calves shown separately. In full implementation of the IEESA account, farm inventories would include only harvested crops.
2. The estimates in all columns result from the valuation method (see text for further discussion of the alternative methods) that produces the low and high estimates of opening stocks.
NOTE.-Leaders indicate an entry is not applicable.

The distinction between fixed assets and inventories is not always clear, and each country will come to its own classifications. One example is the long-standing debate regarding the classification of mineral resources. Proved mineral reserves may seem to be similar to inventories – they are a set number of units waiting to be used up in production. Yet they also fit the classic characteristics of fixed capital – expenditures of materials and labour are needed to produce them, and they yield a stream of product over long periods of time. Further, like a fixed asset such as a machine, the number of units extracted from a new mine or field is uncertain and varies over time and the service life is used up in production. Finally, the treatment of mineral reserves as fixed assets serves equally well as a reminder of the reproducibility of proved reserves.

For these reasons, the IEESAs include these resources, along with structures and equipment, as part of the nation's wealth and give them the same treatment as fixed assets such as structures and equipment in the traditional accounts. This deals with three points of asymmetry between the treatment of mineral reserves and of structures and equipment encountered in traditional accounts. In traditional accounts: (1) depreciation is subtracted from profits to determine true, or sustainable, profits, but depletion is not; (2) depreciation is subtracted from GDP to estimate NDP, but depletion is not; and (3) additions to the stock of plant and equipment are added to GDP as capital formation, but additions to mineral reserves are not.

Detail

In the IEESAs, the standard economic accounting categories are disaggregated to show detail that highlights the interaction of the economy and the environment. For example, the expenditures detail shows spending by households, government and business to maintain or restore the environment. The asset detail shows environmental management (conservation and development, and water supply) and waste management projects (sanitary services, air and water pollution abatement and control) within the standard category of non-residential fixed capital.

The estimating requirements underlying these two main structural features of the IEESAs are apparent in the IEESAs tables, even when they are in skeleton form. Table 1, an asset account and Table 2, a production account, use modified forms of tables presented in the SEEA.

Accounts

Asset accounts

Integrated economic and environmental accounting requires the measurement of stocks and flows related to assets which are presented in an asset account. The IEESAs provide a complete accounting for the relevant assets: they show both stocks and flows associated with changes in those stocks. Table 1 provides for estimates of opening stocks, different kinds of changes in the stock and closing stocks. It also presents the non-financial assets that BEA would try to include in IEESAs asset accounts. These generally follow the subcategories of the 1993 SNA and the SEEA, but some of the subcategories are regrouped to broaden both the production boundary and the definition of assets. Non-financial assets are divided into made assets, developed natural assets and environmental assets. Made assets, which largely replicate the scope

Table 2. IEESA production account, 1987 (billions of dollars).

	Row no.	INDUSTRIES				FINAL USES (GDP)						
						Final consumption						
		Agriculture, forestry, fisheries (1)	Mining, utilities, water, and sanitary services (2)	Other Industries (3)	Total (4)	Household (5)	Government (6)	Gross Domestic Capital Formation (7)	Exports (8)	Imports (9)	GDP (5+6+7+8-9) (10)	Total Commodity Output (4+10) (11)
COMMODITIES												0.0
Made	1							933.0			#	#
Assets	2							933.0			#	#
Fixed assets	3							875.8			#	#
Environmental management	4							10.6			#	#
Pollution abatement and control	5							19.7			#	#
Other	6							845.5			#	#
Inventories	7							57.2			#	#
Government	8							30.1			#	#
Nonfarm	9							32.7			#	#
Farm	10							-5.5			#	#
Other	11	#	#	#	#	#	#		#	#	#	#
Environmental cleanup and waste disposal services	12	n.a.	n.a.	n.a.	n.a.	n.a.	n.a.		n.a.	n.a.	#	#
Other	13	n.a.	n.a.	n.a.	n.a.	n.a.	n.a.		n.a.	n.a.	#	#
Natural and environmental assets	14							n.a.			#	#
Fixed	15							n.a.			#	#
Cultivated biological resources: natural growth	16							n.a.			#	#
Proved subsoil assets	17							16.6 - 64.6			#	#
Developed land	18							n.a.			#	#
Uncultivated biological resources: natural growth	19							n.a.			#	#
Unproved subsoil assets	20							n.a.			#	#
Undeveloped land	21							19.9			#	#
Water	22							38.7			#	#
Air	23							27.1			#	#
Work-in-progress inventories (natural growth products)	24							n.a.				
Total intermediate inputs	25	#	#	#	#							0.0
VALUE ADDED												0.0 0.0
Compensation of employees	26	#	#	#	#							#
Indirect business taxes, etc.	27	#	#	#	#							#
Corporate profits and other property income	28	#	#	#	#							#
												0.0
Depreciation of fixed made assets: structures and equipment	29	n.a.	n.a.	n.a.	-607.9							#
Environmental management	30	n.a.	n.a.	n.a.	-19.2							#
Pollution abatement and control	31	n.a.	n.a.	n.a.	-2.5							#
Other	32	n.a.	n.a.	n.a.	-586.1							#
Depletion and degradation of fixed natural and environmental assets	33	n.a.	n.a.	n.a.	n.a.							#
Growth products: fixed	34	n.a.	n.a.	n.a.	n.a.							#
Proved subsoil assets	35	n.a.	n.a.	n.a.	16.7 - -61.6							#
Developed land	36	n.a.	n.a.	n.a.	n.a.							#
Uncultivated biological resources	37	n.a.	n.a.	n.a.	n.a.							#
Unproved subsoil assets	38	n.a.	n.a.	n.a.	n.a.							#
Undeveloped land	39	n.a.	n.a.	n.a.	n.a.							#
Water	40	n.a.	n.a.	n.a.	-19.9							#
Air	41	n.a.	n.a.	n.a.	-38.7							#
Gross value added (GDP) (rows 26+27+28+29+33)	42	n.a.	n.a.	n.a.	n.a.						#	#
Depreciation, depletion, and degradation (rows 29+33)	43	n.a.	n.a.	n.a.	n.a.							#
Net value added (NDP) (rows 42-43)	44	n.a.	n.a.	n.a.	n.a.							#
												0.0
TOTAL INDUSTRY OUTPUT	45	#	#	#	#	#	#	#	#	#	#	#

n.a. Not available.
These estimates will depend on the integration of the System of National Accounts and the System of Environmental and Economic Accounting as part of the overall modernization of BEA's economic accounts.
NOTE.-Leaders indicate that an entry is not applicable.
GDP Gross domestic product
NDP Net domestic product

of non-financial assets in traditional income and wealth accounts, are subdivided into fixed assets and inventories. Developed natural assets are subdivided into cultivated biological resources (both fixed and work in progress inventories), proved subsoil assets and developed land. Environmental assets are subdivided into uncultivated biological resources, unproved subsoil assets, undeveloped land, water and air (the last two in terms of the economic effects of changes in the stock).

Made and developed natural assets

To better highlight the interaction of the economy and the environment, Table 1 provides more detail on natural resource and environmentally related produced assets than the traditional income and wealth accounts. Within made assets, non-residential fixed capital is disaggregated into environmental management (conservation and development, and water supply) and waste management projects (sanitary services, air and water pollution abatement and control). Detail is also provided on farm inventories of finished goods. Within cultivated biological resources, Table 1 provides detail beyond that contained in the traditional accounts, such as cultivated fixed natural growth assets (for example, livestock), and categories not included in the traditional accounts (for example, trees on timberland). The treatment of proved subsoil assets and cultivated land in Table 1 differs from the SEEA treatment. Proved reserves are generally defined as those reserves that are proved to a high degree of certainty, by test wells or other test data, and are recoverable under current economic conditions and with current technology. In the SEEA, they are classified as non-produced assets. In Table 1, these assets, along with cultivated natural growth assets, are included in the category 'developed natural assets'. As will be illustrated in the production accounts, capital formation that adds to the stock of these assets, both by bringing undeveloped or uncultivated assets into the category of developed natural assets and by adding to their value within that category, is treated in a manner similar to capital formation that adds to the stock of structures and equipment.

This treatment was adopted because it is difficult to rationalize describing proved reserves and cultivated land as 'non-produced' natural assets when expenditures are required to prove or develop them. Agricultural land, for example, must be produced in that expenditures must be undertaken to convert uncultivated land areas into commercially valuable farmland, which yields a return over a number of years. Wetland areas, if they are to become farmland, must be drained and graded and cleared of vegetation. Unproved mineral reserves also require expenditure on test wells, engineering studies, and other exploration and development investments before they are recorded as proved reserves.

Similar treatment of these developed natural assets and made assets facilitates consistent treatment of capital formation of natural assets and more conventional capital formation, such as investment in structures and equipment. Under this treatment, as mineral reserves, for example, are proved, the total value of the produced assets – structures and equipment as well as the proved reserve's value – is included as capital formation. Similarly, as oil field machinery is depreciated, proved reserves associated with the machinery are depleted.

The other major difference between developed assets in Table 1 and in the comparable SEEA presentation is in the treatment of soil. In the SEEA, soil (that is, productive soil on agricultural land) is treated as separate from agricultural land. In Table 1, soil is a subcategory of agricultural land because the value of agricultural land is inseparable from the value of the soil. Available estimates suggest that the effect of soil erosion, or depletion, on agricultural productivity and land values in the USA is quite small. Nevertheless, though soil is not treated separately, it is shown separately because its erosion has a significant effect on environmental quality through its effect on water quality.

Environmental assets

Environmental assets include natural assets with significant economic value that differ from developed natural assets in that they are generally used as raw inputs into production in their natural state, either as intermediate products or as investments. For example, uncultivated biological resources, such as tuna harvested from the ocean, are included as environmental assets, whereas cultivated biological resources, such as rockfish raised on a fish farm, are included in developed assets. Other categories in environmental assets are uncultivated land, unproved subsoil assets, water, and air. The inclusion of unproved subsoil assets broadens the definition of subsoil assets to include reserves that, though unproved, have an economic value over and above that of other undeveloped land because of their location or geological characteristics. As capital expenditures are made to prove these properties, they move from non-produced to produced assets. This broader definition of subsoil resources will facilitate longer term planning and analysis of the use of mineral resources. The stock of proved reserves, like the stock of drill presses, can be expanded by additional investment; hence, firms will keep on hand the stock of reserves dictated by current market prices, finding costs, and interest rates. Thus, complete analysis of mineral resources requires consideration of un-proved, as well as of proved, reserves.

In a distinction similar to that between proved and un-proved subsoil assets, cultivated land such as agricultural land, parkland, and land underlying buildings is included in developed natural assets, whereas uncultivated land such as wetlands and forestland (not included as timberland) is included in environmental assets. The agricultural land must be developed before it can be used as farmland, whereas wetlands are used, for example, for flood control, in their natural state by the economy. Water, which is subdivided by type, and air also provide services to the economy in the form of recreational and waste disposal services.

Although these environmental assets differ from made and developed natural assets, investments that add to the stock of these assets, as noted below in the production accounts, are treated symmetrically with investments that add to the stock of structures and equipment and of developed assets. These investments, for example, include pollution abatement and control to improve the quality and waste disposal capacity of the air and water, or at least to offset the degradation/depletion (which is also recorded in the production account) occurring in the current period. The rationale for treating such expenditures as investment rather than costs is that they represent a decision by the economy to devote its resources to investments that improve air and water quality, rather than investments in structures and equipment.

Production accounts

The next step in integrating economic and environmental accounting is to combine the appropriate flows from the asset account with the flows in a production account. With this integration, the production account of the IEESAs explicitly includes the use of natural resources and environmental services in production through entries for depletion and degradation, and it explicitly includes the additions to the stock of natural and environmental assets through entries for investments that add to stocks of developed natural resources or that restore stocks of environmental assets.

Table 2 combines features of both the supply and use tables in the 1993 SNA. The table has four quadrants (one empty, except for a total) which are separated by double lines, a total column at the far right and a total row at the bottom. The left and right upper quadrants show the use of goods and services (commodities) named at the beginning of the rows, summing to total uses as measured by total commodity output. The left-hand upper and lower quadrants show the use of intermediate inputs and factors of production by the industries named at the top of each column, summing to total supply as measured by total output.

Valuation

The choice of a valuation approach was not a difficult one for BEA. While alternative methods such as maintenance cost and contingent valuation have attractive theoretical characteristics, they are not appropriate for BEA's purpose, and the associated practical difficulties outweigh their charms. In keeping with the goals and criteria stated above, market pricing was the optimal choice for the IEESAs. First, market pricing maintains objectivity by avoiding the bias that may be inherent in 'willingness to pay' surveys. Second, market pricing is consistent with conventional accounts, as well as the SEEA, and facilitates international comparability. Finally, market pricing is consistent with the limits placed on the included interactions because it values those interactions from the perspective of the market. This approach was not problem-free, however. The quality of the estimates released in 1994 varies with regard to the source data available and, as natural and environmental resources are rarely traded, market prices are often not available. Thus, the estimates recorded for 1987 in Table 1 should be regarded as rough order of magnitude, or best available, estimates.

When market prices or other source data were unavailable, BEA used the best available techniques to produce proxies of market values:

(a) The estimates of the value of vineyards and orchards are based on Federal Reserve Board estimates of the value of agricultural land and estimates of the acres of land in vineyards and orchards from the Bureau of the Census.

(b) The values of trees on timberland were estimated based on stumpage value estimates provided by the US Forest Service's Pacific Northwest Research Station. The stumpage value estimates are based on the concept of net rent to the timber stand (as distinct from the land underlying the forest) and are derived mainly from private market data on payments for logging rights. As such, they should correspond to the present discounted value of the timber sales from the tract less the costs of logging, access, transportation and processing. All timber on commercially viable timberland in the United States, public and private, is included in this category.

(c) Sil estimates, obtained from the US Department of Agriculture (USDA), reflect the annual effect of soil depletion in terms of extra fertilizer costs and reduced productivity.

(d) The estimate of capital formation in recreational land is based on Federal Government maintenance and repair expenditures for parks; State and local expenditures are not available. It is assumed that these expenditures exactly offset the degradation/depletion of recreational land; in the case of recreational land, the only estimates available were of maintenance and repair expenditures. This assumption is made only so that both investment and degradation/depletion estimates are illustrated by the table and not to imply any judgment about the true value of degradation/depletion.[5]

(e) For environmental assets, the estimates are more uncertain than even the most uncertain estimates for developed land and proved reserves of subsoil assets. Indeed, most of this section of the table, especially that for renewable natural resources, is shown as 'not available'. No value is available for the stock of undeveloped land and its associated ecosystems, for unproved subsoil assets and for uncultivated biological resources (wild animals and fish, plants and forests).

(f) The SEEA does not recommend that the stock of air, which is truly a global common, or water be valued; instead, it recommends that valuation be limited to changes in these assets – their degradation and investments in their restoration. For these assets, Table 1 includes only aggregate values for the degradation of air and water and for expenditures to restore them or to prevent their degradation. The estimates in Table 1 for degradation of air and water quality, as well as for undeveloped land, are simply place markers which assume that maintenance exactly offsets degradation: they are aggregate estimates of the total costs of pollution of these media. The estimates for air, water and undeveloped land pollution are estimates obtained from the Environmental Protection Agency of the costs of public and private pollution control activities in the USA. Estimates of air pollution include the annualized costs of air pollution and radiation. Water pollution estimates are the annualized costs of maintaining water quality, including drinking water. Estimates of undeveloped land pollution are the annualized costs associated with the Superfund, toxic chemicals and pesticides.

Estimates

When market prices were not available, BEA tried to present a range of estimates reflecting various valuation techniques. In most cases, however, only one estimate, rather than a range, was available and many cells in Table 1 do not contain estimates. In general, the quality and availability of the estimates declines as one moves down the rows from produced to non-produced assets, reflecting the increasing conceptual and empirical difficulties in producing such estimates. The estimates may be best regarded as a measure of the work to be undertaken; they are presented here to serve as a road map for areas in which source data and estimating methods must be developed or improved.

The estimates presented in Table 2 are taken from Table 1. As indicated by 'n.a.', many valuation and measurement issues remain before an IEESAs production account can be completed. Further, work toward filling in the estimates would proceed in tandem with work on modernizing BEA's national accounts in line with the SNA. For example, treating expenditures on government structures, equipment, and inventories as capital formation implements a feature of the SNA. In the table, ' # ' indicates the estimates that would reflect both work toward the IEESAs and SNA-related changes.

The prototype estimates of mineral resources – the focus of the first phase of work released in 1994 – include stocks and flows in accounts that supplement BEA's national wealth accounts and National Income and Product Accounts (NIPAs). These prototype estimates provide a comprehensive picture of the stocks of natural assets and the changes in them. They also allow an examination of the practical consequences of several alternative methods of valuing the stock of resources, additions and depletion. The alternative methods – current rent (of which BEA used two variants), present discounted value, replacement cost and transaction prices – represent the Bureau's technical assessment of the best estimates and framework that are feasible with existing sources and methods. Some of the implications of these estimates are as follows:

(a) The value of additions has tended to exceed depletions. Since 1958, the value of the stocks of proved mineral reserves in the aggregate has grown in current dollars, while showing little change in constant (1987) dollars.
(b) Changes in the stocks of these productive assets over time have largely reflected changes in the resource rents. Increases in resource rents have been accompanied by greater investment in exploration and enhanced recovery technology. Decreases in rents of some resources have been accompanied by reduced exploration activity and the closing of marginal fields and mines.
(c) Proved mineral reserves constitute a significant share of the economy's stock of productive resources. Addition of the value of the stock of these mineral resources to the value of structures, equipment, and inventories for 1991 would raise the total by $471–916 billion, or 3–7%, depending on the valuation method used.
(d) The stocks of proved mineral resources are worth much more than the stocks of invested structures and equipment associated with the resources. In 1991, the value of the stock of subsoil assets was 2–4 times as large as

the value of the associated stock of invested structures and equipment and inventories.

(e) Valuing the effect of depletion and additions, as well as including the value of resource stocks, provides a significantly different picture of returns. Compared with rates of return calculated using income and capital stock as measured in the existing accounts, the IEESAs-based average rates of return on capital in the mining industry for 1958–91 are lower, at 4–5% rather than 23%. Rates of return for all private capital slip from 16% using measures in the existing accounts to 14–15% using IEESAs measures for the mining industries.

(f) Although the trends that emerge from the alternative methods are similar, the range of estimates is large. The highest estimates of stocks, depletion and additions were obtained from the current rent estimates based on capital stock values, and the lowest were from the current rent estimates based on average rates of return to capital.

Uses of the new accounts

The IEESAs will help to identify the use of the various natural and environmental resources, but because of offsetting changes it is difficult to say *a priori* whether there will be a net reduction or increase in their value overall. Indeed, it is not clear whether such a 'bottom line' estimate is even desirable. First, such an estimate may not be very informative. For example, while it is almost certainly true that the economic value of the stocks of some assets in the USA, such as bluefin tuna, is declining, the stocks of other environmental assets, such as timber stocks, have been increasing as planting and growth have more than offset harvests, fire and land conversions. Similarly, while losses of wetlands from development continue to outnumber gains from wetland restorations, increasing rates of investments in cleaner air and water since the mid-1970s appear to have resulted in net improvements in air and water quality; many of the measures of air and water quality, such as the ambient concentrations of air and water pollutants, have shown improvement. It is conceivable that when all entries in Table 2 (or if not all, at least enough more than at present to avoid risks of conclusions based on partial results) have been filled in, the table will show that IEESAs-NDP differs little from traditional NDP.

Second, it is not clear that the information reflected in a bottom line measure is always relevant. For example, as noted above in the case of mineral reserves, the stock of proven reserves varies with the rate of return and remains a fairly stable multiple of annual consumption; the mineral industries make expenditures to prove reserves to ensure a level of supply for a given or projected demand. Changes in proven reserves, therefore, have little to say about long-run sustainability. Instead, as the prototype IEESAs accounts illustrated, the changes in stocks of proven reserves and their prices over time have significant effects on rates of return, in the aggregate and by industry, and the structure of the accounts will allow the analysis of the secondary effects by type of income, type of investment and type of spending. Other effects that could be analysed include the impact of changes in minerals prices on the US terms of trade, the related measure of command-basis GNP (which reflects the resources over which a nation has command as a result of its exports and other income from abroad) as well as a wide range of issues including federal fees for grazing rights, water use, logging, mineral rights, fishing permits, and the impact of environmental regulations and emission taxes.[6] For a developed economy, such as that of the USA, this detailed information about the interaction between the economy and the environment rather than normative measures of our environmental welfare or long-run sustainability will provide the most valuable insights about the implications of different regulations, taxes and consumption patterns.

BEA's plan for natural resource and environmental accounting

BEA's plan calls for work on the IEESAs to be undertaken in conjunction with modernizing its economic accounts. BEA's national accounts are now undergoing the first major redesign since the 1950s. The redesign, which will be along the lines of the 1993 SNA, will feature an integrated set of current and capital accounts, sector by sector. Fully developed capital accounts, along with balance sheets, are essential for a comprehensive set of economic accounts. The conceptual work on these accounts and the more specialized work on natural resources and the environment will be mutually supporting. Further, to make reasoned policy choices involving trade-offs among kinds of capital, one would want a view of the total capital stock, natural and made, consistently covered and appropriately valued.

BEA has developed a three-phase plan for the IEESAs. With the April 1994 issue of the *Survey of Current Business*, BEA completed the first phase of work. The overall IEESA framework is designed to build upon the existing national accounts and is in line with the guidance embodied in the new international SNA about a satellite system and the companion SEEA. In its initial work, BEA focused on mineral resources, consisting of oil and gas, coal, metals and other minerals with a scarcity value. The focus, in accordance with SNA recommendations, is on proved reserves, the basis for valuation is market values, and the treatment given to mineral resources, which require expenditures to prove and which provide 'services' over a long timespan, is similar to the treatment of fixed capital in the existing accounts.

The second phase will incorporate renewable natural resources. Compared with the accounting for proved reserves of non-renewable resources, where the economic literature extends back over 50 years, valuation methods and concepts for many of the renewable resources are less well developed. Renewable natural resources are inherently more difficult to value than non-renewable natural resources for several reasons: renewable resources, such as stocks or schools of wild fish, often have a commercial or production value as well as an amenity or a recreational value; often, ownership rights cannot be established, and they cannot be sold; and they are able to regenerate, so their use does not necessarily result in a net reduction in either their yield or the value of their stock.

These difficulties notwithstanding, there has been rapid progress in environmental benefit valuation for renewable natural resources in recent years as economists have tried to keep pace with regulatory, legal and policy needs for environmental damage and impact measures. Further work by BEA to translate these new concepts and measures into a consistent national framework would need to rely heavily on the expertise of other units within the US Government, for example, the National Oceanic and Atmospheric Administration, the Environmental Protection Agency, USDA, and the Department of Interior. The plan calls for work to extend the accounts to renewable natural resource assets, such as trees on timberland, fish stocks and water resources. Development of these estimates will be more difficult than for mineral resources because they must be based on less refined concepts and less data.

Building on this work, the third phase calls for moving on to issues associated with a broader range of environmental assets, including the economic value of the degradation of clean air and water, or the value of recreational assets such as lakes and national forests. Clearly, significant advances will be required in the underlying environmental and economic data, as well as in concepts and methods, and cooperative efforts with the scientific, statistical and economic communities will be needed to produce such estimates.

Lessons

Several lessons were learned in the process of developing the IEESAs that may be useful cautions for countries about to embark on a similar journey towards integrated economic and environmental accounts.

(a) First, the new accounts, like the conventional accounts, will only be worthwhile if they are used. Thus, consistency with the existing accounts through the use of market prices, and proxies thereof, and the treatment of natural resources and environmental assets symmetrically with economic assets remains an absolute requirement if they are to be used in conjunction with the existing accounts. In addition, if the new accounts are focused and based on the market-related concepts outlined in the existing accounts, many of the concerns related to the subjectivity and conceptual basis of earlier social accounting measures will be avoided.

(b) Throughout the development of the IEESAs, BEA looked outside its own work and built upon the economic and social accounting lessons of the past. BEA also relied on the expertise and data provided by natural resource and environmental and economic accounting experts in the USA and abroad. In this way, it avoided unnecessary 'reinventing' of the wheel.

(c) At the same time, BEA adapted that knowledge to its own needs to make the IEESAs appropriate for analysis of the specific concerns of the USA.

(d) When adequate source data were available, BEA presented a range of estimates to illustrate various methods and to emphasize the uncertainty associated with such accounts.

(e) In order to make its source data, methods and assumptions open and accessible, BEA published detailed information on source data and estimation methods.

(f) The limitation of the accounts to the interaction between the economy and the environment enabled BEA to apply a market approach consistently (though the exact approach and data availability remain issues for future work). If the accounts had extended to the noneconomic functions of natural resources and environmental assets, they would have been forced to incorporate inconsistent valuation methods that would preclude or at least complicate useful aggregation or comparison.

(g) The limitation to interactions between the economy and the environment also enabled BEA to construct its accounts upon a clear and consistent conceptual base. The advantages of this approach have been repeatedly revealed since the release of the prototype accounts. Most recently, the National Academy of Sciences has been able to begin its review of the accounts, as requested by Congress, with a clear understanding of the rationale for its structure and methods.

Notes

1. Copies of the two *Survey of Current Business* articles detailing the accounts are available on the Internet or by calling (202)606-9900.
2. In the last chapter, Kuznets (1996) notes that the result of the restriction of national income estimates to economic activities is that they 'neglect completely any consideration of such costs of economic activity as impinge directly upon consumers' satisfaction or the welfare of the community' and that errors of both omission and commission 'renders national income merely one element in the evaluation of the net welfare assignable to the nation's economic activity'.
3. Jaszi (1971) makes clear his belief that 'the tools we have available to construct a measure of output . . . cannot be used to construct a measure of welfare'.
4. This environmentalist school of thought has a long tradition both within and outside of economic circles. For example, US Vice President Gore (1992) goes beyond economic analysis in his comprehensive evaluation of the environment and society. Daly and Cobb (1989) use a more traditional economic approach in their development of the Index of Sustainable Economic Welfare. Cobb *et al.* (1995) use a similar approach in their calculation of a 'Genuine Progress Indicator' that adjusts GDP for household production, crime, and other welfare effects.
5. Phases II and III of BEA's work plan, described in the next section, include work, building on the damage assessment and recreational valuation literature, to construct estimates of the market value of recreational and environmental amenities.
6. For further discussion of the concept and measurement of command-basis GNP, see Denison (1981) or the *SNA* 1993 (Commission of the European Communities *et al.*, 1993 404–405).

8
USA: Environmental protection activities and their consequences

DEBORAH VAUGHN NESTOR AND CARL A PASURKA, JR.

This study defines the goods and services that constitute environmental protection activities in the United States. The US input–output (I–O) table serves as the basis for the definition of environmental protection activities developed in this study. The US environmental protection I–O tables identify the sectors that receive the revenues associated with purchases of goods and services to comply with environmental regulation as well as the sectors that demand environmental protection goods and services. The I–O framework also allows for the development of a measure of the importance of environmental protection activities relative to the US economy. Employment associated with environmental protection activities is also estimated.

The results show that environmental protection activities constituted between 0.64 and 0.80% of gross national product (GNP) between 1977 and 1991. Value-added for environmental protection was $46,646.6 million in 1991. Employment directly attributable to environmental protection activities increased from 678,000 in 1977 to 741,000 in 1991. Using the I–O framework, it is also possible to estimate employment indirectly attributable to environmental protection activities. If individuals indirectly employed are also counted, employment increased from 1.3 million in 1977 to 2.0 million in 1991. Finally, this study discusses current research to link air emissions with input–output sectors for the USA.

Overview

This chapter has two goals.[1] First, it identifies the production and service activities that constitute environmental protection (EP) activities in the US economy. The identification of these activities is accomplished through the use of an I–O accounting framework. The US I–O table, published by the Bureau of Economic Analysis (BEA) (US Department of Commerce, BEA, 1984; 1991), is adjusted to isolate EP activities from other economic activities. The

---

All views expressed in this paper are the authors' and do not reflect the official position of the US Environmental Protection Agency. We thank Anton Steurer for helpful comments on earlier drafts of this paper. Any remaining errors are strictly our own.

K. Uno and P. Bartelmus (eds.), Environmental Accounting in Theory and Practice, 131–142.

resulting EP I–O tables characterize the sectors whose output is used to comply with environmental regulations as well as the sectors that demand EP goods and services. Second, there is a discussion of efforts to measure the 'output' of EP activities, which is the reduction in the level of pollutants (i.e. undesirable outputs). This study does not attempt to measure or draw conclusions about the net economic impacts of environmental regulation. Rather, it focuses on defining and measuring the amount of resources devoted to EP activities. The composition of EP activities in the USA is assessed, and the size of EP activities relative to the US economy is estimated for 5 years: 1977, 1982, 1985, 1988, and 1991.

The methodological contribution of this study is discussed, followed by an outline of the framework used for defining EP activities. A description of the goods and services that comprise EP activities in the USA is followed by estimates of the size and composition of US EP activities for 1977, 1982, 1985, 1988, and 1991. These estimates are presented both in total and by environmental medium (i.e. air, water and solid waste). Estimates of direct and indirect employment associated with EP activities and an inventory of air emissions data and the potential use of these data are also presented.

Contribution of the study

The EP 'industry'

Policy makers have recently shown interest in quantifying the impacts of environmental regulation on sectors providing EP goods and services, and more generally, in defining an environmental protection 'industry' (Brown *et al.*, 1993). The I–O approach applied in this study is a consistent framework for defining an environmental protection industry as well as for estimating its size and the number of individuals employed in environmental protection activities.

Several approaches to estimating the size of and employment in the EP industry have been taken, and each approach results in a different estimate. For example, the *Environmental Business Journal* (EBJ) measures the size of the EP industry by estimating the revenue received by each of 13 industry segments (Environmental Business Publishing, 1994). In 1990, EBJ estimated that the EP industry was a $122 billion industry. Another measure is given by total annualized costs of EP, as reported by EPA in *Environmental Investments: The Cost of a Clean Environment* (US Environmental Protection Agency, 1990). This cost-based approach yields an estimate of $115 billion in 1990. Finally, the Organization for Economic Co-operation and Development (OECD) estimates total US production of EP goods and services at $80 billion for 1990 (OECD, 1992).[2] This study estimates direct value added of $47 billion in 1991, and direct plus indirect value added of $122 billion.

Discrepancies in estimates of the size of the EP industry stem from at least two factors. First is the general lack of agreement regarding which activities to 'count' as EP. As an example, BEA (see Farber and Rutledge, 1989) counts none of the expenditures for water supply used for water treatment as EP. The EBJ, on the other hand, includes all revenues associated with water supply in its definition. For 1991, EBJ (Environmental Business Publishing, 1994) reports revenues of $21.1 billion for the water utilities segment of the EP industry.[3] The second factor contributing to discrepancies in estimates of the size of the EP industry is more relevant to this study. This factor relates to the method used to calculate the size of the EP industry. As mentioned above, the EBJ estimates revenue received by sectors providing EP goods and services, while the EPA reports total costs of EP. In practice, the revenue-based and the cost-based approaches to defining the EP industry will generate different estimates. First, not all environmental costs involve company to company transactions. Some pollution abatement activities are performed within the polluting industry and costs associated with these activities do not become revenues for companies providing EP goods and services. Also, some environmental control costs do not involve out-of-pocket expenditures (e.g. depreciation).[4] Second, the costs of pollution abatement include expenditures for items that are not part of the EP market (e.g. electricity required for the operation of pollution control equipment). Again, expenditures on these items do not become revenues for companies providing EP goods and services.[5] Finally, companies like engineering firms provide services besides EP. Thus, the total revenues of these companies overstate the amount received for EP goods and services.

Besides leading to different estimates of size of the EP industry, it is questionable whether the size of the EP industry should be measured at all in terms of total EP costs or total EP revenues. Typically, the size of an industry is measured in terms of its contribution to Gross National Product (GNP), which is given by its total value added. Using the I–O framework, it is possible to derive a measure of the EP industry's contribution to national product or its total value added. Computing value-added associated with EP yields a measure of EP activities that is comparable to measures of the size of the national economy and other industries.

Integrated environmental and economic accounting

The EP I–O tables also serve as a foundation for developing integrated environmental and economic accounts. The United Nations has proposed the System for integrated Environmental and Economic Accounting (SEEA) as a special satellite system that is closely related to the core System of National Accounts (SNA).[6] This study represents an application of the SEEA, that is the disaggregation of the US I–O table into environmental and non-environmental components. It is possible to build upon the framework set forth in this study and develop the other parts of the SEEA.

Improved modelling of the economic impacts of environmental regulation

Application of computable general equilibrium (CGE) models to environmental policy has become quite popular. However, when using the models to estimate the impacts of specific environmental regulations, researchers have had to make simplifying assumptions. Without information on the inputs to pollution abatement processes, CGE modellers have made simplifying assumptions about which goods and services are purchased to comply with environmental regulation. These assumptions could influence the accuracy of CGE model results. For example, even though an industry may bear a relatively large regulatory burden, the burden may be offset if its output is used in pollution abatement activities (see Nestor and Pasurka, 1995a). The framework set forth in this study, if institutionalized, can provide information necessary for explicit modelling of pollution abatement processes and lead to improved CGE modelling of environmental policy.

Framework for defining environmental protection activities

EP activities in an I–O framework

This study makes use of an I–O accounting framework for identifying the production and service activities that comprise EP activities in the US economy. EP activities are defined by disaggregating the US I–O tables into EP and non-EP components. This requires a scheme for classifying the various types of EP activities. The United Nations (1993a) provides some guidance, proposing that environmental protection activities be classified into the following five categories: external EP activities, internal EP activities, fixed capital formation for EP, household EP activities and government EP activities.

External EP activities refer to establishments in which EP constitutes the main or secondary production activity. The key identifying characteristic of external EP activities is that they are delivered to other establishments, or a third party. External EP activities are represented as separate rows and columns in an I–O matrix. In Table 1, the entries depicted by the shaded column (n + 1) represent the dollar value of the products purchased as intermediate inputs from other sectors in the economy by the external EP activities sector. The corresponding shaded row in Table 1 represents the dollar value of the external EP activities that other industries purchase for use as an intermediate input.

Internal EP activities are for the establishment in which they are produced. Internal EP activities are ancillary activities analogous to administration or research and development activities. Internal EP activities are measured by inputs purchased for and combined as pollution abatement activity by a polluting industry; they include intermediate inputs and value added. Internal EP

Table 1. I–O Framework modified to display the EP industry.

TO FROM	1	2	...	n	(n + 1)	Y	X
1	$X_{11EP}^{NE} + X_{11}$	$X_{12P}^{NE} + X_{12}^{E}$	...	$X_{1nP}^{NE} + X_{1n}^{E}$	$X_{1(n+1)}$	$Y_{1P}^{NE} + Y_{1}^{E}$	$X_{1EP}^{NE} + X_{1}$
2	$X_{21EP}^{NE} + X_{21}$	$X_{22P}^{NE} + X_{22}^{E}$		$X_{2nP}^{NE} + X_{2n}^{E}$	$X_{2(n+1)}$	$Y_{2P}^{NE} + Y_{2}^{E}$	$X_{2EP}^{NE} + X_{2}$
⋮	⋮			⋮	⋮	⋮	⋮
n	$X_{n1EP}^{NE} + X_{n1}$	$X_{n2P}^{NE} + X_{n2}^{E}$	...	$X_{nnP}^{NE} + X_{nn}^{E}$	$X_{n(n+1)}$	$Y_{nP}^{NE} + Y_{n}^{E}$	$X_{nEP}^{NE} + X_{n}$
(n + 1)	$X_{(n+1)1}$	$X_{(n+1)2}$	...	$X_{(n+1)n}$	$X_{(n+1)(n+1)}$	Y_{n+1}	X_{n+1}
V	$V_{1}^{NE} + V_{1}^{EP}$	$V_{2}^{NE} + V_{2}^{EP}$	...	$V_{n}^{NE} + V_{n}^{EP}$	V_{n+1}		
X	$X_{1EP}^{NE*} + X_{1}^{*}$	$X_{2P}^{NE*} + X_{2}^{*E}$	...	$X_{nP}^{NE*} + X_{n}^{*E}$	X_{n+1}		

activities are not separated from the main activities of an establishment and, in this I–O framework, are accounted for by separating out that portion of total inputs used by polluting industries for pollution abatement. In Table 1 this adjustment is reflected by X_{ij}^{EP} which represents intermediate inputs used for EP activities. The residual, X_{ij}^{NE}, represents intermediate inputs used for non-EP activities. Total value added consists of value added associated with EP activities, V_{ij}^{EP}, and with non-EP activities, V_{ij}^{NE}.

The category fixed capital formation for EP represents the accumulation of fixed assets for EP and corresponds to gross private domestic investment in the I–O format. As an example, the purchase of a scrubber represents the accumulation of capital for air pollution abatement.

In addition, two other types of EP activities are performed in the USA. These are EP activities performed by households and government. Household and government EP activities are like EP investment activities in that they are represented by an adjustment to final demand in the I–O framework. Household, investment and government EP activities are embodied in final demand, depicted by the adjustment Y_j^{EP} in Table 1. Final demand expenditures for non-EP activities are represented by Y_j^{NE}.

Specific activities included and data sources

The BEA I–O table, *The Use of Commodities by Industries* forms the basis of this report. It is assumed that EP activities are embedded in the tables as currently published. Other sources of information (survey data, engineering data and information published by other US federal agencies) are used to disaggregate EP activities from other economic sectors in the I–O tables. To simplify calculations, it is assumed that all inputs purchased for EP purposes are produced domestically.[7]

Limitations

The decomposition of the US I–O tables into EP and non-EP components depends upon a number of simplifying assumptions. For external EP activities, the assumptions required to isolate 'environmental' water supply, sewerage and solid waste management from the I–O tables are crucial to the estimates of EP activities derived in this report. The estimates for both internal EP activities and EP investment are driven by the expenditure patterns used to allocate capital, and operating and maintenance expenditures for air, water and solid waste abatement to specific I–O categories. These expenditure patterns were derived from dated and often incomplete engineering studies. Further, due to data unavailability, results from manufacturing sectors were used to estimate pollution abatement operation and maintenance expenditures for non-manufacturing sectors.

The procedure for decomposing the US I–O tables into EP and non-EP components was applied to the 1977 and 1982 benchmark I–O tables (US Department of Commerce, BEA, 1984; 1991), since at the time this report was prepared these were the most recent economic census years for which benchmark I–O tables had been compiled. Updating the 1982 EP I–O tables to estimate the composition and levels of employment for 1985, 1988 and 1991 required assuming that the expenditure patterns for the various pollution abatement processes remained constant over time.

Finally, the measurement of foreign trade in EP equipment presents an additional set of difficulties. These difficulties and the actual measurement of EP trade are the subject of a separate report *International Trade in Environmental Protection Equipment: An Assessment of Existing Data* (US Environmental Protection Agency, 1993). In this study, the focus is on the domestic portion of the US I–O tables.

Composition of EP activities

It is possible to use the EP I–O tables to identify the sectors that demand EP goods and services.[8] Table 2 shows the distribution of demand for all EP activities as well as by environmental medium for 1991. It is generated by summing down the column of the EP I–O table for each industry.[9] The EP I–O tables also identify which goods and services are purchased to perform EP activities (i.e. the goods and services that serve as inputs to EP activities).

Table 2. Demand for EP activities by sector for 1991 (million of dollars)

Sector	Total	Air	Water	Solid waste
Agriculture, forestry, fisheries	884.5	8.2	103.7	772.7
Utilities and mining	7309.7	2763.8	1993.7	2552.2
Construction	554.8	52.0	128.7	374.1
Food, beverages and tobacco	1235.9	167.0	822.2	246.7
Textiles, leather, wood, paper, and products	2518.5	668.0	1092.2	758.2
Chemicals and allied products	3543.9	698.6	1544.8	1300.4
Petroleum refining	3696.1	1479.7	815.6	1400.8
Rubber, plastic, stone, clay and glass products	1719.2	526.2	499.5	693.4
Primary metals	2083.3	919.6	643.9	519.8
Manufacturing products	1799.6	260.6	692.1	846.9
Machinery and transport equipment	1839.6	342.0	533.0	964.6
Non-EP services	6813.6	358.1	2630.1	3825.4
External EP sector	1931.2	0.0	31.7	1899.5
Other industry	0.0	0.0	0.0	0.0
Households	13,452.9	0.0	9622.0	3830.8
Investment	0.0	0.0	0.0	0.0
Government	284.7	0.0	284.7	0.0
Exports	4.6	0.0	4.6	0.0
Total demand	49,672.1	8244.0	21,442.8	19,985.4

Household, investment and government demand include only purchases from the external EP sector.

Table 3. Inputs purchased for EP activities by sector for 1991 (millions of dollars).

Sector	Total	Air	Water	Solid waste
Agriculture, forestry, fisheries	25.3	0.0	20.1	5.3
Utilities and mining	6364.3	2210.3	2549.4	1604.6
Construction	10,254.5	466.4	9073.4	714.7
Food, beverages and tobacco	3.4	0.0	0.9	2.5
Textiles, leather, wood, paper, and products	429.9	104.3	48.6	277.0
Chemicals and allied products	2138.4	76.7	1014.7	1047.0
Petroleum refining	1570.8	94.4	123.0	1353.5
Rubber, plastic, stone, clay and glass products	1117.7	448.0	621.1	48.7
Primary metals	0.2	0.0	0.0	0.2
Manufacturing products	933.5	0.0	84.7	848.8
Machinery and transport equipment	1121.3	89.0	187.5	844.8
Non-EP services	6257.2	1801.8	2651.1	1804.3
External EP services	1931.2	0.0	21.4	1909.8
Other industry	52.2	0.0	0.2	52.0
Total intermediate inputs	**32,200.2**	**5290.8**	**16,396.3**	**10,513.1**
Labour	11,077.2	1304.0	3036.3	6736.9
Indirect business taxes	1609.8	0.0	49.9	1559.9
Other value added	4785.0	1649.2	1960.3	1175.5
Total value added	**17,472.0**	**2953.2**	**5046.5**	**9472.3**
Total cost	49,672.1	8244.0	21,442.8	19,985.4

Excludes inputs to EP activities purchased as final demand, with the exception of external EP activities.

Table 3 shows the inputs to EP activities for 1991, and is generated by summing across the row of the EP I–O table for each industry. Table 3 illustrates that EP activities are intermediate-input intensive. Construction, utilities and services comprise the bulk of total EP costs.

Using the I–O framework, it is possible to derive a measure of the size of the EP industry relative to the national economy. This measure consists of computing value added associated with EP activities and is analogous to using value added for a specific industry to compute its contribution to national product. The value added for EP activities is simply the sum of value added for internal, external, household, investment and government EP activities. As an alternative measure, one might want to compute the total share of GNP required to support EP activities. This is given by the sum of direct and indirect EP value added. Indirect EP value added is derived using I–O techniques. The derivation of the formula for computing direct and indirect EP value added is taken from Nestor and Pasurka (1995b).

Table 4 lists both direct, and direct plus indirect, EP value added in current dollars, and EP value added as percentage of GNP for 1977, 1982, 1985, 1988 and 1991. Both direct value added and total value added have increased in absolute terms as well as a percentage of GNP.

Table 4. Measures of the size of the EP industry

Year	Direct value added (millions of current dollars)	Direct value added as a percent of GNP	Direct + indirect value added (millions of current dollars)	Direct + indirect value added as a percent of GNP
1977	14,124.9	0.71	30,436.4	1.53
1982	20,593.1	0.64	50,802.5	1.58
1985	28,692.2	0.68	72,404.1	1.74
1988	37,754.5	0.74	96,766.0	1.93
1991	46,646.6	0.80	121,625.2	2.12

Employment associated with EP activities

Computation of employment

The calculation of direct employment is straightforward. Using I–O techniques, it is also possible to estimate employment indirectly attributable to EP activities or, equivalently, the employment that arises as a consequence of 'multiplier effects'. Multiplier effects occur because EP activities require inputs and employment associated with the production of these inputs. The production of the inputs to EP activities also requires inputs and this generates employment and so forth. As an example, consider an industrial plant that has installed a scrubber to abate its emissions of air pollution. The plant will directly employ individuals to operate the scrubber. In addition, the plant will purchase electricity to run the scrubber, and individuals will be employed in the production of electricity. Likewise, the electric power plant will purchase coal to produce electricity, and individuals will be employed to mine coal. The individuals employed in producing the electricity to run the scrubber and the individuals employed in mining the coal used to generate the electricity needed to operate the scrubber constitute indirect EP employment. The multiplier used in this study does not capture employment associated with household income generation and the resulting expenditures (i.e. 'induced' effects). Formal derivation of the multiplier used in this study is provided in Nestor and Pasurka (1995b).

Total employment in EP activities

Estimates for direct, and direct plus indirect employment associated with EP activities in 1982, 1985, 1988 and 1991 are reported in Table 5. The table shows that direct employment increased in absolute terms between 1977 and 1991, fluctuating between 0.64 and 0.79% of total US employment. Total EP employment has increased both absolutely and as a percentage of US employment.[10]

Table 5. Employment in EP activities

Year	Direct employment (number of individuals)	direct employment as a percent of total US employment	Direct + indirect employment (number of individuals)	Direct + indirect employment as a percent of total US employment
1977	678,359	0.79	1,267,082	1.48
1982	640,181	0.69	1,433,502	1.54
1985	659,067	0.65	1,591,940	1.58
1988	698,348	0.64	1,796,027	1.65
1991	741,186	0.66	1,965,818	1.76

Physical emissions

Generating estimates of physical emissions

Currently, estimates of criteria air pollutants are being generated for the USA.[11] Emissions from both production processes and fuel consumption are included in this inventory. Following the National Energy Accounts (NEA) for the USA, estimates are being generated for the period 1970–1985 at approximately the three-digit SIC level of disaggregation. For the period 1986–1990 period, emission estimates are generated for the 20 two-digit SIC manufacturing sectors in the USA. Several considerations have dictated the scope of the effort to generate this inventory of emissions.

First, the NEA ceased to be assembled after 1985. In 1985, 1988 and 1991, the US Department of Energy (DOE, various issues) published the *Manufacturing Energy Consumption Survey* (MECS). The 1985 and 1988 publications report fuel use at approximately a two-digit SIC level of industry disaggregation. The 1991 MECS report presents more disaggregated data, although these are not as disaggregated as the NEA. Since the MECS data exclude non-manufacturing fuel use, these reports do not allow the generation of emission estimates for non-manufacturing sectors from 1986 to 1990. It was necessary to stop the inventory for US criteria emissions in 1990 since EPA changed its methodology for generating emission estimates. Up to 1990, it is possible to use the 'trends' methodology. This is essentially the use of engineering estimates of emissions from production processes and estimated efficiencies of pollution control devices. Since 1991, the US EPA has used the Aerometric Information Retrieval System (AIRS) database to generate its estimates of emissions of criteria air pollutants. The AIRS database can only be used to generate estimates of emissions from 1985 to the present. It will eventually be necessary to reconcile these two approaches in order to maintain a consistent database of emissions, starting in 1970.

A new step will be to generate estimates of greenhouse gas emissions. It will be possible to generate estimates for 1970–1985 for the NEA sectors and

1986–1991 for two-digit SIC manufacturing sectors. Since most greenhouse gas emissions are the result of fuel use, the same data shortcomings of the energy use data that affect the criteria emissions also adversely affect greenhouse gas emissions.

Finally, the US EPA (1987 onward) initiated the *Toxic Release Inventory* (TRI) in 1987. In order to facilitate reporting of the data in a meaningful manner by industry, it will be necessary to develop a methodology to aggregate the 300 + pollutants into a manageable number of undesirable pollutants.

Potential uses of emissions data

The simple collection of data on the emission of pollutants, while a necessary first step, leads to the next question. How are these data to be used? We envision the data being collected being used on several types of projects. First, following Mayer and Stahmer (1989) and Wier (1993) we plan to undertake a decomposition of the factors influencing changes in criteria emissions in the USA between 1972 and 1982. Second, an increase in the level of environmental protection activities has the effect of reducing undesirable outputs (i.e. emissions). There will thus be a decrease in the level of production of desirable output and a decrease in the level of production of undesirable outputs (see Färe *et al.*, 1989). As the relative amount of resources devoted to environmental protection activities increases there is a movement along an existing production possibilities curve.

Traditional measures of productivity focus on the impact of environmental regulations on the level of desirable output while ignoring the productivity of the environmental regulations (i.e. the reduced level of the undesirable outputs). The reduction in the output of the desirable output as a consequence of the reduction in undesirable outputs constitutes the opportunity costs (i.e. shadow price) of the undesirable output (see Färe *et al.*, 1993). This marginal abatement cost of reducing the level of pollution is one manner in which a price can be assigned to the undesirable outputs. Nestor and Pasurka (1994) have made preliminary calculations of adjusted total factor productivity measures for US manufacturing. This report is currently being revised, given new data on desirable and undesirable outputs and inputs.

Finally, it will be possible to specify a production technology with both desirable and undesirable outputs and incorporate it into a CGE model for the USA. This would allow a model to calculate the level of emissions endogenously. Nestor and Pasurka plan to start work on this in 1996.

Notes

1. This study draws heavily on US Environmental Protection Agency (1995a).
2. For a detailed review of estimates of the size of the EP industry, see Parker *et al.* (1993).

3. The EBJ estimate includes revenue of private and publicly owned water utilities for water supply (conversation with Dan Noble on May 18, 1995). The EBJ value excludes revenue of POTWs (publicly owned treatment works).
4. When discussing current account EP activities, the distinction between 'cost' and 'expenditures' warrants clarification. The term expenditure usually refers to tangible out-of-pocket expenses while cost is a broader economic concept. For example, depreciation typically is not referred to as an expenditure because no transaction is associated with this expense. Depreciation is, however, a cost.
5. In addition, environmental fees, taxes, and penalties do not count as EP industry revenues.
6. For an in-depth description of the SEEA, see United Nations (1993a). Eurostat (1994) also contains information on incorporating environmental protection expenditures and physical emissions into a system of national accounts.
7. Additional details are provided in the report, *The US Environmental Protection Industry: The Technical Document* (US Environmental Protection Agency, 1995b).
8. US Environmental Protection Agency (1995a, Appendix A) contains a concordance between the sectors shown in Table 1 and the BEA input–output tables.
9. This table can be generated for 1977, 1982, 1985 and 1988. Due to the simplifying assumptions used in constructing the updated EP I–O tables (e.g. fixed expenditure patterns) it is dangerous to use the tables to make comparisons over time.
10. For caveats, see US Environmental Protection Agency (1995a, p. 18).
11. Criteria air pollutants consist of TSP, SO_x, NO_x, CO, VOCs and lead.

9
Netherlands: What's in a NAMEA? Recent results

STEVEN J KEUNING AND MARK DE HAAN

The National Accounting Matrix including Environmental Accounts (NAMEA) shows environmental pressures in physical units that are consistent with the monetary figures in the national accounts. Based on the expected contribution of each polluting substance to five major environmental problems, emissions are converted into theme equivalents per problem. This results in five summary environmental indicators that are directly comparable with the conventional economic aggregates. In addition, this meso-level information system is increasingly used for integrated analyses and forecasts of economic and environmental changes. This chapter introduces the NAMEA concept, provides some illustrative analyses of the recently completed NAMEA time series and demonstrates that social accounts and social indicators can easily be integrated. This results in a fairly broad, multipurpose statistical information system.

Introduction

At present, integrated environmental and economic accounts are available for the years 1986–1992 in the Netherlands. These accounts are cast into a National Accounting Matrix including Environmental Accounts (NAMEA). In view of the favourable reaction by our users, it is the intention of Statistics Netherlands to continue the annual compilation of these NAMEAs, as well as their presentation as an integral part of the national accounts.[1] This chapter provides a description of the NAMEA framework as it is applied in practice, analyses some results from a recently completed time series of NAMEAs and demonstrates that social accounts and social indicators can easily be integrated into this framework.

The NAMEA approach

In a paper for the 1991 Special IARIW Conference on environmental accounting, an illustrative NAMEA was presented for the first time (de Boo *et al.*, 1993), following the conceptual design by Keuning (1993). The original design

---

The results shown in this paper are based on NAMEA time series compiled jointly by staff of the Environment Statistics and the National Accounts Department of Statistics Netherlands. The views expressed in this paper are those of the authors and do not necessarily reflect the views of Statistics Netherlands.

K. Uno and P. Bartelmus (eds.), Environmental Accounting in Theory and Practice, 143–156.

contains a complete system of national flow accounts, including a full set of income distribution and use accounts, accumulation accounts and changes in balance sheet accounts. Not only emissions of pollutants and extraction of natural resources are represented, but also their effects. A distinction is made between effects of current emissions that are absorbed in the current period (noise, stench etc.), current effects of emissions in the past (e.g. leakage from a waste dump), net capital losses due to natural causes (e.g. a severe drought), referrable damage to economic assets and to other (natural) assets due to environmental effects and non-referrable degradation to non-economic, natural assets. All these transactions are summarized in additional balancing items, culminating in a new total for the changes in net worth. It soon became clear that insufficient data were available for an immediate operationalization of this conceptual framework. It was therefore decided to compile a more modest pilot NAMEA, making use of the work undertaken at the Netherlands Ministry of Housing, Spatial Planning and the Environment (1989). This Ministry had developed a so-called national environment policy plan in which a number of environmental themes were distinguished. For each of these themes, a single indicator had been designed by weighing together the emissions that contributed to each theme (Adriaanse, 1993). The conversion of emissions into theme equivalents was based on the expected contribution of each polluting substance to a particular environmental problem. Target values for each theme in the year 2000 were also formulated. By the time the first NAMEA was compiled this environment policy plan had been approved by the Dutch Parliament. The first NAMEA became available in 1993 (de Haan *et al.*, 1994), and the present NAMEAs largely maintained this format (de Haan and Keuning, 1996). The 1994 NAMEA paper also contains an appendix on compilation issues, while this year a more extended methodological publication was published (CBS, 1996).

Table 1 presents an aggregate NAMEA for 1991.[2] It consists of a conventional national accounting matrix extended by three accounts on the environment: a substances account (account 11), an account for global environmental themes (account 12) and an account for national environmental themes (account 13). These accounts do not express transactions in money terms but include information on the environment as it is observed in reality, that is, in physical units. In order to emphasize that currency units and physical units cannot be added up, the physical units are positioned higher in the rows, and more to the left in the columns of accounts 2, 3, 6 and 9.

The rest of the NAMEA presents the regular national accounts flows in a matrix format (cf. Keuning and de Gijt, 1992). Some transactions that are relevant to an environmental concern are singled out (cf. account 1a). As usual in a matrix presentation, receipts are shown in the rows and outlays in the columns. Most of the accounts contain a balancing item in the column (the doubly framed cells in Table 1), defined as the difference between total receipts, in the row, and the (other) outlays, in the column. In this way, row and column totals are equal for each account, which guarantees the consistency of the complete system.

The first account presents intermediate and final uses of goods and services in the row, and domestic and foreign supply in the column. Supply at market prices consists of output at basic prices plus trade and transport margins plus taxes on products. Supply and use of environmental cleansing services are shown separately. In order to provide a complete picture of the industries' expenses on behalf of the environment in the NAMEA, these services are inclusive of the so-called internal environmental cleansing. Internal environmental cleansing services are produced by the same establishment that uses them within its own production process (cf. de Boo, 1995). As intra-establishment deliveries are not considered output in the standard statistics, both production and intermediate consumption are higher in the NAMEA than in the conventional national accounts. Evidently, this affects neither net domestic product (NDP) nor any other balancing item.

The second account is a specific consumption account, which re-allocates consumption expenditures (matrix 1,2) to consumption purposes (vector 2,5). The latter are connected to specific pollution patterns (2,11). Consumer goods that are purchased in order to protect the environment are presented separately (cell 2a,5). This concerns, for example, the extra costs of cars fitted with catalytic converters. Pollution generated by the government is connected to government production and not to government consumption in the NAMEA.

The third account shows in the row the production of goods and services, and in the column intermediate use and value added. 'Other taxes on production' are recorded on a separate tax account (cell 8,3), and consumption of fixed capital is directly put on the capital account (cell 6,3), so that the balancing item in cell (4,3) equals NDP at factor cost. In row 3, the production of goods and services is expanded with the concomitant emissions of pollutants (row-vector 3,11). Vector (11,3) shows the amounts for several inputs which have not been explicitly paid for. In fact, their depletion may not even be incorporated in the output price, and thus also not be contained in regular NDP. This concerns, for instance, natural resources and the waste that is re-processed in incineration plants. Note that the emissions of waste incineration plants are in fact taken into account, namely in row-vector (3,11). In principle, re-cycling of materials should also be shown here, but at present insufficient data are available.

The fourth row shows the components of NDP (wages and salaries, employers' social contributions and operating surplus), as well as wages and salaries from abroad. Cell (4,3) reflects value added tax invoiced by sellers, but not handed over to the government for various reasons. In the column of account 4, primary income is allocated to the institutional sectors (corporations, households, government, etc.) and to the rest of the world. In the fifth account, income is (re-)distributed and used for consumption and saving.

Table 1 A national accounting matrix including environmental accounts (NAMEA) for the Netherlands, 1991 (million guilders).

ACCOUNT (classification)		Goods and services (Product-groups)		Consumption of households (Purposes)		Production (Branches of industry)	Income generation (primary input categories)	Income distribution and use (Sectors)	Capital	Taxes (Tax categories)		Rest of the world Current
				Environment	Other purposes					Environmental taxes	Other taxes	
		1a	1b	2a	2b	3	4	5	6	8a	8b	9
Goods and services (Product groups)		Trade and transport margins		Household consumption		Intermediate consumption		Government consumption	Gross capital formation			Exports (fob)
Environmental cleansing services	1a			24	-	6305		1410				
Other goods and services	1b	-	-	710	321727	501763		76837	114818			293086
Consumption of households (Purposes)								Consumption of households				
Environment	2a							734				
Other purposes	2b							321727				
Production (Branches of industry)	3	Output, basic prices 7627	994861									
Income generation (Primary input categories)	4					Net Domestic Product, factor cost 429118				VAT not handed over to the government	1880	Wages from the rest of the world 820
Income distribution and use (Sectors)	5						Net National Generated Income, factor cost 430650	Property income and current transfers 573920		Taxes 3982	137518	Property income and current transfers from the rest of the world 60190
Capital	6					Consumption of fixed capital 61560		Net saving 72960				
Financial balance	7								Net lending to the rest of the world 17340			
Taxes (Tax categories)		Taxes less subsidies on products				Other taxes less subsidies on production		Taxes on income and wealth	VAT on land and taxes on investment			Taxes from the rest of the world
Environmental taxes	8a		907			855		2220				
Other taxes	8b	112	45787			2887		88730	992			1050
Rest of the world Current	9	Imports (cif)	267386				Wages to the rest of the world 1170	Property income and current transfers to the rest of the world 67720		Taxes to the rest of the world	160	
Rest of the world Capital	10								Capital transfers to the rest of the world 2350			Current external balance -18710
Substances (CFCs and halons in 1000 kg, gas and oil in pj and other substances in million kg)						Absorption of substances in production						Trans-border pollution to the rest of the world
CO2	11a											
N2O	11b											
CH4	11c											
CFCs and halons	11d											
NOx	11e											488
SO2	11f											159
NH3	11g											113
P	11h											24
N	11i											581
Waste	11j					2645						
Gas	11k					2595						
Oil	11l					138						
Global environmental themes									Environmental indicators			
Greenhouse effects (GWP)	12a								188890			
Ozone layer depletion (ODP)	12b								3816			
National environmental themes												
Acidification (AEQ)	13a								156			
Eutrophication (EEQ)	13b								287			
Waste (KG)	13c								23761			
Loss of Natural resources (PJ)	13d								-759			
TOTAL		Supply, market prices 7739	1308941	Consumption of households 734	321727	Costs, basic prices 1002488	Allocation of generated income 431820	Current outlays 1206258	Capital outlays 135500	Tax receipts 3982	139558	Current receipts from the rest of the world 336436

Due to rounding totals do not always add up.

In account 6, gross saving is allocated to gross capital formation and to net lending and capital transfers to the rest of the world. In the present NAMEAs, accounts 6 and 7 are not subdivided by institutional sector, so that inter-institutional capital transfers and financial flows cancel out. Account 7 presents the financial balances (net lending) of the total economy and the rest of the world, respectively. These balances add up to zero by definition. Therefore,

Rest of the world Capital	Substances (CFCs and halons in 1000 kg, gas and oil in pj and other substances in million kg)												Global environmental themes		National environmental themes				TOTAL
	CO2	N2O	CH4	CFCs & halons	NOx	SO2	NH3	P	N	Waste	Gas	Oil	Greenhouse effect	Ozone layer depletion	Acidification	Eutrophication	Waste	Loss of Natural Resources	
10	11a	11b	11c	11d	11e	11f	11g	11h	11i	11j	11k	11l	12a	12b	13a	13b	13c	13d	
																			Commodity use 7739 1308941
	Emission of pollutants from households 36372	2	4	656	156	5	-	15	115	6663									Consumption of households 734 321727
	Emission of pollutants from industries 128040	59	724	4375	397	191	220	155	1257	19742									Production, basic prices 1002488
																			Generated income 431820
																			Current receipts 1206258
Capital transfers from the rest ofthe world 980	Other changes in natural resources										1836	138							Capital receipts 135500
Net lending from the rest of the world -17340																			0
																			Tax payments 3982 139558
	Trans-border pollution from the rest of the world				93	99	27	20	415	.									Current payments to the rest of the world 336436
																			Capital payments to the rest of the world -16360
													Allocation to global environmental themes 164412 61 728	5031	Allocation to national environmental themes (accumulation of substances) 158 136 134	166 1206	23760	-759 -	Destination of substances 164412 61 728 5031 646 295 247 190 1787 26405 1836 138
																			Theme-equivalent global 188890 3816
																			Theme-equivalent national 156 287 23760 -759
Capital receipts from the rest of the world -16360	Origin of substances 164412	61	728	5031	646	295	247	190	1787	26405	1836	138	Theme-equivalents, Global 188890	3816	Theme-equivalents, National 156	287	23760	-759	

the (empty) column of this account has been deleted.

Account 8 of the NAMEA is a separate tax account, in which a variety of taxes are presented: taxes (less subsidies) on products in sub-matrix (8,1), other taxes on production in sub-matrix (8,3), taxes on income and wealth in sub-matrix (8,5), and so forth. In the detailed NAMEA, all kinds of environmental taxes such as energy levies, levies on pollution of surface waters and levies on waste water drain-offs are presented separately. The column of this account shows the collection of these taxes by the government (row-vector 5,8) and the rest of the world (cell 9,8b).

Accounts 9 and 10 represent all transactions with the rest of the world. The row of the current account (9) presents imports, not only of goods and services, but also of less wanted substances. For the time being, data are only available for the entry of pollutants via the rivers or through the air. The column contains exports, including the outflow of pollutants to other countries. Cell (10,9) reflects the current external balance of the rest of the world with the Netherlands. The figures show that the Netherlands managed to create a surplus for goods and services as well as for pollutants. Of course, this 'pollution trade balance' only refers to 'direct' flows of pollutants. However, estimates on pollutants embodied in imported products, or invoked by foreign demand, can easily be made with the help of a NAMEA-based analysis (discussed below).

Account 11 registers in the column the origin of 10 types of pollutants. This pollution is caused by producers (vector 3,11), consumers (vector 2,11) and the rest of the world (vector 9,11). This column also registers additions to proven reserves and other changes in natural resources (vector 6,11). The row of this account shows in vector (11,3) the extraction of natural resources (crude oil, natural gas and wood) as well as the re-absorption of pollutants into the economic process. This concerns, for instance, part of solid waste that is incinerated. The rest of the pollutants is exported to other countries (vector 11,9), or is re-allocated to five environmental themes (sub-matrices 11,12 and 11,13). The use of natural resources is allocated to a sixth theme: loss of natural resources. Account 11 is expressed in kilograms or in petajoules. Of course, the row and column totals of account 11 are equal.

The so-called 'environmental themes' presented in accounts 12 and 13 are adopted from the second Netherlands' National Environmental Policy Plan (Netherlands Ministry of Housing, Spatial Planning and the Environment, 1993). Environmental themes are used as an inventory framework of current environmental problems in the Netherlands. The column totals of account 12 and 13 reflect a weighted aggregation procedure. The weights reflect for each theme the potential relative stress on the environment of each substance. These aggregation methods have been accepted by the Dutch Parliament and are for the major part based on international research on the effects of different substances on environmental quality (Adriaanse, 1993). The conversion from the substance units, as given in the heading of account 11, into the corresponding theme-related stress equivalents is shown in Table 2.[3]

This method yields a limited number of physical environmental indicators, shown in vector (12-13,6). Account 12 monitors two global environmental problems: the greenhouse effect and the depletion of the ozone layer. The corresponding indicators reflect the Netherlands' contribution to these global problems. If, however, the environmental damage within the country is at

Table 2. Environmental themes and their corresponding substances (in parentheses, the weights used to convert substance units into theme-equivalents)

Environmental themes	Substances	Theme-equivalents
Greenhouse effect	CO_2 (1), N_2O (270) and CH_4 (11)	Global warming potentials
Depletion of the ozone layer	CFCs 11, 12, 13, 112 (1), CFC 113 (0.8), CFCs 114, 115 (0.6), trichloroethane (0.1) tetrachloride (1) and halon 1301 (10)	Ozone depletion potentials
Acidification	NOx (0.22), SO_2 (0.31) and NH_3 (0.59)	Acid-equivalents
Eutrophication	P (1) and N (0.1)	Eutrophication-equivalents
Accumulation of waste	Waste	Million kilograms
Waste water	Waste water	Million kilograms
Natural resources	Gas (1) and oil (1)	Petajoules

stake (cf. the themes in account 13), summary indicators on the national accumulation of pollutants are directly comparable with the economic indicators from the national accounts. For these themes, domestic pollution is augmented with the import, and reduced by the export, of pollutants.

The aggregate NAMEA in Table 1 presents the interrelation between macro-indicators for the economy (NDP, net saving, external balance) and the environment (environmental theme indicators). However, a much more detailed information system is underlying this table. This is illustrated below.

A preliminary analysis of the 1986–1992 NAMEAs

Thus far, NAMEAs have been compiled for 1986 through 1992 (CBS, 1996) for a detailed review of results and compilation methods. This section presents a selection of these results, using the kind of tables that are typically also shown in the press releases by Statistics Netherlands on this subject.

Table 3 shows the 1992 contribution of each industry to GDP and to five major environmental problems in the Netherlands. As expected, the share of services in GDP largely surpasses its share in each of the environmental problems. Note that the contribution of manufacturing (excluding food processing, chemicals and related industries) to GDP exceeds its contribution to all environmental problems except one (ozone layer depletion). The environment intensity of construction is also relatively low, apart from waste generation. On the other hand, the contribution of agriculture to most of the problems concerned is significantly higher than its share in GDP. The same applies to electricity generation, the chemical industry and, somewhat surprisingly, to the food, beverages and tobacco industry. In all cases, more than half the pollution is generated by two or at most three of the 20 industries. More than half the greenhouse effect, the acidification and the eutrophication is caused by industries that generate less than 10% of GDP.

Table 4 presents an overview of the annual changes in economic and

Table 3. Contribution of each industry to GDP and to five major environmental problems in the Netherlands, 1992 (%).

Industry	Gross domestic product (factor cost)	Greenhouse effect	Ozone layer depletion	Acidification	Eutrophication	Waste
Agriculture and fishery	4	15	2	46	82	6
Crude petroleum and natural gas production and exploration	3	2	0	1	0	0
Other mining and quarrying	0	0	0	0	0	0
Food, beverages and tobacco	3	4	5	1	3	11
Textiles, clothing, leather and leatherware	1	0	0	0	0	0
Wood and furniture	1	0	0	0	0	1
Paper, paperware, printing and related industries	2	1	1	0	1	3
Petroleum	1	7	0	11	0	0
Chemicals	2	15	26	7	2	14
Rubber and plastic-processing	1	1	12	0	0	0
Manufacture of building materials, earthenware and glass	1	2	0	2	0	3
Basic metal industry	1	3	0	3	0	1
Manufacture of metal products and machinery	3	2	9	0	0	2
Other manufacturing	4	1	4	0	0	1
Electricity generation	1	24	0	10	1	2
Other public utilities	1	1	0	0	0	0
Construction and installation on construction projects	6	2	6	1	0	28
Transport and storage	6	8	5	12	1	2
Environmental cleaning services	0	2	5	4	5	6
Other services	61	9	24	1	3	19
Total	100	100	100	100	100	100

environmental indicators for the period 1986–1992. It is quite remarkable that all environmental indicators decrease, or increase considerably less than the GDP and consumption volume. So, at least for the problems covered here, the pollution per unit of production and consumption has fallen. The same applies to virtually every consumption category and major industry distinguished in this table. Not surprisingly, the results are most impressive for the ozone layer depletion problem. In manufacturing, for instance, the emissions that cause this problem have been reduced by almost 10% per year, on average, despite an annual value added volume growth of almost 2%. In addition, acidifying emissions from agriculture have been reduced by 4% per year, despite an annual value added increase of 4%. In general, the relatively less harmful services industry has been somewhat less successful in the reduction of its pollution. Finally, the waste problem appears to have been most difficult to tackle in this period: all categories, except the waste caused by own transport of households (car wrecks and such), still showed an absolute increase in environmental pressure. In general, the most polluting industries have been most successful in abating emissions, perhaps under the influence of (anticipated) government measures. For instance, the chemical industry has realized a substantial annual reduction in the emission of substances that cause the ozone layer depletion, acidification, eutrophication and generate waste, while the volume growth of value added in this industry amounted to 2.1% over the same period. Only the emission of gases that caused the greenhouse effect continued to increase. A similar picture emerges in electricity generation. On the other hand, the score was somewhat less positive in, for instance, the paper, paperware, printing and related industries. Such results may be of use in the continuing dialogue between the government and representatives of the industries concerned.

So far, the discussion has been limited to the direct contributions of production activities to environmental indicators. This, though, does not provide a complete picture, as some production processes generate a relatively modest pollution but use more harmful intermediate inputs. These backward linkages can be taken into account when computing the pollution intensities of final products. For instance, the demand for food products generates pollution not only in food processing but also in agriculture.

This 'total' environmental pressure caused by the final demand for all kinds of products can also be analysed with the help of the NAMEA framework. Considering the 'total' pollution caused by one unit of final demand for a major product category, relative to the average 'total' pollution caused by one unit of final demand shows that one guilder of final demand for services generates less than half the pollution of an average guilder of final demand for Dutch products. The opposite applies to a guilder of final demand for agricultural and manufacturing products. An impression of the 'direct' relative pollution per guilder is obtained by dividing the last five columns in Table 3 by the first column. For instance, the direct contribution of agriculture to the

Table 4. Annual volume change of economic and environmental indicators in the Netherlands, 1986–1992 (% volume change)

	Economic indicators		Environmental indicators				
	GDP at factor cost	Household consumption	Greenhouse effect	Ozone layer depletion	Acidification	Eutrophication	Waste
Total	2.9	2.8	0.1	−8.7	−3.1	−1.7	1.8
Consumption of households		2.8	−0.8	−3.3	−1.0	−2.3	0.6
Own transport		0.7	1.4	–	−0.7	−0.6	−7.8
Other purposes		3.0	−1.9	−3.3	−2.8	−2.7	0.8
Production	2.9		0.2	−8.8	−3.4	−1.7	2.3
Agriculture and fisheries	4.2		2.0	−2.8	−4.0	−1.8	–
Manufacturing industries	1.9		0.0	−9.7	−4.1	−1.3	1.5
Services	3.3		0.1	−7.5	0.9	−0.8	2.9

eutrophication problem is more than 20 times its GDP contribution. If all backward linkages are taken into account, this ratio drops to almost 14. In general, the relatively large direct contribution of agriculture to the environmental problems concerned is somewhat mitigated if indirect effects are incorporated. All in all, a guilder of demand for food contributes most to the greenhouse effect, acidification and eutrophication, while a guilder of demand for manufactured products generates most waste and ozone layer depletion. However, when drawing policy conclusions from these figures, it should also be kept in mind that foreign substitutes for domestic products are not necessarily less harmful to the environment. In this regard, the theory of comparative advantage may also be of use in assessing the policy implications of the negative side-effects of production and international trade.

If the 'total' pollution per unit of final demand in 1992 is compared with that in 1987, all products generated less pollution per guilder in the later year. The total reduction varied from 7% for waste to not less than 48% for the ozone-depleting gases. Pollution has been reduced most in the production of the most harmful products (agricultural products in the case of eutrophication, acidification and the greenhouse effect, and manufactured products in the case of ozone layer depletion and waste generation). Analogously, the smallest decrease in pollution has been achieved in services, with the exception of the ozone-depleting emissions. Finally, even the spread of the pollution reduction is positively correlated with the original spread of pollution intensities. For example, the biggest difference in pollution reduction between agricultural products, on the one hand, and services, on the other hand, occurs with respect to the themes of acidification and eutrophication.

The above analyses have also been made at a more detailed level. It appears, for instance, that not only the 'direct' pollution by the chemical industry, but also the 'total' environment intensity of the final demand for chemical products were still substantially above average, with the exception of their contribution to the eutrophication problem. However, the 'distance' to the average was reduced by about half when pollution embodied in intermediate inputs was also taken into account. A similar result was found for electricity. On the contrary, the position of the food, beverage and tobacco industry worsened considerably when backward linkages were incorporated, because of their reliance on environment-intensive agricultural inputs. The very favourable score of services was also diminished when indirect effects were included.

For the different final demand categories, most waste is generated by exports, although per guilder of final demand capital formation is the greater culprit. The increase in waste emission between 1987 and 1992 was entirely due to the growth of final demand; per guilder of final demand, the 'total' waste production declined substantially. All in all, the purpose of such figures is to demonstrate that the same problem can be viewed from different perspectives.

In summary, the NAMEA can be used for several Leontief-type analyses. Clearly, the usefulness of such analyses would be enhanced if the social dimension was also incorporated.

Integrating social accounts and indicators into the NAMEA[4]

The 1993 System of National Accounts (SNA) contains a chapter on Social Accounting Matrices (SAM) (Commission of the European Communities *et al.*, 1993). This chapter indicates that a further extension, in the direction of a so-called System of Economic and Social Accounting Matrices and Extensions (SESAME), is both useful and feasible. In this context, it states: 'This approach could equally well be followed when dealing with environmental issues' (para. 20.33).

At present, both a 1990 SAM and a 1990 NAMEA are available for the Netherlands (Timmerman and van de Ven, 1994; de Haan and Keuning, 1995). These two frameworks have been integrated into a so-called Social Accounting Matrix including Environmental Accounts (SAMEA). At the aggregate level, the only difference of a SAMEA from the NAMEA, shown in Table 1, is a breakdown of the distribution-and-use-of-income account into three subaccounts. This serves to provide a better insight into the income distribution and use by household subsector. Such subsectors are not distinguished in the NAMEA.

The additional information contained in the SAMEA, in comparison with the NAMEA, is only revealed in the detailed tables. For instance, wages and salaries by branch of industry, as shown in the generation of income account, are broken down by sex and by seven educational levels. Concomitantly, full-time equivalent employment by branch of industry has also been subdivided by these labour categories. This enables an analysis of the relationship between the remuneration of each labour category and the pollution that is caused by the economic activities in which they are employed. The contribution of each labour category to major environmental problems can be computed by allocating, first, the pollution equivalents per environmental problem to the substances that cause these problems, secondly, the pollution equivalents by substance to the economic activities that emit these substances and, finally, the pollution equivalents by economic activity to the primary input categories that generate the value added in these activities. This then provides an indication of the type of labour that might benefit or suffer from a shift in economic structure towards less polluting activities. For instance, women typically work in industries that burden the environment less. They make up 24% of the wage bill and 31% of employment, but account for only 11–14% of the environmental stress equivalents. To some extent, the same applies to men with lower secondary, vocational or general education (accounting for 7% of wages and employment, and 5–7% of the pollution) and men with a university degree (accounting for 8% of the wage bill, 5% of total employment and 4–7%

of the problems). On the other hand, a relatively large share of the environmental problems is caused by activities that employ relatively many men with middle-level education (higher general secondary, and middle vocational education). Their share in the wage bill equals 40%, in employment 41%, and in the stress equivalents by theme 50–59%. Evidently, these differences are closely related to the representation of each labour category in services, which contribute relatively much to GDP and relatively little to the environmental themes incorporated in the SAMEA.

For operating surplus and mixed income, it is found that their contribution to GDP is lower than their contribution to most of the environmental problems. In particular, this applies to eutrophication, which is predominantly caused by an activity with many self-employed, namely agriculture. An exception to this rule is ozone layer depletion, which is associated with two manufacturing industries with a comparatively high share of compensation of employees in value added.

The SAMEA also contains the allocation of these value added categories to institutional subsectors, including 10 household groups. This means that the above allocation of pollution units to primary input categories can be taken one step further, so that the contribution of each subsector is revealed, at least as far as their supply of production factors is concerned. On the other hand, households also directly generate pollution through their consumption. The contribution of the elderly to all environmental problems surpasses their population share. The opposite applies to households that mainly depend on other transfer income (unemployment benefits and such). In both subgroups, the pollution caused per guilder of consumption is below average, except for the ozone layer depletion and the waste problem. SAMEA can also be used for an analysis of the 'total', cumulative pollution generated by the consumption pattern of each household group (cf. the previous section of this paper). Analogously, the 'total' employment, of each skill level and sex, involved in one unit of pollution might be computed for all environmental problems concerned.

Summary and conclusions

The conceptual and statistical development of the NAMEA is continuing and the number of environmental themes will be expanded when new information becomes available. This relates, for instance, to other themes from the Netherlands' Environmental Policy Plan (the dispersion of toxic substances, stench and noise nuisance, and excessive use of ground water). Another expansion of the system refers to a decomposition of supply and use data in the NAMEA into physical units and average prices. A direct connection can then be made between the use of natural resources and the emissions of pollutants (Konijn *et al.*, 1995). In turn, the research on material flows, for instance on energy balances, may lead to more detailed emission estimates in the future.

In order to identify the incidence of environmental pressure, the present

NAMEAs already distinguish between global and national environmental themes. For the global themes, it may be assumed that each citizen is equally affected, so that the benefits of any reduction in environmental pressure would be distributed in the same way. For the other themes, it is necessary, and perhaps also sufficient, to subdivide the environmental theme equivalents and the labour and household groups by region. The classification of regions should be such that they are fairly homogeneous regarding environmental burden. Any subsequent analysis will then show both the immediate costs, in terms of less income or higher unemployment, and the benefits, in terms of less environmental pressure, of all kinds of environmental policies. For an analysis of the likely longer-term economic benefits of more stringent environmental policies, the NAMEA should provide more details on the origin and destination of capital formation. This can be achieved by merging it with the National Accounting Matrix that is already available for the Netherlands (CBS, 1995; Keuning and de Gijt, 1992).

Finally, research is carried out on the integration of national accounts, environmental accounts and sociodemographic accounts in a single information system that also yields the core economic, social and environmental indicators for monitoring human development (Keuning, 1996b). This SESAME system allows for an extension of regular macro-analyses that are now based on purely monetary national accounts. In turn, that would mean that the social and environmental consequences of all intended (monetary and fiscal) policies are automatically taken into account. This can only be achieved if the summary indicators that result from economic analysis are: (1) derived from an underlying analytical framework (SESAME) and (2) not themselves based on a large number of modelling assumptions ('greened economic aggregates'). In fact, a sustainable national income could very well be the result of a very important policy scenario that can be calculated with a SESAME-based model, namely the scenario in which all emissions and depletion have been reduced to acceptable norms.

Notes

1. For a review of the present status of the NAMEA approach and of its actual and potential uses, and for comparison with the System of integrated Environmental and Economic Accounting (SEEA), see Keuning (1996a).
2. Readers who are already familiar with the NAMEA framework may want to skip the rest of this section, as it resembles earlier explanations (e.g. in de Haan *et al.*, 1994).
3. Refer to de Haan *et al.* (1994) for a more detailed exposition of each theme.
4. Refer to Keuning and Timmerman (1995) for a more extensive presentation of these results.

Part III

Physical and spatial accounting

10

Measuring Canada's natural wealth: why we need both physical and monetary accounts

ALICE BORN

If sustainable development is going to be more than a slogan, it requires definition, quantitative objectives and indicators for measuring any progress in preserving the environment for future generations. Expanded national wealth accounts are seen as one of the tools in measuring sustainability whether it is a measure of wealth per capita or a measure of a nation's total capital over time. Fundamental to the concept of sustainable development is capital maintenance (produced, natural, human and social capital). However, does the concept of capital maintenance refer to the value of natural resources or to the physical volume of natural resource stocks? As shown in the paper there is a need for both physical and monetary accounts in order to show whether or not Canada is maintaining its mineral stocks in both monetary and physical terms.

Introduction

There is growing interest in sustainable development. Expanded wealth accounts are seen as one of the tools in its measurement (World Bank, 1995; Hamilton, 1994). Whether it is a measure of wealth per capita or a measure of a nation's total capital over time, wealth accounts and national balance sheets will include the value of natural capital as well as the value of produced capital in the future. Estimation of the value of human capital is still in the early stages of development but the World Bank has recently provided some estimates (World Bank, 1995). The development of natural resource accounts has focused on placing a monetary value on known physical quantities of the resource in order to obtain a wealth value for natural capital.

At Statistics Canada, natural resource stock accounts, which are aimed at measuring Canada's natural wealth, are one of four environmental accounts being currently developed. The other three accounts are natural resource use accounts, waste output accounts and environmental protection expenditure accounts.[1] The natural resource use and waste output accounts are recorded in physical terms, and the environmental protection expenditure accounts in monetary terms. Only the natural resource stock accounts are recorded in both physical and monetary terms.

K. Uno and P. Bartelmus (eds.), Environmental Accounting in Theory and Practice, 159–170.

This paper reviews some of the preliminary results of the natural resource stock accounts in the context of Canada's wealth accounts, which currently only include the values of produced capital and land. In accordance with the 1993 international System of National Accounts (Commission of the European Communities *et al.*, 1993),[2] Canada plans to add the value of natural wealth (e.g. subsoil and timber assets) to the national balance sheet accounts beginning in 1997. The results presented here are limited to subsoil assets.

The first part of this chapter discusses the uncertainty of measuring Canada's resource base and its effects on physical accounting. The second part focuses on the linkages between the physical and monetary accounts, and the 1993 SNA and the System of Economic and Environmental Accounts (United Nations, 1993a). The final section examines the usefulness of the expanded wealth accounts with respect to the concept of sustainable development.[3] Partial results of Canada's wealth accounts are also presented.

Physical accounts: resource base and uncertainty

Physical accounting of subsoil assets shows that the assumption of a 'fixed and non-renewable' mineral stock can be limiting in determining the size of a nation's resource base. What we observe are 'inventories' of proven reserves which are being constantly renewed by investment in exploration and development. Furthermore, in the case of metals, the rate of depletion is affected by the rate of recycling. The size of the resource that is considered 'economic' and, therefore, currently recoverable depends on costs and prices, including the prices of substitutes (Adelman *et al.*, 1991). This suggests that we cannot know how much of our mineral resources should be shared with future generations or what is the optimal depletion rate by simply estimating when supplies will be exhausted at current rates of depletion (Mikesell, 1994). Since resource base and reserve estimates are revised from year to year as extraction continues, new discoveries are made and the overall understanding of the resource base improves, physical accounts of fuel and non-fuel minerals should reflect these changes.

Physical accounts should present a range of available resources and should not be restricted to 'proven reserves', as suggested by the 1993 SNA. For example, mineral resources can be divided into two components: discovered or known recoverable resources and undiscovered recoverable resources. Discovered recoverable (or known) resources are those resources that are estimated to be recoverable from known deposits using current technology and under current economic conditions. Included in this category are cumulative production,[4] remaining established (or developed) reserves and yet-to-be established (or undeveloped)[5] reserves. Undiscovered recoverable resources are those that are estimated to be recoverable from resources that are believed to exist on the basis of available geological and geophysical evidence but not yet shown to exist by drilling, testing or production. The sum of discovered and undiscovered mineral resources is referred to as ultimately recoverable resources (Table 1).

For the purpose of monetary valuation, we focus on that part of the resource base which is estimated to be recoverable using existing machinery and equipment and structures in place, and under current economic conditions. These are defined as remaining established (or recoverable) reserves in Table 1. Established reserves represent a conservative estimate of the available stocks, given changing technology and markets. Nevertheless, based on this definition, it is apparent that rapid depletion has occurred over a short time period (Figures 1–4), particularly for oil and nickel reserves. However, these represent only a small portion of Canada's known reserves and ultimately recoverable resources of fossil fuels and metals. In the case of crude oil, remaining established reserves represent 38% of the known reserves and 11% of the total ultimately recoverable resources. For crude bitumen, we are placing a monetary value on only 1% of known reserves and for natural gas, on 62% of known reserves and 13% of the total ultimately recoverable resources.

Table 1. Energy resource estimates, 1992.

	Discovered recoverable resources				Undiscovered	Ultimately
	Cumulative production	Remaining established reserves	Yet-to-be established reserves	Total[1]	recoverable resources	recoverable resources[2]
Crude oil (millions of m^3)	2 394	680	1 113	4 207 [3]	4 361	8 568 [3]
Crude bitumen[4] (millions of m^3)						
Surface mineable	210	434	9 356	10 000	-	10 000
In situ	59	48	38 893	39 000	-	39 000
Total crude bitumen	269	482	48 249	49 000	-	49 000
Natural gas (billions of m^3)						
Western Canada	2 163	1 909	-	4 072	3 144	7 216
Frontier areas	32	9	1 187	1 228	8 055	9 283
Total	2 195	1 918	1 187	5 300	11 199	16 499
Coal[5,6] (megatonnes)						
Lignite	..	2 236	..	..	..	44 360 [7]
Subbituminous coal	..	871	..	..	..	
Bituminous coal	..	3 471	..	..	..	34 245
Total	..	6 578	..	..	..	78 875
Uranium (kilotonnes)	155 [8]	309	163	627	..	..

Notes:

1. Sum of cumulative production, remaining established reserves and yet-to-be established reserves.
2. Sum of total discovered reserves and undiscovered recoverable resources.
3. Totals in source data.
4. Occurs in oil sands.
5. Reserve definition used is *recoverable* rather than *established*, however the definitions are similar.
6. As of 1985.
7. Lignite plus subbituminous coal.
8. Approximated.

Sources:

Natural Resources Canada (1993a and 1993b).
National Energy Board (1994).

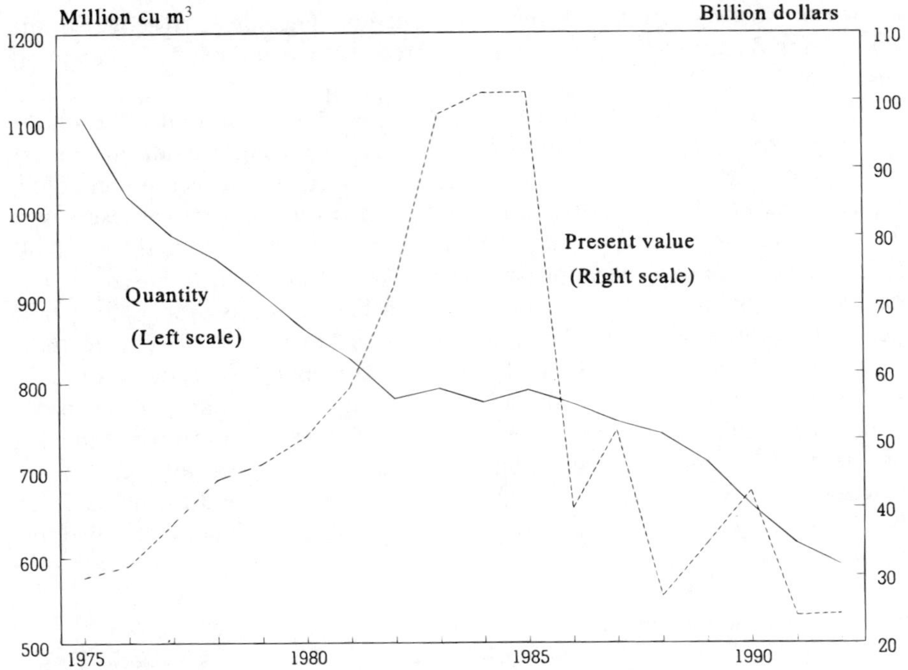

Figure 1. Quantity and present value of established oil reserves of Canada.

The importance of physical accounts is demonstrated here. While we are placing a monetary value on only a small part of Canada's known reserves, a large part of the resource remains unvalued. 'Option values' based on transaction values might be used to value undeveloped reserves to reflect more closely the total wealth of a nation.[6] A broader definition of resources in the physical accounts shows the resources that are available in the long term.

Physical and monetary accounting

A comprehensive description of the inter-relationship between the environment and the economy is not possible without using physical data: in many cases, these are more suitable than monetary data. This is especially true for the flow of materials within the natural environment, and from the natural environment to the economy and back into the environment as wastes (United Nations, 1993a). Thus, Statistics Canada's Natural Resource Use and Waste Output Accounts, which are based on input–output models, are expressed in physical terms. Linking those data to the monetary data of the national accounts, however, is necessary for the development of other environmental–economic accounting systems. At Statistics Canada, both developed and unde-

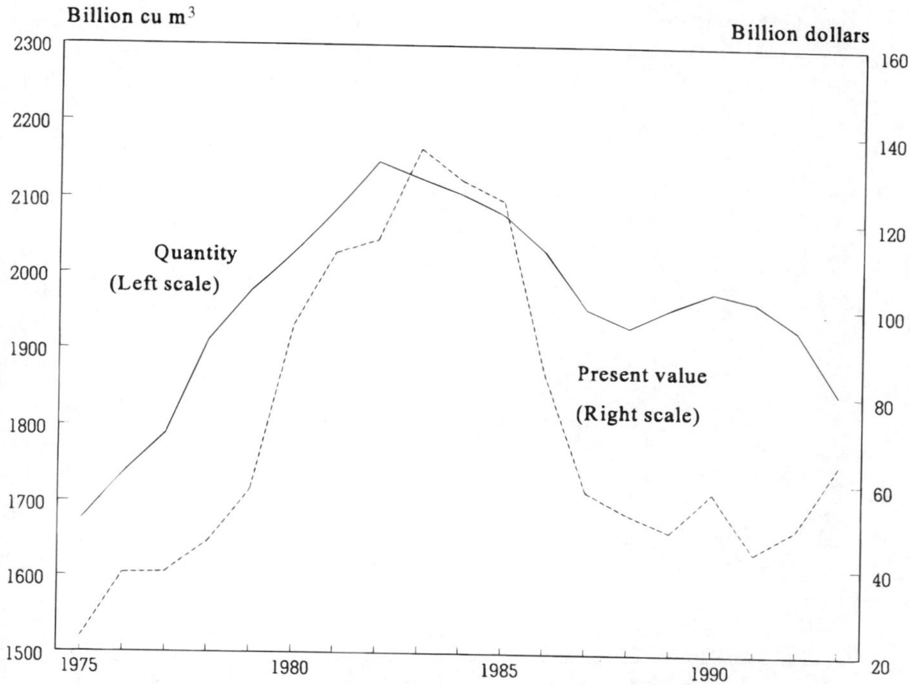

Figure 2. Quantity and present value of established natural gas reserves of Canada.

veloped reserves appear in the physical accounts while only developed reserves are valued in the corresponding monetary accounts.

Physical data are difficult or impossible to aggregate across the range of natural resources because of the different statistical units used. Trying to assess the sustainability of an economy or a set of human activities based on inventories of a stock of natural capital also raises the problem of incomparability of physical units. For example, if a stock of timber increases and a stock of natural gas decreases at the same time, how can it be determined whether the total stock of natural capital has fallen or risen, or remained the same (Victor, 1990)?

In many cases, the only way to obtain comparable results is to make a monetary valuation. Such a valuation is the norm for produced assets and is necessary for including natural (as well as human) capital in the balance sheet. Using current prices simplifies the valuation of natural capital but can give a distorted view of equilibrium prices in the longer term. In the case of subsoil assets, negative externalities generated by the production of marketed goods, valued in terms of hypothetical environmental protection and restoration costs, may not be adequately reflected in market prices.[7] Furthermore, if the price or the net price rises as the resource stock declines, the value of the

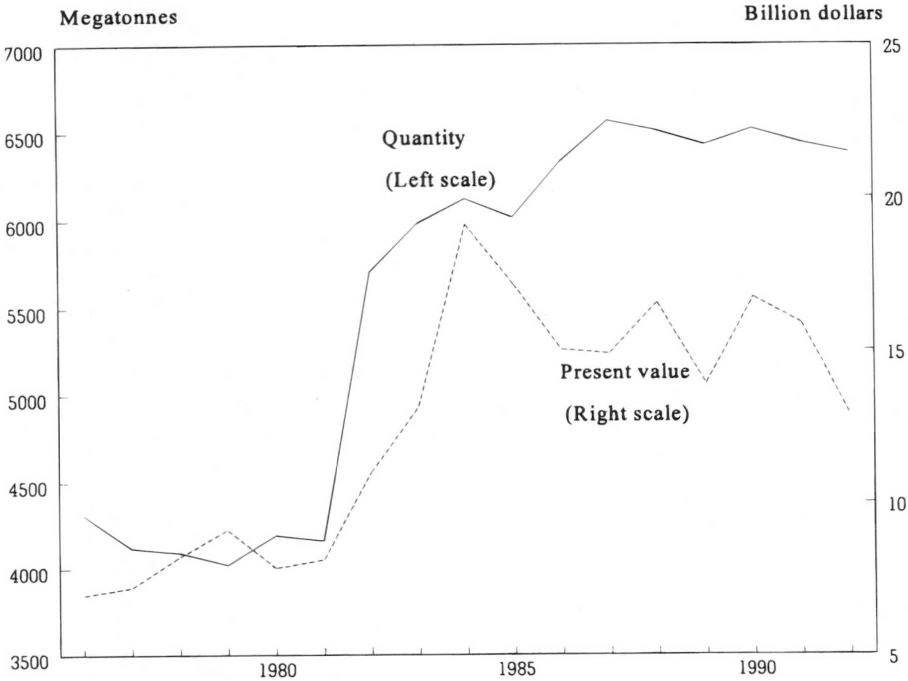

Figure 3. Quantity and present value of recoverable coal reserves of Canada.

resource stock as an indicator of sustainability may give the wrong policy signal to the government.

Figures 1–4 show the physical quantity of remaining reserves of oil, natural gas, coal and uranium[8] and the present value of the stocks from 1975 to 1992. While monetary values provide a common numéraire with which to compare these commodities, the physical quantities sometimes reflect a different picture of the remaining reserves. Physical quantities of oil reserves have declined almost by half since 1975 while values of the stock increased until 1986 and declined thereafter, reflecting the major drop in world oil prices in that year. Natural gas shows a similar trend to oil; however, physical depletion of the stock is less dramatic. Since 1982, coal reserves have been maintained between 6.0 and 6.5 billion tonnes with their present value peaking at $19 billion in 1984 and declining to $13 billion in 1992.[9] Physical reserves of uranium have declined by 200 kilotonnes since 1979 reflecting a price drop from $135/kg[10] in 1980 to $71/kg by 1990. Recent increases in reserves are due to the opening of high quality mines in Saskatchewan. The changes in present value over time reflect both the price changes discussed above and long-term contracts above world prices for Ontario uranium in the 1980s. Quantities of nickel have been declining since 1979. The value shows declines from 1979 to 1986, increases in

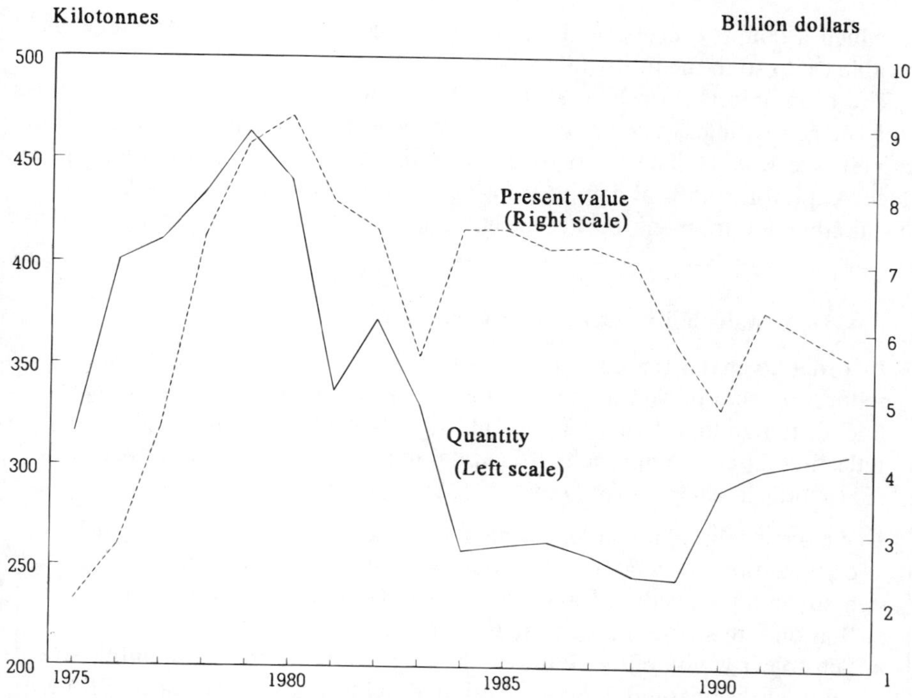

Figure 4. Quantity and present value of recoverable uranium reserves of Canada.

the late 1980s, followed by a strong decline in 1989 due to large deliveries of nickel from the former USSR into the market and the effects of the recession.

While monetary valuation of developed mineral reserves is necessary in order to calculate wealth, these values alone do not clearly indicate trends in the remaining reserves from which income is to be derived, nor do they show whether 'resource rents' are being consumed or reinvested elsewhere in the economy. Physical accounting provides a measure of resource availability in the medium or long term, depending on the reserve definition used. Monetary valuation is only one aspect of natural resource accounting, and it is just as useful to measure a resource in physical terms as in monetary ones. From a national accounts' perspective, a natural resource is as much an economic as a physical concept.

Sustainable development and mineral resources

The most familiar definition of sustainable development is that of the Brundtland Commission: 'development that meets the needs of the present without compromising the ability of future generations to meet their own needs'. It follows that the national accounts should provide an indication of the extent to which a country meets the basic material needs of its inhabitants and will be able to do so in the future (Schrecker, 1995).

The economic approach to sustainable development is based on the principle of generating a maximum income flow while maintaining the stock of capital (see Box 1). The controversy over the degree of substitution of produced capital for natural capital translates into a continuum of conditions for sustainable development, between what is known as weak and strong sustain-

Box 1 Sustainability and capital maintenance

In order to make the concept of sustainability operational, it is usually reduced to that of capital maintenance as "a reasonable indicator of baseline sustainability" (see Chapter 16). Several definitions of sustainability reflecting different approaches to capital maintenance have been proposed (Bartelmus in Chapter 16; Daly, 1995; Serageldin and Steer, 1994):

- Weak sustainability seeks to maintain income flows without regard to composition of capital (e.g. produced, natural, human and social). It assumes that produced and natural capital are substitutable and thus that only their sum needs to remain intact.
- Sensible sustainability requires that, in addition to maintaining the total level of capital, its composition should be taken into account (i.e. ensuring that the levels of the different types of capital are maintained). Oil may be depleted if the receipts it yields are invested in other forms of capital. According to this definition, produced and natural capital are substitutable but also complementary.
- Strong sustainability requires the full preservation of the natural resource, or at least that the different types of capital remain intact, that is, receipts from oil extraction should be invested in renewable types of energy. It assumes that produced and natural capital are complementary, which requires each to be maintained intact separately.

The focus of weak sustainability is income maintenance, by reinvesting returns in other production processes rather than looking for substitutes of the resource being depleted. The question is how much income derived from resource extraction should go to consumption, saving or investment in other forms of capital (weak sustainability), or in substitutes for the resource (sensible sustainability). For natural assets that are non-renewable, such as subsoil assets, the criterion of strong sustainability must be relaxed. In its most extreme form, strong sustainability is not a viable approach, since the non-use of a nation's natural resource can be considered squandering part of a country's economic potential. In this chapter, indicators reflecting both sensible and weak sustainability are examined.

ability. However, it is this maintenance of capital or a nation's wealth, both produced and natural, that is at the heart of sustainable development, no matter how it is defined (Hamilton, 1994).

If it is accepted that rents from natural capital depletion (both renewable and non-renewable) should be shared with future generations, then sufficient assets should be maintained to ensure a non-decreasing flow of income or welfare (per capita). Total capital stock, in monetary terms, may be preserved if revenues from the resources being depleted are invested in other forms of productive capital. However, it is difficult to identify the types of capital to be maintained (e.g. produced, natural, intangible or human capital) and to value some of these assets (Munashinghe, 1993; Born, 1992).

Sustainable development requires that consumption by present generations should not be at the expense of future generations. For subsoil assets, this imposes two conditions. First, investments should be made in alternative wealth-generating assets as a substitute for the depleting mineral asset. Second, the environmental damage caused by refining and smelting and by consumption should be minimized (Auty and Warhurst, 1993). For subsoil assets, especially fossil fuels, the limited carrying capacity of the environment implies that the second constraint is more important than the first (the depletion of the resources).

The sustainable development of subsoil assets raises a number of questions such as the valuation of annual depletion, the substitutability of produced capital for depleted mineral resources and the mechanism for transferring the 'capital value' of the depletion to future generations. Some environmentalists suggest that sustainable development requires limiting the rate of mineral extraction to the rate at which their renewable substitutes are developed. Since the exact physical volume of mineral stocks can vary from year to year with discoveries, technological change, depletion and resource evaluation, we do not know the optimal depletion path for these resources (Mikesell, 1994). The problem of sharing mineral resources with future generations lies in estimating when supplies will be exhausted at current rates of depletion. Physical stocks of minerals are depleted by each generation even though additional reserves are added to the 'recoverable' stock. However, extraction does not represent a loss of 'capital value' if an equivalent value of additional reserves is discovered. It would suffice to maintain the capital value of mineral reserves for future generations to be assured an income equal to that earned from the reserves by the present generation. However, this assumes that produced capital financed by the rent from mineral depletion is a substitute for the depleted mineral. It also raises the issue of whether sustainable development requires a constant stock of both produced capital and natural capital or just total capital stock. If there is a high degree of substitutability between produced capital and natural capital such as minerals, the reinvestment of the rent from the extraction of subsoil assets should prevent future economic output from being constrained by depletion of minerals (Mikesell, 1994). The assumption of strong sustainability, on the other hand, implies the need to maintain both produced and natural capital.

Wealth accounts are one measure of weak sustainable development; they could be complemented by other accounts. Such accounts would help to answer the question of how human activities impact on the potential of natural resources to generate income in the future (Schrecker, 1995). Table 2 presents wealth accounts and wealth per capita for Canada, including preliminary estimates of subsoil assets. Increases in the value of wealth per capita are mainly attributable to increases in value of produced assets and land. The value of subsoil assets declined slightly, reflecting the value of depletion being greater than reserve additions. The physical accounts show a more obvious rate of resource depletion. All recoverable reserves apart from uranium are being depleted and not being replaced (Table 3). At this time, we cannot determine whether the owners of subsoil assets, mainly provincial and federal governments, employ the rent derived from their extraction to produce other assets. However, Canada's overall wealth and wealth per capita are being maintained and are increasing over time.

Government revenues from the extraction of natural resources represented approximately 3.5–4.0% of total provincial and territorial government revenues between 1990 and 1995. For a province such as Alberta, which produces about 80% of Canada's oil and natural gas, the percentage increases to about 20%. Since most natural resources in Canada are owned by provincial and federal governments, it is likely that most of the returns from the extraction/harvesting of these resources are invested in health care or education and therefore human capital. In Alberta, 30% of oil and natural gas

Table 2. Expanded wealth accounts, 1992[1].

		Non-produced assets		Wealth per capita	
	Produced assets	Land[3]	Subsoil assets	Excluding subsoil assets	Including subsoil assets
	billion dollars			dollars per capita	
Opening stock (December, 1991)	**1 897.7**	**559.9**	**118.7**	**86 788**	**90 979**
Depletion of non-produced assets	...	..	6.4	...	...
Use of produced assets	..	...	...	...	...
Other volume changes[2]	..	..	5.6	...	...
Revaluation due to market price changes	..	..	-	...	...
Closing stock (December, 1992)	**1 940.1**	**587.5**	**117.9**	**87 942**	**92 044**

Notes:
1. Includes the value of Canada's oil, natural gas, crude bitumen, coal, uranium, and nickel reserves. Excluded are the value of iron, potash, gold, silver, copper, zinc, lead, and molybdenum.The results presented here are preliminary.
2. Includes changes in technological progress, prices, costs and estimation methods.
3. The value of land includes residential, non-residential and agricultural land.
Source:
Statistics Canada, National Accounts and Environment Division.

Table 3. Physical accounts of subsoil assets, 1992.

	Crude oil	Natural gas	Crude bitumen	Coal	Uranium	Nickel
	million m^3	billion m^3	million m^3	megatonnes	kilotonnes	kilotonnes
Opening stock (December, 1991)	**614.1**	**1 965.3**	**501.7**	**6 434.2**	**305.0**	**5 691**
Depletion of sub-soil assets	69.8	125.7	23.8	79.0	9.1	231
Other volume changes[1]	44.4	89.7	4.3	23.0	13.1	145
Closing stock (December, 1992)	**589.8**	**1 929.5**	**482.2**	**6 378.3**	**309.0**	**5 605**

Notes:
Figures may not add due to rounding.
1. Includes changes in technological progress, prices, costs and estimation methods.
Source: Statistics Canada, National Accounts and Environment Division.

revenues were put into its Heritage Savings Trust Fund for a number of years and reinvested in other forms of capital. In Alberta's case, some of the resource rents were converted to other forms of assets through investments in public infrastructure, public enterprises and human capital (health and education expenditures). However, in recent times, no revenues have been placed in the fund. Calculations by Smith (1992) suggest that from 1963 to 1988, the Alberta government was consuming revenues from oil and gas extraction at a rate that may have resulted in a decrease in wealth over time.

Conclusion

If sustainability is going to be more than a slogan it requires a definition, some quantitative objectives, and some indicators for measuring progress or regression. Sustainable development is a macroeconomic issue since it concerns all resources: produced, natural and human capital. The aim is to sustain welfare, i.e. the consumption of economic and environmental goods and services, for present and future generations. A more accurate picture of the capital base on which future income and consumption depend requires the inclusion of natural capital in wealth accounts.

The availability of natural capital and the absorptive capacity of the environment will pose limits to sustainable economic growth. Developing alternative growth indicators and including measures of natural capital and its depletion in economic growth models could provide tools to determine the nature of sustainable economic growth (Bartelmus, 1994a). The definition of sustainability in terms of capital maintenance requires that the owners of natural capital invest the rent in the production of other assets. Does the concept of capital maintenance refer to the 'capital value' of mineral resources or to the physical volume of natural resource stocks? As suggested by Peskin (see Chapter 21), the distinction is important in the case of natural capital, since there may be little association between an asset's physical condition and its value. As shown in this paper, there is a need for both physical and monetary accounts, but if one talks about economic sustainability, it is economic value that is used to measure total national wealth (produced, human and natural capital).

If the measure of national wealth is expanded to include the value of subsoil assets and timber resources, then total wealth per capita may become a useful indicator of sustainability. If wealth per capita does not decline, then (economic) development is considered to be sustainable. Whether or not GDP or an adjusted 'green' GDP per capita adequately measures short-run economic performance, wealth per capita does give a picture of how wealth creation and population growth influences future income per capita.[11] This is, for Statistics Canada, an area of future work in natural resource and environmental accounting.

Notes

1. For a view of natural resource and environmental accounting at Statistics Canada, see Smith (1994).
2. Major modifications were made to the System of National Accounts in 1993 (Commission of the European Communities *et al.*, 1993). The balance sheet accounts now include the value of subsoil and other natural assets. The change in the value of these assets is recorded as a reconciliation item and therefore does not affect net product aggregates.
3. Many concepts and definitions of sustainable development are currently being explored and discussed. The concept of sustainability used in this paper suggests that 'each generation should maintain the capital value of the natural resources it inherits'. In the case of non-renewable resources, this condition can be approximated by saving and reinvesting resource rents from mineral extraction into other forms of capital (Mikesell, 1994).
4. The total amount of hydrocarbon or metal extracted up to a given date.
5. Developed reserves are defined as recoverable through produced capital already in place (e.g. machinery and equipment and structures). In case of undeveloped reserves, additional exploratory and development work would be required in order to classify them as developed reserves.
6. This was suggested by the US Department of Commerce, Bureau of Economic Analysis (1994).
7. For a more complete discussion of mineral prices and their impact on the value of subsoil assets, see Born (1995).
8. In Canada, the term established reserves is used for oil, crude bitumen and natural gas reserves, recoverable reserves for coal and uranium reserves, and proven and probable reserves for nickel. While these definitions reflect reporting conventions in Canada, they are considered to be similar.
9. For a more complete discussion of the quantity and value of Canada's coal reserves, see Born *et al.* (1995).
10. The price quoted here is per kilogramme U (uranium).
11. Hamilton (1994) describes the policy implications of the 'greening' of national accounts and of the inclusion of natural capital among the assets. He shows how the growth in wealth can be separated from the growth in population.

11
Building physical resource accounts for Namibia: depletion of water, minerals and fish stocks, and loss of biodiversity

GLENN-MARIE LANGE

Natural Resource Accounts (NRA) need to be viewed as problem-defined and, from their inception, designed to address the pressing environmental–economic issues for a given country, the region to which it belongs, and the world. The key elements of a framework for NRA include classification, units of measurement and methods of estimating data. To provide a framework that is appropriate to a specific country's needs, the classification for NRA must be jointly determined by the most important resource variables as well as the critical economic and institutional characteristics associated with the management and use of resources in that country. A useful framework for NRA will also be forward-looking, not just reflecting conditions in the past but anticipating conditions likely to arise in the future that result from a country's development strategy. Naturally, these NRA may differ significantly from one country to the next.

Namibia, like many developing countries, is highly dependent on its resource base. It faces competing demands for the use of resources from a growing population and economy, as well as the threat of resource degradation. In Namibia, the critical natural resources include water, land and livestock, wildlife, minerals and fisheries. This chapter describes the construction of NRA for Namibia in terms of the choice of classification and units of measurement for these resources and the basis for this choice in the environmental, economic, and institutional conditions unique to that country, and the development strategies of Namibia.

Introduction: environment, economy and institutions in Namibia

With a current population of nearly 1.6 million, Namibia is one of the world's most sparsely populated countries. It is sub-Saharan Africa's driest country; roughly 80% of its 842,000 km^2 consists of desert, arid and semi-arid land. Only 8% of the country receives more than 500 mm of rain a year, which is considered the minimum for dryland cropping (Brown, 1994). Rainfall is not only low but extremely variable over much of the country; periodic droughts

K. Uno and P. Bartelmus (eds.), Environmental Accounting in Theory and Practice, 171–186.

are common. Much of the land is suited only to grazing livestock and the carrying capacity varies a great deal with annual rainfall, requiring considerable flexibility in stock management to prevent land degradation. However, Namibia is richly endowed with other resources, notably minerals such as diamonds and uranium, as well as one of the world's most productive fisheries. In the past, many of these resources have been exploited with little thought for the provision of future income. The construction of NRA represents a step away from past practices toward sustainable resource management.

Namibia's resource-based economy is highly capital-intensive with a modern extractive sector that accounts for much of GDP, government revenues and foreign exchange earnings. The four mainstays of Namibia's economy all depend on the resource base – mining, agriculture (livestock), fisheries, and tourism (Table 1). Backward and forward linkages to other sectors of the economy are poorly developed. Much of Namibia's requirements for food and manufactured goods are met through imports mainly from South Africa. The small size of the internal market, dispersed over large distances, has proven an obstacle to the development of industry in Namibia, especially in the face of competition from neighbouring South Africa.

Namibia's uneven economic development has been exacerbated by its pre-independence legacy. Under South African rule until 1990, two very different

Table 1. The Namibian economy in 1994.

Population	1.5 million
Urban	29%
Rural	71%
GDP	N$ 10,243 million
	US$ 2885 million
GDP by sector or origin (% of total)	
Agriculture & fishing	14
Mining	16
Manufacturing	9
Trade	8
FIRE	12
Government	26
Other	15
Total	100
Principal exports (% of total)	
Diamonds	31
Processed fish	22
Other minerals	19
Live cattle, sheep and goats	8
Meat and meat products	8
Fish, semi-processed and unprocessed	6
Other	5
Total value of exports	US$ 1321 million

Source: Central Statistical Office (1995a, 1995b); Economist Intelligence Unit (1996).

and separate economies developed: the so-called communal areas where the majority of the people were restricted to a disproportionately small land area and practised mostly subsistence agriculture, and a commercial economy based on export-oriented mining and agriculture controlled by a minority. The highly skewed distribution of income and access to resources has resulted in uneven population pressure on land and water; the pressure is most severe in communal areas where population is growing very rapidly and there are relatively few alternative sources of employment (Brown, 1994; National Planning Commission, Namibia, 1995).

Namibia now faces the enormous task of integrating the two economies. Rapid economic growth is a primary objective of virtually all countries, but this goal has added significance in Namibia because of the urgency of reducing the great social and economic inequalities. Namibia faces competing demands for the use of resources from a growing population and economy, as well as the threat of resource degradation. Opportunities for industrialization are limited and economic development will continue to remain highly dependent on the natural resource-based sectors.

This chapter describes the construction and use of natural resource accounts (NRA) for Namibia, focusing on the physical accounts (the accounts will also be partially monetized, where appropriate). It begins with a description of the objectives of natural resource accounting in Namibia and how the NRA will be used by policy makers. The paper then describes the choice of classification and units of measurement for natural resources (in physical terms) and the basis for this choice in the environmental, economic and institutional conditions unique to Namibia. This is followed by a discussion of the accounts themselves. I conclude with an assessment of the progress to date and the research agenda for the future.

Objectives for natural resource accounting in Namibia

Natural resource accounting is an activity of the Government of Namibia's Directorate of Environmental Affairs, Ministry of Environment and Tourism. It is part of a seven-point programme to develop environmental and resource economics within the ministry and is intended both to draw on and to contribute to other aspects of the programme. The first phase of this project, from September 1995 to September 1996, is a pilot project to construct preliminary accounts and demonstrate how they can be of use to policy-makers. In addition, the project is establishing a regional network of environmental and resource economists in Southern Africa who are pursuing similar activities, or are interested in doing so in the future. Since many countries in Southern Africa share similar resource constraints and management issues, we expect this to be a very productive collaboration.

The NRA proposed in the United Nations (1993a) SEEA can be used for three purposes:

(a) to provide an improved indicator of macroeconomic performance, the environmentally adjusted domestic product (EDP), which more accurately indicates whether we are living off our (produced plus natural) capital;
(b) to monitor the state of the environment and economy, linking specific resource variables to economic activities which affect or are affected by them;
(c) to provide the environmental and natural resource input to economic tools and models for policy analysis at the sectoral, regional and national levels.

In Namibia, NRA will be constructed and used for all of three of these purposes. For policy analysis, NRA will be linked to a simple economic model based on the 1993 Social Accounting Matrix (SAM). This simple model will be used to explore the resource implications of the development strategies outlined in the First National Development Plan (National Planning Commission, Namibia, 1995). Several resource management issues have been identified for in-depth treatment: these include livestock and factors affecting annual variation in rangeland carrying capacity, alternative strategies for water supply, demand management, cost-benefit analysis of alternative uses of water, and an evaluation of alternative uses for land, including tourism.

The Namibian NRA will be based generally on the guidelines presented in the SEEA (United Nations, 1993a). Both stock and use accounts are constructed where appropriate. The benchmark is 1993, the year for which the Social Accounting Matrix (SAM) for Namibia has been constructed. Accounts for earlier years will be constructed when the data are readily available and when historical data have a significant bearing on scenario analysis about the future. While NRA are generally considered a macroeconomic policy tool, the NRA for Namibia maintain a great deal of regional (sub-national) information in order to reflect the local variation in environmental conditions necessary for meaningful policy analysis.

A framework for natural resource accounting in Namibia

NRA need to be viewed as problem-defined and, from their inception, designed to address the pressing environmental–economic issues for a given country, the region to which it belongs, and the world. The key elements of a framework for NRA include classification, units of measurement and methods of estimating data. The publication by the UN Statistics Division of a handbook on integrated environmental and economic accounting has made considerable progress toward standardizing the construction of the accounts. However, the classification system for natural resources is defined at a level of generality that is often too broad for in-depth policy analysis. Relatively little guidance is given regarding measurement issues or the construction of more detailed accounts tailored to a particular country's needs. Consequently, countries have had to define their own classification systems within the overall framework established by the UN.

To provide a framework that is appropriate to a specific country's needs, the classification for NRA must be determined jointly by the most important resource variables as well as the critical economic and institutional characteristics associated with the management and use of resources in that country. A useful framework for NRA will also be forward-looking, not just reflecting conditions in the past, but anticipating conditions likely to arise in the future. Naturally, these NRA may differ significantly from one country to the next.

The accounts presented here include water, fisheries, land, livestock, wildlife and minerals. Both stock and use accounts are constructed where appropriate. The proposed framework for NRA in Namibia also includes accounts for forests and energy; these accounts will not be discussed here since work on them will only commence in the next phase of the project. Accounts for air pollution and solid waste are not included in the NRA because of the relative insignificance of these problems in Namibia.

Land tenure is the most important institutional characteristic affecting resource management and is a defining attribute for both land and livestock accounts. Agricultural land is divided into commercial areas where title to land is assigned to individuals and communal areas where ownership resides in the state, or local community, and use is determined under traditional practices. Roughly half the land is under commercial farm management, owned by less than 5000 farmers; the remaining 50% of the land lies in communal areas where it supports over 900,000 people. Other institutional factors are explained in the discussion of specific accounts.

The physical accounts and analysis reported here parallel (future) monetary accounts and analysis at all points, in part because the physical accounts underlie the monetary accounts. More importantly, physical accounts provide a basis for collaboration between economists and natural scientists by presenting information in units which can be utilized by natural scientists. This collaboration is necessary to develop a better understanding of the interaction between economic systems and natural systems in Namibia, which will provide a more sound basis for policy making.

Accounts for water and water-based resources

Water

Water is the single most important constraint to development in Namibia. Water is obtained from three natural sources: groundwater, perennial surface water supplied by the rivers that form the northern and southern boundaries of Namibia, and ephemeral surface water resulting from rainfall which, if heavy enough, produces temporary rivers. These sources of water vary in terms of location, renewability and reliability. More than half of Namibia's water is supplied from groundwater with the remainder equally split between perennial and ephemeral surface water.

The heavy reliance on groundwater in Namibia results from a combination of several factors:

(a) a sparse population spread over large distances;
(b) predominance of extensive livestock ranching (drinking water must be within daily travelling distance for cattle);
(c) high costs for transportation of water (the economically productive, non-coastal areas rise more than 1500 m above sea level);
(d) high annual variability of ephemeral water supply;
(e) distance of perennial water sources from much of Namibia's economic activity.

Two different institutions were established to deliver water to end users. The bulk water supply, managed by the Directorate of Water Affairs (DWA), relies on relatively large-scale dams, transport, and storage technology, to supply urban areas and commercial agriculture (about 40% of supply). The rural water supply (until recently managed by the Ministry of Rural Development) utilizes relatively small-scale, localized means for collection (mostly boreholes in rural areas) and does not generally transport water over large distances (roughly 60% of supply). Established prior to Independence to serve well-off urban areas, mining, and commercial farming areas, bulk water supply has always been more closely monitored than rural water supply. Consequently, more information is available about bulk water resources than rural water supply, and the accounts based on bulk water supply are probably more reliable at this point in time.

Stock accounts for water are constructed for each of the three natural sources. Full stock accounts would require information about annual extractions as well as natural gains and losses. While the former is available, the latter is not. Consequently, only closing stocks in each year are reported. In order to represent both natural and socioeconomic characteristics associated with delivery and use of water, the use accounts are disaggregated by both natural source (three types) and institutional source (two types). Accounts representing the annual use of the six categories of water are constructed for 19 economic sectors, two categories of households, and a government sector.

Stock accounts for water: groundwater

Groundwater is the cheapest and most reliable source of water for much of Namibia's dispersed population since it can be tapped at the point of use and is not directly dependent on annual rainfall. Determination of sustainable rates of extraction is based largely on monitoring of water tables of boreholes. No comprehensive information about the total volume of groundwater is available. Estimates have been undertaken in the central area of Namibia for selected aquifers under the administration of the bulk water supply in 1992.

Since groundwater supplies cover half of all Namibia's water, the possibility of depletion is a serious concern. There is, however, a great deal of uncertainty concerning the severity of depletion of groundwater stocks. Namibia's highly variable rainfall can result in long periods of a continuously falling water table, followed by a single rainfall which recharges an aquifer. Under such circumstances, it may be more appropriate to assess depletion not on an annual basis, but on the basis of the rainfall cycle normal for the area, which could be 5 years, 20 years, or even longer. Continuous extraction of groundwater over this time may be justified, subject to an evaluation of other factors, notably the environmental impact on ecosystems of a falling water table and the potential collapse of certain aquifers from excessive extractions. There is not enough information at this time to determine the appropriate cycle for most aquifers.

Of the 88 million cubic metres of bulk water supplied in 1993, 14% came from groundwater stocks considered to be experiencing long-term depletion (as evidenced by continuously falling water levels in boreholes without recovery for 5 years or more). Much of the coastal region is affected, which includes important urban and tourist centres (Swakopmund and Walvis Bay), mining (uranium) and industry (fish processing). Groundwater provided by rural water supply is also experiencing depletion in some areas. However, it is not yet possible to quantify this phenomenon since systematic monitoring of boreholes under the management of rural water supply was not undertaken in the past and is just now beginning.

Stock accounts for water: ephemeral surface water

The amount of ephemeral water in a given year depends on annual rainfall; in some instances there may be carry-over from a previous year. For the NRA, the stock is measured in two ways: annual runoff from the major ephemeral rivers (a measure of gross potential stock), and annual volume of water stored in a dam each year at the beginning of April (a measure of available stock). April marks the end of rainy season. The stock will vary greatly from one area to the next based on the annual distribution of rainfall (Figure 1). Since the local availability of water is a critical factor for water supply and there is little infrastructure for moving water from surplus to deficit areas, the stock accounts are compiled separately for each of the major dams. Only the aggregate figure is reported here.

Stock accounts for water: perennial surface water

Perennial rivers are those which never dry out. Nevertheless, the volume of all perennial rivers is subject to considerable variation over time. The stock is measured as annual runoff for each of the perennial rivers. All perennial rivers originate outside of Namibia and form international boundaries between Namibia and its neighbours; consequently, the amount of water actually available to Namibia each year is subject to international agreements among the

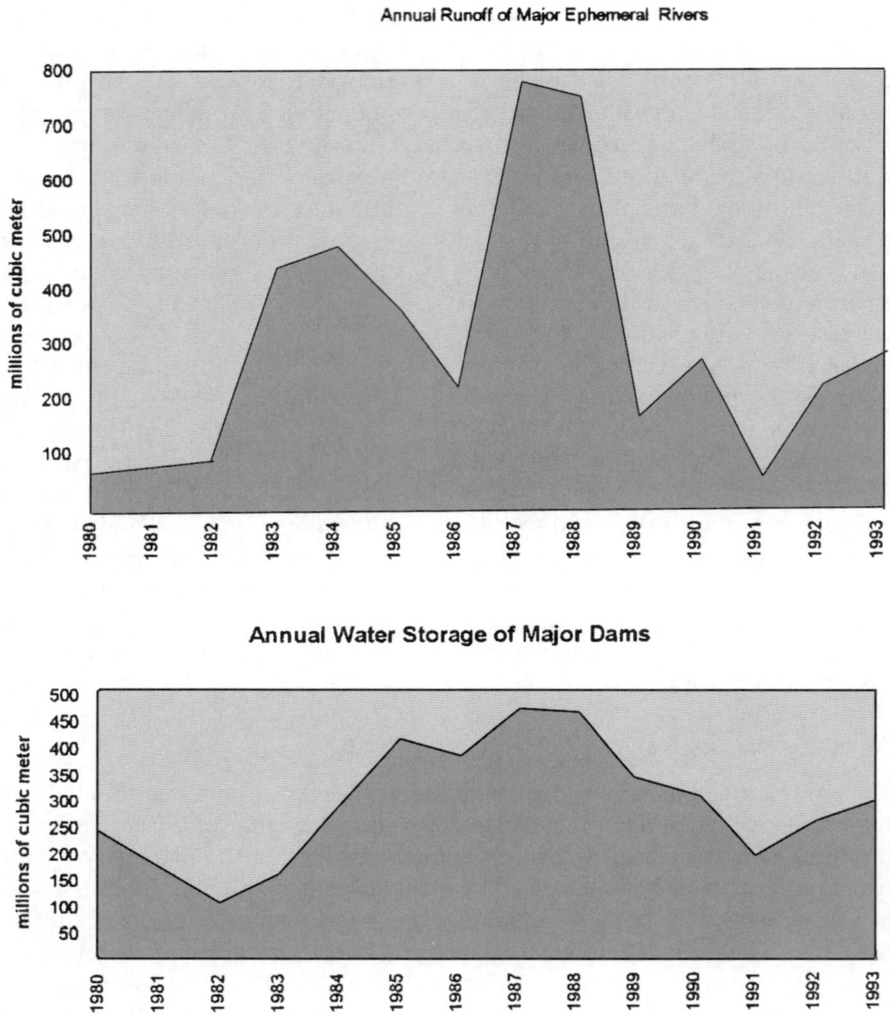

Figure 1. Stock of ephemeral water, 1980–1993. *Source*: Lange (1996).

countries sharing the perennial rivers. Such agreements have been concluded with South Africa and are under negotiation with other neighbouring countries.

Use accounts for water: annual use in 1993

Of the 212 million m^3 of water used in 1993, agriculture accounted for 70%, households for another 21%, and the remaining 9% was divided among all other sectors (Figure 2). Though urban households account for only 29% of

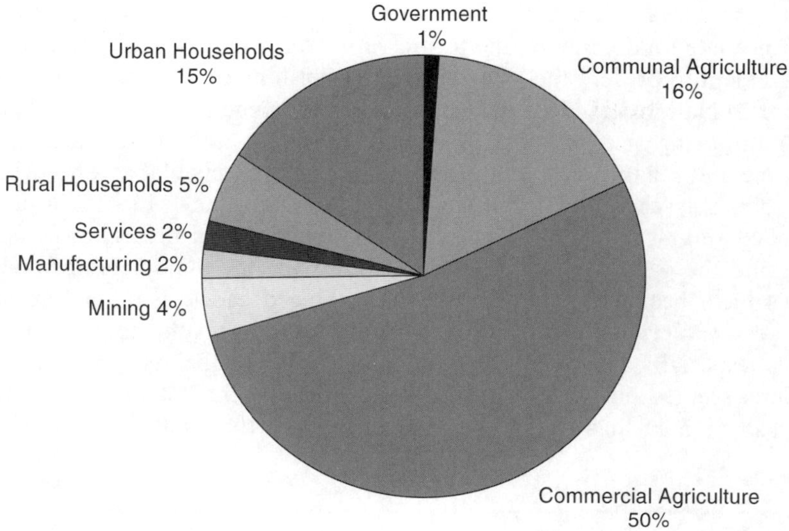

Figure 2. Distribution of water use by sector, 1993. *Source*: Lange (1996).

the population, they consume more than three times as much water as rural households.

Crop irrigation alone accounts for 50% of all water used in Namibia. Water supplied for livestock accounts for only 27% of all water used for agriculture, but accounted for 76% of the value of agricultural output in 1993. Commercial agriculture uses a great deal more water than communal agriculture (77% of all water used for agriculture). Much of the water used by commercial farmers is for irrigated crop production. Spray irrigation is the predominant technology, even though it is the least water-efficient method for Namibia's climate. Furthermore, the major crops grown under irrigation are low-value crops such as maize. It is unlikely that these crops would continue to be raised without the substantial price protection from international competition given by the government.

The use of ephemeral water for commercial crops is possible because of past government investment in dam and water transportation infrastructure to support these minority farmers. In communal areas irrigated crop farming is only possible in regions close to the perennial rivers since comparable investments were not made in the past to supply water to these farmers.

Marine Fisheries

The coastal waters of Namibia provide a rich habitat for many fish. The fishing industry constitutes a significant share of GDP and exports, and provides a large number of jobs in fish processing. It is currently the only manufacturing industry in which Namibia has a clear comparative advantage.

Future growth of this industry may also provide an opportunity for development of related industries in the future. However, fisheries will require careful management to prevent the recurrence of over-fishing experienced in the past.

Prior to Namibia's independence in 1990, the fishing industry was subject to little monitoring or regulation. Many foreign operators fished the area during that time and reliable data about fish landings are not available. The Ministry of Fisheries is currently trying to obtain these results, which will provide expanded and more reliable data for the construction of fisheries accounts.

Despite the paucity of data, stock accounts for the major commercial fish species have been estimated for the past three decades. Namibia's fisheries have been subject to significant over-fishing resulting in the collapse of some species, especially pilchard (Figures 3 and 4). While the less valuable horse mackerel seemed to expand to fill the niche left by the collapse of pilchards, total fish biomass has declined significantly since the 1960s, and even horse

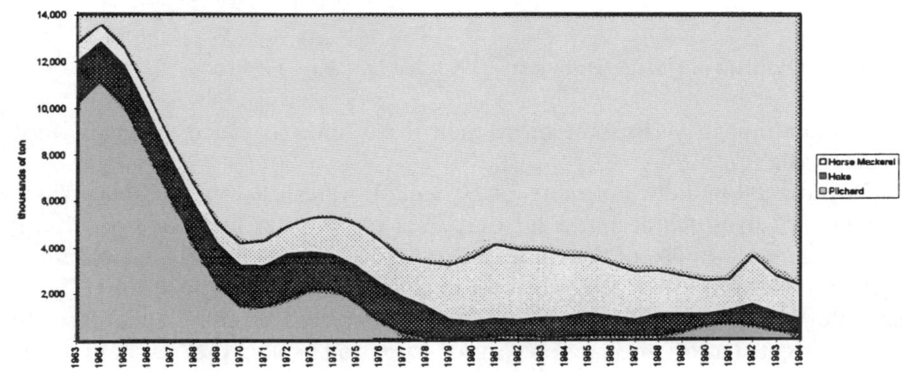

Source: unpublished data from Ministry of Fisheries and Marine Resources and author's calculations

Figure 3. Biomass of pilchard, hake and horse mackerel, 1963–1994.

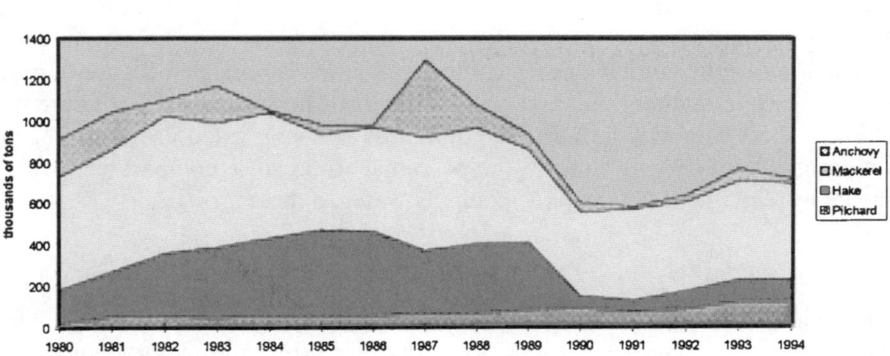

Source: unpublished data from Ministry of Fisheries and Marine Resources

Figure 4. Landings of major fish species.

mackerel has declined over the last decade. The collapse of the fish stock has resulted in reduced landings of fish and a long-term loss in income to Namibia. After 1990, the government of Namibia established a 200-mile exclusive economic zone and instituted strict restrictions on the allowable catch to prevent further over-fishing. It is uncertain whether the fish stock will recover. Even at reduced stock levels, fisheries continue to contribute significantly to national income.

Accounts for land and land-based resources

Land

Land accounts are being constructed in order to assist in land use planning and to provide input to the public discussion about land reform. Most of the land in Namibia with agricultural potential is best suited for grazing livestock and is currently used for this purpose. Relatively little land is suitable for crop agriculture. Some land has been set aside for national parks or protected areas; in some cases, this land has considerable potential for livestock. A careful assessment of the relative economic value of the land in use for different purposes is essential, especially as the population grows and pressure for land increases. In addition, there is some evidence that mixed livestock and game ranching may be more valuable than just livestock from both an ecological and economic point of view.

Several sets of land accounts will be constructed for Namibia based on carrying capacity for livestock, suitability for tourism and susceptibility to degradation. Only the first set of accounts has been completed and is discussed here. Because of the critical role of land tenure in land management and the socio-economic issues raised by past inequality of access to resources, the stock accounts for land are compiled by combinations of tenure attributes (commercial and communal areas) as well as geographic attributes.

Land is disaggregated into 39 veterinary districts, each of which can be assigned to communal or commercial areas, within 14 veterinary regions. Since the economic importance of land is primarily its ability to support livestock (carrying capacity), the stock of land for each district is classified according to eight ranges of long-term carrying capacity (in ha per Large Stock Unit, LSU; see Figure 5). Carrying capacity is not always a well-defined concept, particularly in regions like Namibia with extremely low and erratic rainfall. For lack of a better means to classify land in relation to livestock management, the carrying capacity concept is used. Since carrying capacity in a given year is determined primarily by rainfall, significant changes in carrying capacity and in the stock accounts for land will occur from year to year as a result of these natural conditions. This variability is especially important for land use planning because some uses of land (e.g. livestock ranching) are more sensitive to annual variation in rainfall than others (e.g. tourism). In determining the

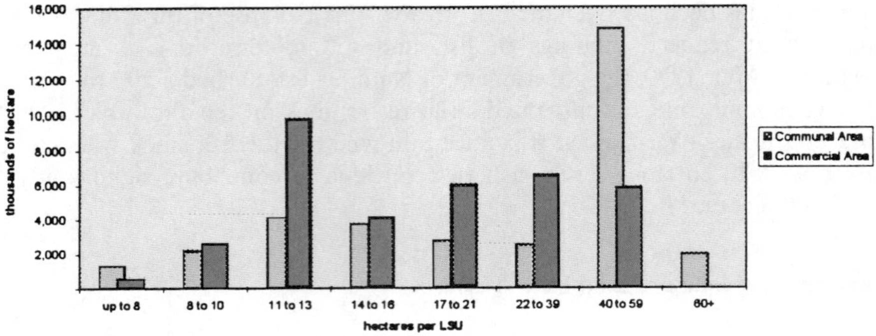

Source: estimates by Jon I. Barnes and author based on Adams et al. (1990).

Figure 5. Stock of land by long-term carrying capacity and system of tenure.

most economically valuable use of land, this variability needs to be taken into account.

Based on analyses which will be conducted in the coming year, the accounts will be constructed to represent annual changes in carrying capacity. These annual land accounts will be based primarily on annual rainfall and other environmental factors such as the quality of groundwater and the access to boreholes. Carrying capacity is also affected by human actions. There is anecdotal evidence of rangeland degradation attributed to human actions associated with overgrazing and poor management of livestock, but few actual measurements of degradation exist. Whether annual variation in vegetation cover actually represents degradation and what the source of this degradation might be are controversial questions and not well understood.

Bush encroachment in the commercial areas is a relatively well-defined phenomenon and has been subject to some measurement. Consequently, the NRA will include estimates of bush encroachment, but will probably not include other environmental impacts until the issues are more fully understood.

Livestock

Agriculture, primarily livestock, provides the primary source of livelihood for most Namibians even though output may be close to subsistence levels in the communal areas. Livestock constitutes the major capital asset for many people in Namibia. Accounts for livestock include cattle, goats, sheep, horses and donkeys. Accounts have been compiled for both commercial and communal areas according to the 39 veterinary districts classification used for the land accounts (shown at the national level in Figure 6). To construct full stock accounts, information about annual off-take as well as natural gains and losses is needed. Information about annual off-take of livestock is available only for the formal market, and only at the national level, not by veterinary district. It

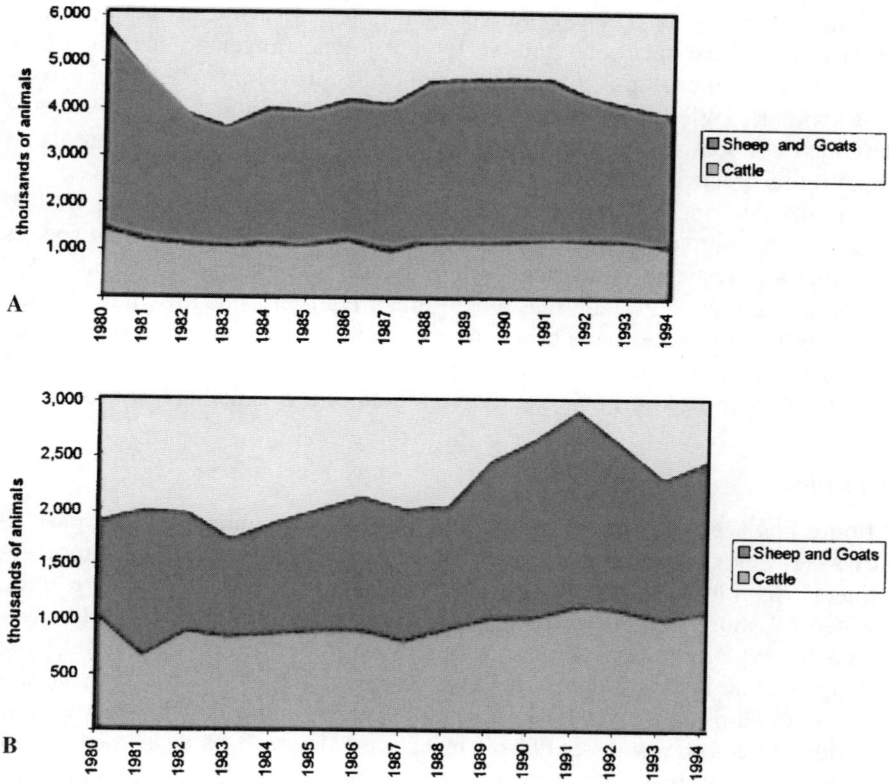

Source: Directorate of Veterinary Services, annual.

Figure 6. (A) Numbers of livestock in commercial areas, 1980–1994. (B) Numbers of livestock in communal areas, 1980–1994.

is estimated that in the communal areas, informal market off-take is at least as great, if not greater than formal market slaughter. However, no reliable data about informal market off-take are currently available. In the future, the accounts will attempt to estimate these data.

Biodiversity: accounts for wildlife

Maintaining biodiversity is important both for its intrinsic value and because of the economic importance of wildlife-based tourism. Tourism is one of the few industries with major growth potential in Namibia. It can also provide employment in rural areas, stemming migration to urban areas and relieving pressure on the land. However, setting aside a vast land area to preserve wildlife habitat for tourism can conflict with the use of land for agriculture by the largely poor rural population.

Game ranching has been proposed as an alternative use of the land; it is often more environmentally suitable for the fragile rangeland than livestock and can be more profitable for farmers while also providing a means of preserving wildlife. Wildlife tourism through game ranching is already being exploited on some commercial farms. There are a number of proposals to expand this practice on both commercial and communal areas.

For tourism and the preservation of biodiversity, both the mix of wildlife and its density are important factors. Accounts for wildlife (key animal species in a region) are being constructed which distinguish commercial areas, communal areas, and protected areas (government land set aside for wildlife). Currently, only a few areas have been surveyed, for intermittent years. In virtually all regions there has been a decline in numbers of wildlife over the past two decades. Closer monitoring is clearly warranted of this valuable resource.

Minerals

Mining has accounted for roughly 75% of export earnings over the past 20 years. However, the easily accessible reserves of Namibia's most important mineral, diamonds, will be depleted in the near future. New diamond reserves, located off-shore, are being developed, but these require large capital investments for exploitation. Uranium, the next most important mineral, is suffering from a decline in world demand. Figure 7 shows trends of extraction of diamonds and uranium. Newly discovered off-shore natural gas reserves may provide the country with significant revenues if they can be developed. Other important minerals include metals (mainly gold, copper, lead, zinc), industrial minerals and semi-precious stones.

Stock accounts for the major minerals have been constructed. Use accounts are not constructed since there is virtually no domestic processing or use of minerals. Data about annual reserves have been provided by the Ministry of

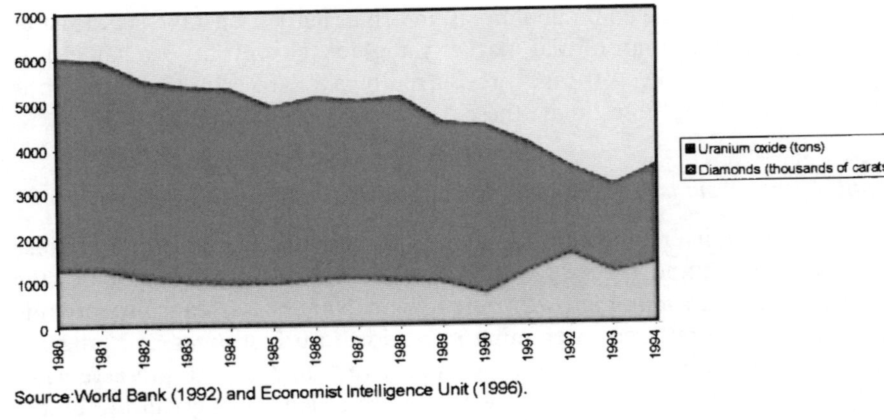

Source: World Bank (1992) and Economist Intelligence Unit (1996).

Figure 7. Production of diamonds and uranium, 1980–1994.

Mines and Energy, but are considered confidential and cannot be publicly reported. The accounts will be used to calculate resource rents in order to demonstrate the importance of a policy in which a portion of the revenues from mining could be invested in activities that will generate income after depletion of current mineral reserves.

Conclusions

To date, a fairly extensive set of accounts has been constructed in Namibia. In the remaining stage of this pilot project, we will focus on improving selected accounts, developing indicators (physical and monetary) to evaluate the accounts, and linking the accounts with an economic model to assess the demands on the resource base of Namibia's development strategy. The major policy questions to be addressed using the NRA include:

(a) Minerals and Fisheries. How successful has the government of Namibia been in recovering resource rents from these sectors?

(b) Water. How is water currently used in the economy?
What is the economic contribution of water use in each sector?
To what extent is water use subsidized in each sector?
What are future water demands likely to be and what will this cost?
Does the current pattern of water use and infrastructure development represent the best use of scarce water and financial resources?

(c) Land and livestock. What is the cost of rangeland degradation measured in terms of lost productivity?
How has land degradation been counteracted by changes in herd management practices?
What are the implications of land degradation for long term carrying capacity, especially in the context of global warming, which is expected to increase the variability of already erratic rainfall?
What is the value of land used for extensive livestock ranching compared to mixed livestock and game ranching, or ecotourism, taking into account the different degrees of variability of economic returns to each activity due to drought?

(d) First National Development Plan. What is the impact of the development strategies described in Namibia's First National Development Plan on the natural resource base (assessed through an integration of the NRA with an input-output model of Namibia)?

Several areas of the NRA which have been targeted for data improvement and research are expected to prove useful for policy-making in other government ministries and for researchers as well. For example, better estimates of informal-market livestock off-take would provide a more accurate picture of livestock management and real incomes in communal areas. More reliable data about fish landing in the past would provide an improved basis for assessing the status of the fish stock and understanding its response to overexploitation. Very little is known about wildlife populations, making management of wildlife difficult and fragmented. Research will be undertaken to determine the annual change in carrying capacity and the resultant annual change in economic value of the land for the NRA. A complete set of land accounts, perhaps linked with GIS capability, which assesses the value of land in alternative uses, will provide an important input to land use planning and the debate about land reform.

The groundwork for expanding this work to other countries in southern Africa is also being prepared. Several countries, notably South Africa, Botswana, Swaziland, and Zimbabwe, have expressed interest in constructing NRA to assist in macroeconomic environmental policy analysis. Future work will include construction of detailed water accounts by several countries which then can be used to address a regional management of this resource.

12
Material and energy flow analysis in Germany – accounting framework, information system, applications

WALTER RADERMACHER AND CARSTEN STAHMER

Material and energy flows – part of environmental–economic accounting

Introduction

It is no longer a matter of dispute that material and energy flows are important factors in environmental policy, both for the diagnosis of environmental problems and for the development of corresponding corrective measures. We are therefore required to observe these flows and to quantify their scope, their trends, etc. in order to manage them better.[1] There is thus agreement on the principle. However, the proposals differ vastly in detail. The broad spectrum of material balance sheets, product line analyses, eco-balance sheets, etc. indicates that various analyses have very different subjects and purposes. It is not the purpose of this chapter to give a complete overview of the different approaches. Most of the activities in this field are geared to individual cases, be they products, activities or regions. We are, however, faced with the question which material and energy flows are important when taking, as in German Environmental–Economic Accounting (GEEA), a macroscopic view of the inter-relationship between the environment and the economy. A brief review reveals the great variety of distinguishing criteria:

(a) The fundamental assessment of what is a problem and what is not has a great deal of influence over the method. In this field opinions differ substantially. The potential damage done by material flows is traditionally assessed according to criteria of (eco-)toxicity, i.e. qualitatively. More recently, however, there has been growing demand for greater account to be taken of quantitative aspects, even putting them to the fore. Both approaches have their strong points, but also their limits. With the qualitative approach, it is doubtful whether detailed information required can be available, or become available soon enough. On the other hand, using the quantity approach it is difficult to draw the line between trouble-free and critical quantity movements. The different approaches are based on specific appraisal or reference systems (e.g. the 'geogenic material flow' in the

K. Uno and P. Bartelmus (eds.), Environmental Accounting in Theory and Practice, 187–211.

approach of Baccini and Brunner, 1991) which serve to put the general objective (e.g. sustainability) into concrete terms, but which are also crucial for the choice and delimitation of the units observed and the collection of related data.

(b) Material flows are observed at very different hierarchical levels between 'macro' and 'micro'. These tags can be applied in a number of ways:
- materially, i.e. whether the balance is taken of individual elements, materials or goods or of a totality of them;
- regionally, i.e. according to the size of the observation area (regional system limits);
- economically, i.e. whether concerned with individual processes or undertakings or with the economy as a whole.

The GEEA field of observation must be classified as macro. It is a counterpart of the corresponding activities at the level of local units and products.

(c) Flow accounts may be incoming (what material flows or material accumulations etc. affect the region?) or outgoing (what material flows go out from the activities of a region or period?).

(d) A distinction can also be made between actual physical material flows and imputed material flows ('ecological rucksacks', e.g. raw material consumption and emissions contained in inputs, unused part of extracted raw materials, etc.).

(e) The following breakdown is now used internationally:
- driving forces = human/economic activities that cause environmental burdens;
- pressures = output side of environmental burdens;
- state = impact side of environmental burdens, environmental condition;
- response = environmental protection activities.

Most treatments of material flows to date have concentrated on the relationship between pressures and state. In the material flow analysis of the GEEA, however, it is the relationship between driving forces and pressures that is to the fore.

These five distinguishing criteria can be used to produce a requirement profile for flow accounts in the GEEA. Both quantitative and qualitative assessments should be possible; the macro viewpoint is taken (from production branches to the whole economy, from larger regions to the area of the Federal Republic, from materials/goods to total material/energy); an 'outgoing' account is to the fore; indirect effects must therefore be imputed; the relationship between driving forces and pressures takes priority.

Scarcity of raw materials

It is still customary to cite the scarcity of raw materials as a problem in its own right alongside those discussed above. In this connection, it is recommended to begin by recording in physical units, to construct a 'natural resource account' on the Norwegian model, and finally to follow one of the suggested valuation methods. This division into input side (scarcity of raw materials) and output side (environmental pollution) of the material flows through the economy may in some respects be helpful and provide a structure. A complete methodological and organizational separation of the two issues would, however, be pointless, and fails to address the problem. To begin with, it would make it impossible to record the relationship between the consumption of raw materials and the production of residues, but it is precisely this relationship that is increasingly becoming the centre of attention. In addition there would (as experience shows) be the risk that, since the scarcity of resources is more accessible to the economic and statistical method, it would unjustifiably dominate the preparation of GEEA in practice.

This brings us to a point that has not yet been addressed in the discussions. Apart from the mere observation of flow variables, there are two kinds of stock variables that have a part to play: those which are drawn on (scarcity of raw materials) and those which are added to (accumulation of goods and residues).

In a physical balance sheet, the law of the conservation of matter and energy applies, i.e. material flows are basically only displacements between the various (e.g. regional) stocks. Both withdrawals and additions are however associated with specific environmental problems or risks.

An integrative concept

Aims

It is clear from what has been said above that only an integrative approach is capable of doing justice to the various requirements. 'Integrative' should be understood as meaning the aim of taking a broad look at various problems and how they are interrelated. It implies a method that differs from a selective and additive approach in many respects:

(a) Valuations/objectives etc. should be independent of the data as far as possible. In particular and specifically, this means that material flows are recorded in such a way that they can be valued by both quantitative and qualitative criteria.
(b) Material inputs and outputs of the national economy must be examined jointly and combined.
(c) The main focus is the recording of flow variables. However, stock variables must also be included in the account if necessary.
(d) Data on quantity structures and technical coefficients must be combined.

The economic metabolism

The plan is based on a simple view of the material flow from the sources of raw materials at home and abroad by way of the production processes and consumption to reductions for emissions, waste, overburden, etc. at home and abroad. Figure 1 distinguishes three levels for data collection and analysis:

(a) Territorial flow account (for a region, e.g. Germany)
- Imports of raw materials (3), goods and residues (2), inflow of material by water and air (1);
- Exports of raw materials, goods and residues (7 and 8), outflow of material by water and air (6);
- Accumulation of materials on the territory (balance +), outflow of surplus from the territory (balance −);
- Scarcity of international primary stocks;
- Imports/exports of indirect effects (imputed international material flows).

(b) Economic flow account
- Imports and exports (2 + 3 and 7 + 8);
- Withdrawal of raw materials (used/unused) from domestic nature (4);
- Discharge of residues and pollutants into domestic nature (5);
- Accumulation of materials in the economy;
- Scarcity of national primary stocks.

(c) Activity flow account (activities of the domestic economy)
- Output: withdrawals from domestic nature, imports, domestic economic output;
- Input: exports, domestic economic use, discharges into domestic nature;
- Accumulation of materials in production activities.

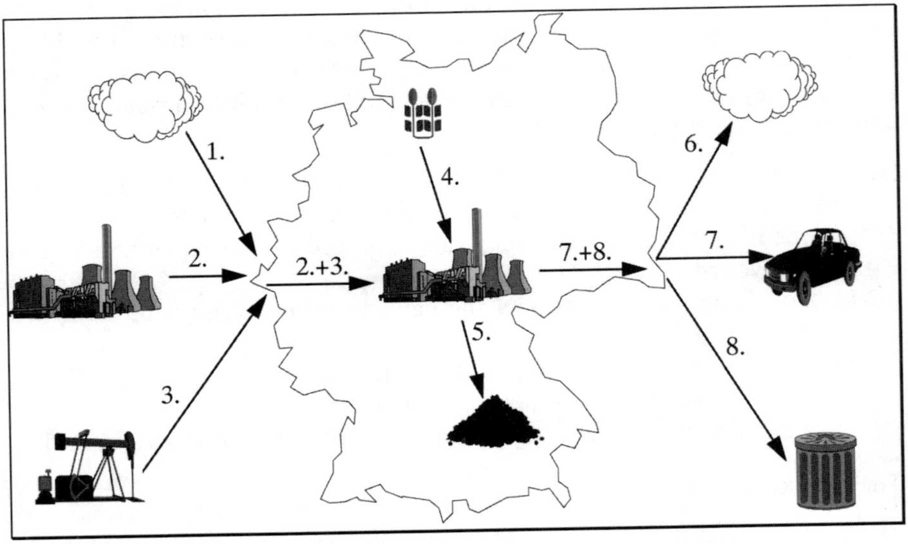

Figure 1. Economic metabolism.

Macro–meso

The plan's second important structural characteristic is its breakdown into the following levels:

(a) inclusive view of all quantitatively relevant flows;
(b) problem-related view of selected flows, production branches or product classes.

Level (a) is therefore primarily geared to quantitative analysis and evaluation, while at level (b) the advantage of focusing more on particular problems with greater attention to quality comes at the price of having to be selective. At level (b) a selection of materials, goods, etc. must be made for which a flow account is drawn up. The recommendations to this effect by commissions of enquiry of the German Bundestag are a first pointer, but at this level, too, the priority objective in every case is to produce as complete a flow account as possible.

Flow accounts presuppose that there are adequate statistical records of the individual stations in the total flow. This requirement is only partly satisfied. There is a major weakness at the cross-over between economy and environment. In particular, we have only an incomplete picture of the sectoral breakdown of the amounts of residues produced. If we are to be sure that the sectoral breakdown is consistent with the total burden of emissions, etc., a separate methodical working model needs to be introduced. In the GEEA this model is called the 'emitter structure'. It is designed as a sectoral structuring of material flows and therefore concentrates on the output of residues in the production branches of the economy.

Finally, attention should be drawn at this point to the limits of the field. These are set more or less of necessity by the data available in the official statistics. This means that it will normally be impossible to distinguish between individual chemical elements. The flow account in the GEEA will as a rule concentrate on materials/goods.

'Incorporation of technical coefficients, consistency test'

Experience shows that the lack of complete, high-quality direct data on the volume of individual flows is a major obstacle to the production of such material and energy flow accounts. This obstacle is overcome by taking data from a variety of sources and linking them appropriately. For example, technical coefficients, which give the average emissions of production processes, are multiplied by the quantitative data about those processes (e.g. use of energy sources) in order to estimate the sectoral output of emissions (=emitter structure). This method naturally presupposes that the coefficients and quantity structures match and that the linkage model used is stable and significant.

Valuations

The question of the valuation of material and energy flows must be understood in a broad sense. When taking an overview of this complex subject matter, it is helpful to keep weightings and appraisals separately. Weightings give different things a common denominator so that quantitative comparisons can be made. Appraisals themselves also contain a comparison, normally with a defined objective, a 'standard'. Some of the current suggestions for appraisal are set out below.

(a) WEIGHT OF QUANTITY FLOW AS A YARDSTICK

An approach that has become known as material intensity per service unit (MIPS; Schmidt-Bleek, 1994) does without an (unequal) valuation and adds and interprets quantity flows as they are 'weighed'. Among other things, this conceals a conflict with the limits of knowledge, which basically militates against a more accurate valuation, e.g. according to the requirements of eco-toxicology.

(b) WEIGHTING WITH SCIENTIFIC COEFFICIENTS

As with the flow accounts, so with valuations, too, a distinction can be made between integral and selective, problem-related approaches. The former include, for example, conversions into exergy[2] or emergy (see Ulgiati *et al.*, 1994) flows, where the sum of potential physical work or the sum of energy expended are used as a common unit of account. The second category includes the problem-related weighting of emissions of individual materials and their addition into total burden potentials, as have been estimated in detail for some problem fields by the Netherlands Central Statistical Office (De Haan *et al.*, 1994).

(c) MONETARY WEIGHTING

Characteristics with different units/scales are reduced to a common denominator by direct or derived valuation (market prices, calculated scarcity premiums etc.). We are thinking here primarily of the proposals from resource economics for the valuation of quantitative raw materials shortages (ratio of stocks to withdrawals); these approaches are doubtless less suitable for valuing material flows.

(d) WEIGHTING WITH SOCIOLOGICAL COEFFICIENTS

A topical example of such an approach, where different physical series are given weightings based on a poll of experts, is Eurostat's 'pressure index'.

(e) APPRAISAL BY PERFORMANCE INDICATORS

This measures the extent to which predetermined targets have been met.[3] The specific scaling of the indicator is retained.

(f) MONETARY APPRAISAL

On the one hand, an estimate can be made of the costs involved in reaching the targets (standards) (avoidance costs).[4] On the other hand, macroeconomic models can be used to simulate a modified economy in which the desired set of objectives was achieved, in other words where material and energy flows did not exceed their thresholds (model of an 'environmentally sustainable economy' and its results; see Radermacher, 1995).

(g) SCIENTIFIC APPRAISAL

Baccini and Brunner (1991) looked for regional sustainability, which they defined on the basis of the 'geogenic material flows', that is the volume of flows that would be normal without human intervention. By their method, comparing anthropogenic with geogenic flows provides an indicator of whether the target has been achieved or not.

It is often said that material flows must be valued before they can be combined so as to even out the different impacts of the various materials. Without such adjustment, it would be like adding chalk and cheese. Taken precisely, this position would of course mean that chemical elements and compounds would have to be looked at separately and that we would in fact have to know enough about the effect they have on ecological acceptors in various quantities. Since neither condition is adequately fulfilled, however, we are obliged to analyse aggregates (e.g. in the form of goods) that may already in themselves be quite heterogeneous. There is also the problem that the specific effect of substances is highly dependent on quantity and can therefore change from completely unproblematic in small quantities to destructive in extremely high quantity flows (e.g. soil excavation/spoil).

All this speaks in favour of regarding the quantity matrix of flows and weighting/valuation schemes as independent methodical elements. This gives the necessary flexibility for demonstrating the effects of different weightings on one and the same quantity structure rather than tying oneself exclusively to the same system of objectives in advance.

From the valuation list referred to, a selection can be made of the processes that come into play in the GEEA. As already explained, the method of quantity recording mentioned under (a) is at any rate a statistically feasible starting point. The balanced aggregation of quantities into total flows also provides indicators on the basis of which an initial approximation can be obtained of the structure of material flows and how they change over time (Kuhn *et al.*,

1994). Weightings with scientific coefficients and monetary appraisals are likely to be the most suitable methods of valuation. In addition, monetary weightings may be used for quantitative scarcity of raw materials. On the other hand, the GEEA plan makes no provision for weightings with sociological coefficients and scientific appraisals in the material and energy flow accounts.

Concept of a Material and Energy Flow Information System (MEFIS)

The methodology of material flow analysis can be applied to extremely different observation levels: Figure 2 demonstrates this by breaking down the units into smaller sub-units, a process that can be started at the national level and continued to the micro-level of chemical processes. Consequently, an information system has to be located on a specific scale within this hierarchy. General accounting units are economic sectors (production and consumption activities). It is the ultimate objective of the concept to achieve as far as possible a comprehensive and detailed balance of the flows into, through and out of the sectors. However, the level of detail might be not sufficient in the starting situation, since economic goods, which are already accounted in traditional statistics, do not represent the entire set of material flows which are relevant from an environmental perspective. Hence, missing flows have to be estimated by linking known flows with corresponding coefficients.

The first experiences from single case studies indicate an urgent need for a more systematic organization of data processing. Three targets can be set that should be reached in this context:

(a) The performance of data processing has to be improved (standardized processes, use of information technology etc.).
(b) The data quality has to be documented in a transparent way (metadata-like reliability, 'pedigree').
(c) The access of (external) users has to be defined (user support, confidentiality etc.).

The variety of data, data sources and data processing leads to a complexity which calls for the aid of information technology. Consequently, the introduction of an information system that can bridge the gap between the 'chaos' of basic data and the 'order' of an accounting framework is essential.

After having explained the metabolism concept the fundamental lines and structures of a material and energy flow information system (MEFIS) are almost obvious: the core of the system is a virtual 'cube' which contains the space for each kind of material flow on a sectoral scale. In comparison to the well known structure of make and use or input–output tables two kinds of enlargements are necessary: first, two new accounts/sectors are introduced, one for the economic cleansing industries (environmental protection, recycling etc.) and a second for nature as a stock account. Second, the vector of com-

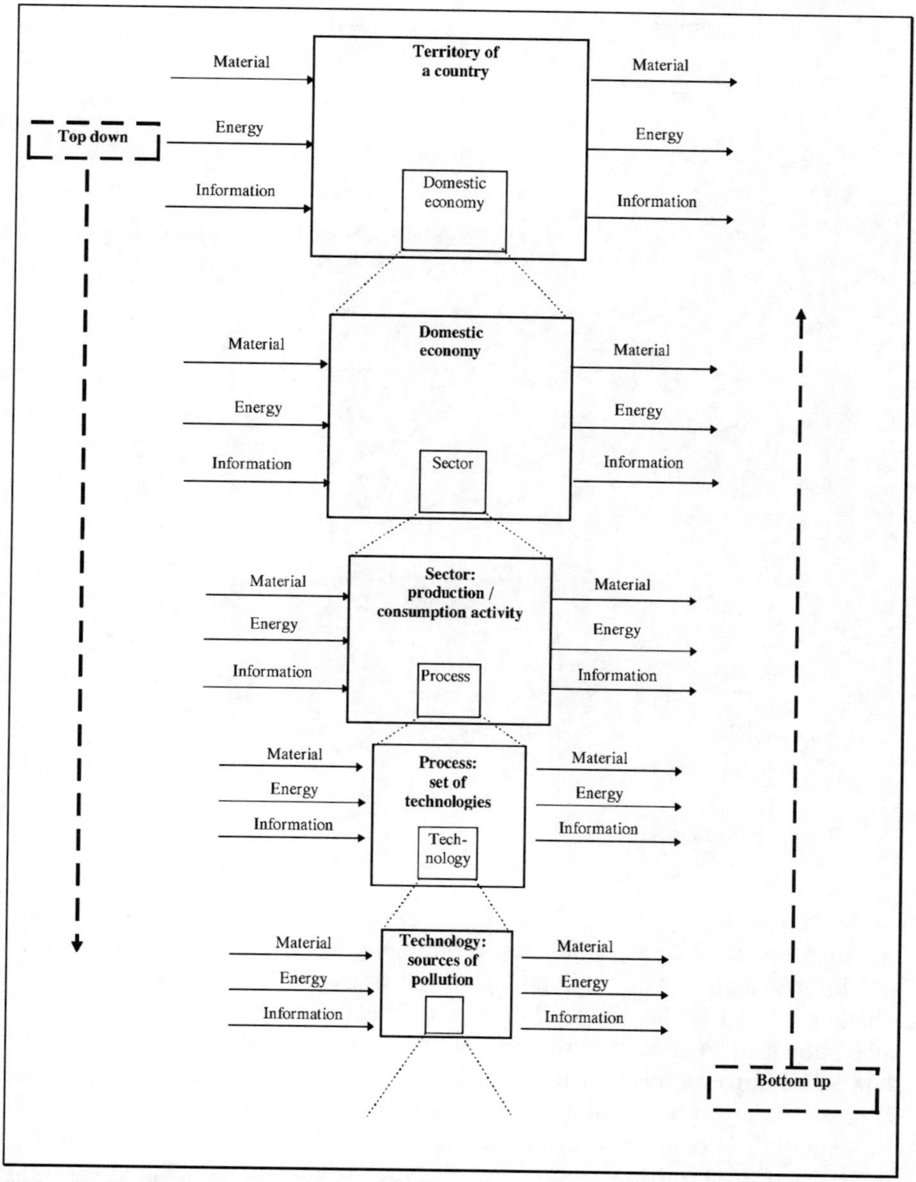

Figure 2. Metabolism: scales of reporting.

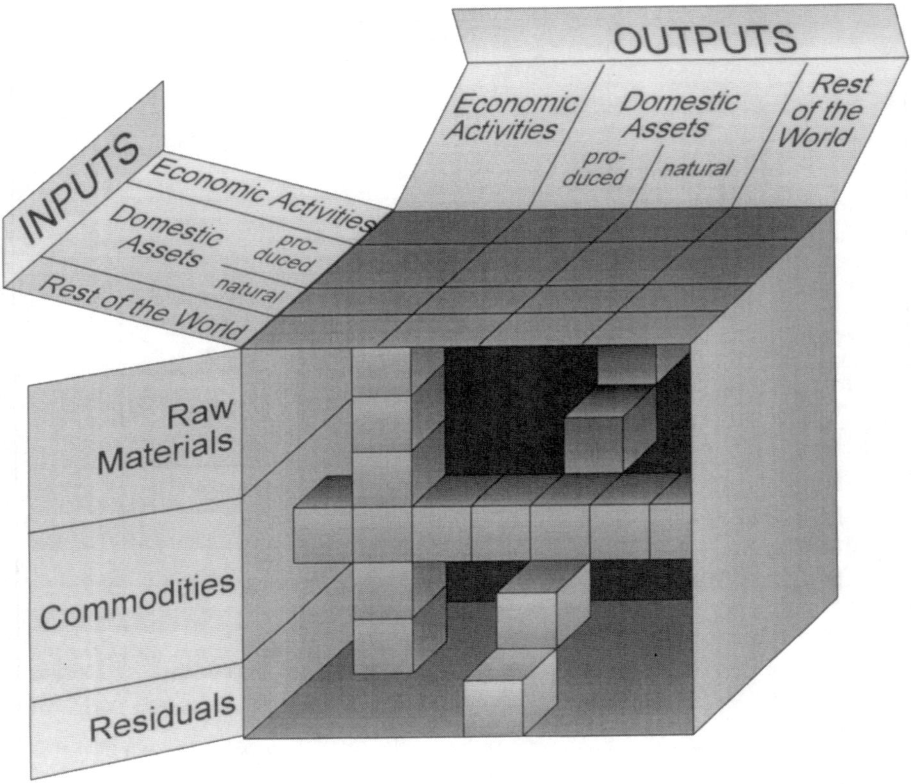

Figure 3. The MEFIS cube.

modities has to be enlarged for non-economic raw materials on the input side and for residuals (waste, pollutants) on the output side. For each substance, whether it is an economic good or not, this leads to one level of the MEFIS cube shown in Figure 3. Normally only the edges of this cube are filled with data – this corresponds to separate make and use tables for material flows. In some cases, however, the input and the output side of flows can be linked, e.g. the 'supply' of waste in economic sectors and their 'use' by cleansing industries. Furthermore the cube provides the virtual space for the integration and storage of flow analysis for specific substances. While physical input–output tables are organized from a sectoral perspective (separated tables for inputs and outputs of the sectors) the flow analysis for specific substances (e.g. nutrients, heavy metals) have to follow their specific path from one to another stage of production and consumption. This leads to the need of linking outputs and inputs of the substance which can be organized within one single sheet of the material flow cube.

In addition to this 'cube', matrices which contain the standardized basic data and the procedures of estimation of flows have to be stored within the information system. Finally (and this will be the most relevant part for the application of an information system) standardized evaluations have to be provided. Those applications require standardized evaluation software (e.g. input–output analysis) and the availability of 'metadata' describing the data quality with respect to the genesis of data and their suitability for specific applications. A small selection of applications from the current work programme should demonstrate the scope and ultimate objective of the information system.

Applications

Material flow accounts: the macroeconomic perspective

This describes the relationships between the national economy on the one hand and nature and the rest of the world on the other. Withdrawals from nature for economic utilization and imports of goods are shown as material withdrawals. The term 'material discharges' is used for materials that leave the national economy or are discharged into nature as overburden, waste, air pollution, waste water or exports of goods. However, this material flow account does not include every process within the national economy (national economy is taken in the national accounts sense, including general government institutions and private households as final consumers).

What is new, and at first unaccustomed and therefore perhaps hard to understand, about this method of analysis is the straightforward addition of quantities. It is important to remember what was said above about the origin of environmental pressures and the essential components 'quantity' and 'toxicity'. As a first approximation it is worthwhile and permissible to prepare a pure quantity balance sheet in which all materials moved are counted in tonnes and combined into totals. This approach does not (yet) make a differentiation or weighting according to specific ecological risks of individual materials. Gravel goes into the total with its own weight just as do uranium, gold or water. The background to this is, admittedly, also that complete information about all or at least the major ecological risks of the various materials is not available.

What distinguishes the table shown here from conventional raw materials balances, however, is that this approach requires the quantities of unused raw material withdrawals (overburden, slag, etc.) to be counted as well. The quantity ratio between used and unused withdrawals is sometimes extraordinarily high, e.g. 1:100,000 in the case of gold. Once they have been mined, these quantities tend to be immediately dumped or deposited back into nature in some other way (sometimes recultivated), thereby counting as both withdrawals and discharges of material as a self-balancing item. A similar situation

Table 1. **Material flows, 1960/1990 (former territory of the Federal Republic of Germany)[1].**

Materials	1960	1990		Materials	1960	1990	
	mn t		1960 = 100		mn t		1960 = 100
Withdrawal				**Discharge**			
Solids, fuels 2)							
Raw material withdrawal	1 253	2 072	*165*	Material application	227	252	*111*
Raw materials, extraction used	780	995	*128*	Fertilizer	226	251	*111*
Biotic raw material	133	188	*142*	Farm manure	224	246	*110*
Abiotic raw material	647	807	*125*	Commercial fertilizer (nutrient)	3	5	*161*
Fuel	248	193	*78*	Pesticides (active substances)	0.01	0.03	*300*
Other	399	614	*154*	Sewage sludge	0.7	0.9	*129*
Not used raw materials, extraction(incl. soil excavation, excavated materials a. rubble)	474	1 077	*227*	Not used raw materials, extraction (incl. flate)	415	982	*237*
Imported goods	136	387	*285*	Exported goods	75	207	*274*
Biotic goods	25	65	*254*	Biotic goods	6	46	*825*
Abiotic goods	110	323	*292*	Abiotic goods	70	160	*230*
Fuel	45	169	*373*	Fuel	35	24	*70*
Other	65	153	*235*	Other	35	136	*388*
				Waste (incl. soil excavation, rubble)	113	164	*145*
Recycling	16	58	*363*	Recycling	16	58	*363*
Total	1 405	2 517	*179*	Total	846	1 662	*196*
Oxygen consumption and air pollutants							
Oxygen input processes (CO2; CO)	409	533	*130*	Emission to air	568	738	*130*
				Carbon dioxide	555	727	*131*
				Carbon monoxide	9.8	7.3	*74*
				Nitrogen dioxide	1.6	2.6	*163*
				Sulphur dioxide	3.3	1.0	*30*
Total	1 814	3 050	*168*	Total	1 414	2 400	*170*
				Balance (material retained)	400	650	*162*
Water in million m3							
Water withdrawal	18 880	42 970	*228*	Waste water discharges	16 510	43.570	*264*
Precipitation and impurity water 3)	1 270	3 470	*273*	Discharged untreated	11 480	34 130	*297*
				Discharged after treatment 4)	5 030	9 440	*188*
				Water discharged unused	1 290	815	*63*
				Water consumption 5)	2 350	2 055	*87*
Total	20 150	46 440	*230*	Total	20 150	46 440	*230*

1) Discrepancies in totals by rounding of figures. - 2) Including useful gases and certain liquid materials. -
3) Precipitation and impurity water accruing in waste water treatment plants. - 4) Including precipitation and impurity water accruing in waste water treatment plants. - 5) Mainly water lost by evaporation or seepage during use.

exists with soil excavation: soil excavated when houses are built is withdrawn from nature and deposited elsewhere unchanged. It starts as a withdrawal from nature and is then a discharge into it.

Table 1 contains the material flow account for the former territory of the Federal Republic of Germany in the years 1960 and 1990. The left-hand side describes the material withdrawn by the economy from nature and the rest of the world, and the right-hand side describes the material discharged by the economy into nature and the rest of the world. The balancing item comprises materials remaining in the economy (and statistical differences).

Withdrawals of material in the form of solids and fuels (and to a lesser extent other gases and certain liquid materials), namely extraction of (used and unused) raw materials, soil excavation and imports have already been mentioned briefly above, as have some headings of materials discharged into nature. Apart from unused withdrawals, soil excavation and exports, the discharge side shows fertilizers, pesticides and waste. A significant quantity of fertilizer takes the form of farm manure (dung and slurry), which could also be interpreted as an input for agricultural products. In the case of agricultural products it is especially difficult to draw the line between nature and the economy. Nature (soil, rain, sun) is used on the one hand, with labour, capital and intermediate consumption (seed, fertilizers, pesticides) on the other. In principle, the withdrawal of nutrients from the soil, of rain and of solar energy ought also to be quantified. As this is virtually impossible, the (statistically well recorded) plant products of agriculture, as the raw material withdrawal of biotic goods, are taken as an indicator of the utilization of nature. Fertilization must, therefore, be regarded as an application of material into nature, which is then in part contained in the withdrawal of biotic raw materials from the environment. The fact that a large proportion of farm manure is not processed in plants but enters the ground water as nitrate pollution supports this approach.

The waste shown in the material flow account includes waste discharged into nature which (after refuse incineration) has to be dumped or disposed of in some other way. The question of where nature begins and the economy ends arises with waste treatment, too. Table 1 shows the waste output (without weight loss in refuse incineration) as a material discharge. It would also be possible to regard waste dumps as stores within the economy. The reutilization of residues has no place in the scheme described above, since recycling takes place within the economy. Figures on residues have, nevertheless, been included in the table in order to give an impression of the magnitude of these environment-preserving activities. This is only an extension of the balance sheet that has no effect on the balancing item and which even in 1990 constitutes only a small proportion of the total material flows.

Apart from the discharges into nature previously described, selected air pollutants are also covered: these are largely attributable to the combustion of fuels (including refuse incineration) and the oxygen consumed in the process.

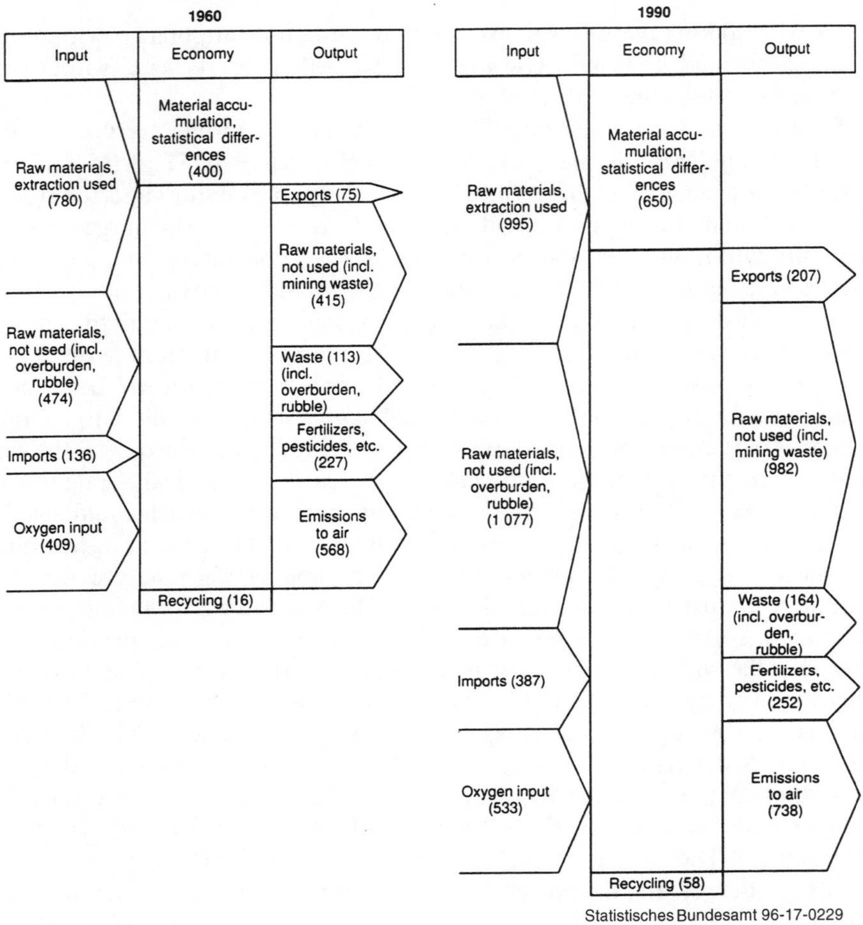

Figure 4. Material flows, 1960/1990, former territory of the Federal Republic of Germany (Million tonnes).

The oxygen consumed in combustion processes is shown as a withdrawal of material; the fuels used already appear in usable raw material withdrawals and imports of goods.

The final sub-item of the material flow account is the use of water, both withdrawal from and discharge into nature. It should be pointed out that it was not possible to take account of possible pollutant loads in water when converting cubic metres into tonnes; water and waste water were therefore each entered in the account at 1 tonne/m^3.

The main findings of Table 1 are shown in graphic form in Figure 4. In the period from 1960 to 1990 there were considerable increases in quantity throughputs through the economy. However, for most withdrawals and discharges of materials the increase is less than that of the gross domestic product in real terms, which in 1990 was two and a half times the 1960 figure. Under withdrawals from nature, the increase in withdrawals of water was particularly great (see Table 1). There were appreciable rates of increase for other materials, too. Utilization via imports of goods (fuels in particular) and overburden as an unused joint product of domestic raw material extraction increased above average. A basically similar trend can be seen on the discharge side, where the quantitative focus was on waste water and exports of goods. However, the discharge side also has a few developments that are insignificant in quantity terms but which are nevertheless of interest for the environment: the pattern of fertilizer/pesticide use shows a rapid increase in pesticides and an average increase in commercial fertilizers (those that are sold commercially). This is where a quantity balance like the material flow account reaches the limits of what it is able to express; qualitative indicators are also needed that will allow us to say something about the ecological risks of these substances. Over the three decades the percentage increase in the quantity of farm manure (dung and slurry) applied was small, although in 1960 the amount of dung and slurry was already very large. What is also lacking here is a further breakdown according to load areas that will show the geographical concentration processes with the corresponding overloads in particular regions. This, too, would go beyond the limits of accounts based on the national economy as a whole.

The balancing item of this material flow account shows how much material is additionally bound in the national economy in a year. This may be in the form of such economically used installations as houses, roads, plant and machinery, or of e.g. increasing stocks of consumption goods and private durables. The balance is calculated for solid, liquid and gaseous materials, not including water. There is no aggregation of all material withdrawals and discharges because the water flow is many times greater than that of other materials and would, therefore, have too great an effect on the overall result. A balance for water is, therefore, not shown: it is assumed that water withdrawn from nature will over a period return to the natural cycle by discharge, evaporation or seepage.

Let us now make a brief critical examination of the balance for materials. In both 1960 and 1990 the balance amounted to around one-fifth of the material withdrawal. This means that between 1960 and 1990 the stock of products and raw materials in the national economy grew from 400 to 650 million tonnes. The quantities that are withdrawn from nature but do not enter the economy, more or less passing it by and being deposited again elsewhere (mining waste, overburden), are even greater. The quantity of waste (113 and 164 million tonnes) seems small by comparison. These calculated balances should not however be overinterpreted. Phase transitions between the various states of aggregation and a large number of margins of error and statistical consolidations oblige us to be very cautious about what the balances seem to say. It should therefore be assumed that the respective orders of magnitude at the most provide a basis for broad interpretations.

All in all, as already mentioned, the material flow account presented here is the first visible result of a new kind of resource accounting which in some respects has not yet been fully worked out. What is missing is primarily some light on the indirect material flows abroad occasioned by the import and export of goods. Until it is clear how big the invisible 'rucksack' is that an imported finished product, for example, carries with it (how much overburden accrued during its manufacture and how many emissions and how much waste were produced), it will be impossible to say for certain whether the use of natural resources becomes more intensive, i.e. economically and ecologically better, in the course of time. Initial estimates have been prepared by the Wuppertal Institute (see Bringezu *et al.*, Chapter 13); these will have to be examined in the future and improved by more refined methods.

Physical input–output tables (PIOT)

Concepts of PIOT

An important part of the work on material and energy flow accounting in Germany is the calculation of PIOT for the year 1990 (see Table 2). This work is based on the methods and concepts explained in Chapter III of the SEEA Handbook (System for Integrated Environmental and Economic Accounting; United Nations, 1993a). The physical stock and flow accounting constitutes a very important part of the SEEA. Production and consumption activities with their different kinds of stress on the natural environment are in the centre of attention. PIOT are the next logical step after establishing macroeconomic material accounts: they are conceptionally and empirically very similar. As mentioned, material and energy accounts include all incoming materials from the domestic nature or from the rest of the world ('withdrawals' and 'inflows') on the input side and all outgoing materials to the domestic nature or to the rest of the world ('discharges' and 'outflows') on the output side. As a consequence, all materials are included which flow into or out of the German economy as a whole. These material flows are presented in more detail in PIOT and material flows between the different economic activities are also described. Therefore, PIOT show, by analogy with the monetary input–output tables, the domestic production, the whole supply of products (which summarizes the domestically produced and the imported products) and the use of these products in physical units (e.g. in tonnes). PIOT also incorporate raw material extraction (including water and oxygen) from nature by producers or households and show the origin, the treatment and final destination of residuals (recycling, treatment of waste and waste water, storage, discharge of residuals to nature). The most important addition to traditional monetary input–output tables is the introduction of an additional 'production factor nature'.

The integration of data about raw material extraction, discharge of residuals and flows of commodities leads to complete material balances for each production and consumption activity. For each activity account the law of conservation of matter and energy is valid, i.e. the total material inputs are balanced with material output. For each domestic activity (that is, for production branches and households) the following equation holds:

extraction of raw materials (including water and oxygen) + use of domestic and imported products + use of residuals in order to protect the environment (treatment or storage of residuals) = products of domestic production + supply of residuals by domestic activities

If these material balances, or the PIOT as a whole, are supplemented with physical stock accounts, which incorporate at least produced assets (including natural assets in agriculture, forestry and fishing and consumer durables), the concepts for material flow accounts are largely implemented as described in the SEEA. As far as possible, such PIOT should be supplemented with qualitative data for both physical stocks and flows. For example, that could be pressure indices which are necessary for data systems such as NAMEA and ECSTASY.

A linkage of physical and monetary input–output tables would be very useful. As a first step it is more important to set up a comprehensive, consistent and complete physical data system which might have a higher degree of independence of traditional (monetary) input–output tables. In the MEFIS-cube, the PIOT could be interpreted as two outside edges of the cube.

Table 2. **Physical input–output table, 1990 (1000 t).**

Ser. No.	Branches Final uses	Input of materials						Output of materials	
		from nature			man made		Total input	man used	
		Oxygen	Water	Raw material	Re-cycling material	Commo-dities		Commo-dities	Recyclin g material
1	Agriculture	8 481	203 133	137 707	0	98 092	447 413	132 396	0
2	Forestry and fishing etc.	2 631	0	50 104	0	5 383	58 118	50 149	0
3	Electricity, steam, hot water supply	233 791	28 695 417	0	0	207 366	29 136 574	0	6 606
4	Gas supply	609	0	0	0	46 834	47 443	46 569	6
5	Water supply	125	4 634 300	0	0	89 049	4 723 474	4 642 028	97
6	Coal mining	5 538	1 895 549	177 758	0	42 222	2 121 067	208 422	19
7	Other mining (excl. coal mining, crude petroleum, natural gas)	1 149	138 681	24 841	0	7 604	172 276	27 603	174
8	Extraction of crude petroleum, natural gas	1 596	18 889	15 337	0	1 306	37 129	14 109	4
9	Man. of chemical products (incl. nuclear fuel)	26 037	3 911 504	0	0	203 155	4 140 697	100 526	5 453
10	Man. of refined petroleum products	16 255	267 600	0	0	103 396	387 250	93 094	67
11	Man. of plastic products	1 949	53 106	0	0	30 648	85 703	6 630	259
12	Man. of rubber products	1 272	31 731	0	0	6 516	39 519	1 276	142
13	Quarrying of stones and clays, man. of building material, etc.	17 555	312 504	589 389	0	301 084	1 220 532	794 558	1 200
14	Man. of ceramic products	1 685	3 645	0	0	4 197	9 527	1 098	66
15	Man. of glass and glass products	5 380	16 901	0	0	25 777	48 058	6 640	172
16	Man. of iron and steel	41 386	1 079 806	0	0	264 896	1 386 088	129 992	16 672
17	Man. of non-ferrous metals, semifinished products thereof	3 858	208 695	0	0	52 583	265 135	6 246	342
18	Man. of foundry products	2 713	23 198	0	0	21 015	46 925	4 005	359
19	Man. of drawing plants products, cold rolling mills etc.	2 302	30 985	0	0	29 551	62 839	10 390	4 160
20	Man. of structural metal products, rolling stock	1 091	6 669	0	0	5 334	13 095	3 312	1 349
21	Man. of machinery and equipment (excl. electrical)	4 868	40 264	0	0	41 681	86 813	8 211	2 120
22	Man. of office machinery, autom. data processing equipment	242	4 474	0	0	6 642	11 357	90	48
23	Man. of road vehicles	5 137	134 131	0	0	74 274	213 542	11 625	2 960
24	Building of ships, boats and floating structures	188	4 805	0	0	4 484	9 477	541	163
25	Man. of aircraft and spacecraft	252	2 651	0	0	3 717	6 620	15	25
26	Man. of electrical machinery, equipment and appliances	3 947	64 622	0	0	36 390	104 958	4 935	889
27	Man. of precision and optical instruments, clocks and watches	494	3 065	0	0	5 601	9 160	163	69
28	Man. of tools and finished metal products	2 376	14 498	0	0	25 050	41 924	8 129	1 509
29	Man. of musical instruments, games and toys, sports goods, etc.	199	663	0	0	4 484	5 346	293	39
30	Man. of wood	1 236	12 427	0	0	43 224	56 888	18 072	4 852
31	Man. of wood products	1 589	3 977	0	0	22 188	27 754	5 468	969
32	Man. of pulp, paper and -board	7 225	590 336	0	0	70 612	668 173	13 619	764
33	Man. of paper and -board products	1 525	59 154	0	0	13 330	74 009	9 265	777
34	Printing and duplicating	982	6 296	0	0	14 284	21 562	4 147	1 055
35	Man. of leather and leather products, footwear	247	6 048	0	0	933	7 228	133	105
36	Man. of textiles	3 542	185 498	0	0	72 566	261 606	2 375	141
37	Man. of wearing apparel	555	2 651	0	0	3 694	6 901	264	60
38	Man. of food products (excl. beverages)	12 680	227 833	0	0	199 113	439 625	70 079	7 396
39	Man. of beverages	3 352	79 451	0	0	65 617	148 421	25 963	2 173
40	Man. of tobacco products	153	1 243	0	0	1 531	2 926	242	18
41	Construction (excl. installation and building completion)	5 755	0	0	0	548 058	553 812	1 869	22 746
42	Installation and building completion	2 256	0	0	0	26 952	29 208	0	266
43	Wholesale trade, etc., recovery	11 953	0	0	101 888	10 604	124 445	57 441	810
44	Retail trade	11 689	0	0	0	14 673	26 362	0	867
45	Railway transport	2 148	0	0	0	10 888	13 036	0	37
46	Water transport, ports	2 894	0	0	0	3 486	6 380	0	10
47	Post and telecommunication	1 286	0	0	0	2 709	3 995	0	74
48	Transport activities n.e.c.	27 447	0	0	0	13 939	41 386	0	29
49	Banking	1 250	0	0	0	3 108	4 358	0	50
50	Insurance (excl. social security funds)	648	0	0	0	1 127	1 776	0	55
51	Renting of real estate	233	0	0	0	8 017	8 250	0	1
52	Hotels and restaurants, homes and hostels	3 238	0	0	0	40 548	43 786	0	158
53	Education, research, cultural services and publishing	1 178	0	0	0	4 211	5 389	0	297
54	Health and veterinary market service activities	1 614	0	0	0	45 590	47 204	0	76
55	Other market service activities, etc.	8 062	0	0	0	41 223	49 284	0	386
56	Central and local government	22 031	0	0	0	392 246	414 277	0	741
57	Social security funds	512	0	0	0	1 436	1 948	0	77
58	Private non-profit institutions, private households	2 591	0	0	0	34 460	37 051	0	54
59	**Branches, total**	**532 976**	**42 976 400**	**995 136**	**101 888**	**3 458 697**	**48 065 098**	**6 521 981**	**90 009**
60	Private consumption	204 963	n.a.	0	0	3 227 455	3 432 417	0	11 133
61	Investments					14 474	14 474	0	746
62	Exports					207 904	207 904		
63	Changes of stocks in commodities, statistical discrepancies					4 460	4 460		
64	**Final uses, total**	**204 963**	**0**	**0**	**0**	**3 454 293**	**3 659 255**	**0**	**11 879**
65	**Total**	**737 939**	**42 976 400**	**995 136**	**101 888**	**6 912 990**	**51 724 353**	**6 521 981**	**101 888**

Calculation of PIOT

The calculation of PIOT comprises three different steps:

(a) Calculation of physical commodity flows without the 'production factor nature', in line with the compilation of traditional monetary input–output tables;

(b) Incorporation of the 'production factor nature', therefore extension on the input side with raw materials spent by nature and on the output side with residuals absorbed by nature;

(c) Introduction of the analyses of treatment or storage of residuals within the economy and of stock accounts.

Table 2. (Continued)

Ser. No.	Branches / Final uses	Output of materials				Balance of materials			
		to nature			Total Output	Water	Energy	Re-maining balance	Total balance
		Air emis-sions	Waste water	Waste					
1	Agriculture	8 176	208 750	n.a.	349 322	n.a.	103	97988	98091
2	Forestry and fishing etc.	2 337	3 450	n.a.	55 936	n.a.	19	2163	2182
3	Electricity, steam, hot water supply	254 883	28 356 873	9 737	28 628 099	388644	75563	44367	508475
4	Gas supply	456	0	295	47 326	0	-8	125	117
5	Water supply	118	0	374	4 642 617	80899	3	-44	80858
6	Coal mining	6 272	1 134 950	518	1 350 181	760599	11679	-1392	770886
7	Other mining (excl. coal mining, crude petroleum, natural gas)	825	124 585	172	153 359	14096	31	4789	18916
8	Extraction of crude petroleum, natural gas	1 085	3 020	25	18 243	17047	-1101	2940	18886
9	Man. of chemical products (incl. nuclear fuel)	21 873	3 881 143	5 453	4 014 448	110361	14869	1019	126249
10	Man. of refined petroleum products	13 362	211 662	111	318 296	64351	-3600	8203	68954
11	Man. of plastic products	1 543	68 974	659	78 065	7583	69	-14	7638
12	Man. of rubber products	968	32 036	316	34 738	3906	85	790	4781
13	Quarrying of stones and clays, man. of building material, etc.	17 396	309 463	5 709	1 128 326	85965	799	5441	92206
14	Man. of ceramic products	1 232	4 759	197	7 352	1086	35	1054	2174
15	Man. of glass and glass products	4 258	22 752	710	34 532	4231	26	9268	13526
16	Man. of iron and steel	48063	1 020 778	2 519	1 218 023	139028	444	28593	168064
17	Man. of non-ferrous metals, semifinished products thereof	3266	208 472	600	218 926	39417	13	6780	46210
18	Man. of foundry products	2805	25 254	1 920	34 343	6923	49	5610	12583
19	Man. of drawing plants products, cold rolling mills etc.	1688	40 954	1 181	58 373	3937	88	440	4465
20	Man. of structural metal products, rolling stock	929	6 313	281	12 184	356	23	532	911
21	Man. of machinery and equipment (excl. electrical)	4088	66 242	1 190	81 851	3916	149	898	4962
22	Man. of office machinery, autom. data processing equipment	192	9 949	37	10 316	633	6	401	1041
23	Man. of road vehicles	4003	179 648	1 529	199 765	12533	337	907	13777
24	Building of ships, boats and floating structures	146	7 084	82	8 016	781	5	675	1461
25	Man. of aircraft and spacecraft	189	5 586	58	5 873	315	8	424	747
26	Man. of electrical machinery, equipment and appliances	3337	86 141	845	96 147	5485	181	3145	8811
27	Man. of precision and optical instruments, clocks and watches	412	6 957	117	7 718	540	22	880	1442
28	Man. of tools and finished metal products	1928	28 287	452	40 305	2695	77	-1152	1620
29	Man. of musical instruments, games and toys, sports goods, etc.	166	3 408	64	3 970	166	5	1204	1376
30	Man. of wood	1103	16 441	809	41 277	2614	-6	13003	15611
31	Man. of wood products	1475	6 510	1 219	15 641	1421	-24	10716	12113
32	Man. of pulp, paper and -board	6197	610 872	2 085	633 537	28123	383	6130	34636
33	Man. of paper and -board products	1174	33 184	438	44 838	31566	31	-2426	29171
34	Printing and duplicating	752	13 550	266	19 770	1911	19	-137	1793
35	Man. of leather and leather products, footwear	219	5 858	100	6 415	560	5	248	813
36	Man. of textiles	2835	237 894	357	243 602	16161	89	1754	18004
37	Man. of wearing apparel	490	4 867	121	5 802	384	10	704	1099
38	Man. of food products (excl. beverages)	10634	302 200	2 961	393 270	25572	266	20517	46355
39	Man. of beverages	2826	95 648	224	126 834	18905	41	2641	21587
40	Man. of tobacco products	131	1 690	31	2 112	583	6	225	814
41	Construction (excl. installation and building completion)	5448	18 841	104 941	153 845	-416	2404	397979	399967
42	Installation and building completion	2131	2 531	2 050	6 978	-56	-10	22296	22230
43	Wholesale trade, etc., recovery	10930	5 445	200	74 826	-120	-14	49753	49619
44	Retail trade	10088	9 280	267	20 502	-205	32	6033	5860
45	Railway transport	2003	8 462	215	10 717	-187	15	2491	2319
46	Water transport, ports	2754	102	14	2 881	-2	2166	1335	3499
47	Post and telecommunication	1173	2 020	44	3 311	-45	0	729	684
48	Transport activities n.e.c.	25629	3 119	87	28 864	-69	1825	10765	12522
49	Banking	1058	2 608	26	3 741	-58	7	668	617
50	Insurance (excl. social security funds)	562	818	16	1 451	-18	2	341	324
51	Renting of real estate	219	0	879	1 099	0	4	7148	7151
52	Hotels and restaurants, homes and hostels	2809	29 425	549	32 941	-650	22	11473	10845
53	Education, research, cultural services and publishing	1074	2 020	261	3 652	-45	2	1780	1737
54	Health and veterinary market service activities	1489	45 480	140	47 185	-1005	-5	1029	19
55	Other market service activities, etc.	7352	38 296	174	46 208	-846	-56	3979	3076
56	Central and local government	17998	360 236	4 337	383 312	-7961	-82	39008	30965
57	Social security funds	475	997	62	1 611	-22	-3	362	337
58	Private non-profit institutions, private households	2374	33 976	323	36 726	-751	2	1074	325
59	**Branches, total**	**529 398**	**37 949 860**	**158 348**	**45 249 597**	**1870740**	**107110**	**837651**	**2815501**
60	Private consumption	181 000	3 162 898	15 322	3 370 353	-69898	-993	132955	62065
61	Investments	0	0	0	746	0	0	13728	13728
62	Exports				0				0
63	Changes of stocks in commodities, statistical discrepancies				0				0
64	**Final uses, total**	**181 000**	**3 162 898**	**15 322**	**3 371 099**	**-69898**	**-993**	**146683**	**75792**
65	**Total**	**710 398**	**41 112 758**	**173 670**	**48 620 695**	**1800842**	**106117**	**984334**	**2891293**

Physical data available in the basic statistics, such as production and foreign trade statistics, are the starting point. These statistics are used to compile the vectors of domestic production, imports and exports. The above mentioned production statistics report on about 60% of all domestically produced goods in weight units (tons); other domestic products have to be converted from specific physical units (such as piece, litre, square metre and so on) into tons. This has to be done at a very detailed level in order to keep the margin of faults as small as possible – at the six-digit level of the classification for input–output tables about 1500 different goods are specified. Further estimates are necessary for the production of small and medium-sized enterprises, because the statistics of producing industries record only the establishments of corporations with more than 19 employees (as a rule). Foreign trade statistics report on all imports and exports in a very detailed classification in tons. Thus, it is possible to set up the following equation for nearly 1500 different kinds of goods:

domestic production + imports = domestic use + exports = total resources

Applying partly special information in physical units and partly information from the monetary input–output table, the sum of domestically produced and imported goods (at the six-digit level) in tons has been distributed among 58 branches and categories of final uses (private consumption, investments or exports). The result of this first step of work is the base structure of commodity flows between different branches and from branches to categories of final uses in tons.

The second step involved in compiling PIOT is to add the 'production factor nature'. That means adding the withdrawal of raw materials, water and oxygen (for burning purposes) from domestic nature on the input side. On the output side, we have to add the discharges of residuals (air emissions, waste and waste water) to domestic nature. Results for raw materials and air emissions from emitter structure analysis can be integrated, to the extent that they fit into the classification of the PIOT. For waste and waste water special investigations were necessary to achieve results which fit into the PIOT classification. In the future, this part of the work will become easier as soon as the material and energy flow information system has been established.

Our current activities focus on this part of the PIOT: we incorporate in particular the data of energy input–output tables (Statistiches Bundesamt, 1994) and results of research projects carried out by the Wuppertal Institute for Climate, Environment, Energy and the Eco-Institute (Darmstadt). A very important tool for incorporating these results and checking the PIOT is the material and energy balances by branch and household sector. These balances are accounted as the difference between inputs and outputs. To obtain a useful tool we distinguish three partial balances:

(a) water balances (water input, waste water, water incorporated in products), including still water evaporated or oozed away etc., and statistical discrepancies;
(b) energy balances (energy inputs + oxygen − energy outputs − related emissions), which should be approximately zero but which still contain statistical discrepancies;
(c) remaining balances (inputs of products, without water, energy and oxygen, output of products, waste) including apart from statistical discrepancies especially material accumulation.

If all information about inputs and outputs were available, the balances would only show the material accumulation or changes in material stocks. Balances normally have to be positive because of the law of conservation of matter and energy. Nevertheless flows from stocks can influence the balance.

Another important problem which has not yet been solved is the treatment of services. From a conceptual point of view, we want to represent all material flows which are actually connected with providing a service by one production activity to another or to a category of final uses. Restaurants are a good example: they provide services to households which are connected for example with consuming a meal. Thus, the weight of the meal consumed by households has to be shown in the PIOT as a material flow from restaurants to private consumption.

In a third step, the economic treatment of residuals has to be analysed. This work comprises the calculation of recycling loops and detailed accounts for the treatment of waste and waste water. The recycling loop shows, for example, the supply of waste paper, the use of this paper in the recycling industry and the supply and use of recycled paper as a secondary raw material. For waste and waste water we intend to show supply, treatment and discharge of these residuals. In a last phase of future work, stock accounts might complete the description of material flows.

Analysis of the outputs: emitter structure

Within the emitter structure, matrices of residue outputs are linked with input–output tables. By combination of these two matrices it is possible to calculate 'indirect' emissions in addition to the 'direct' emissions of the economic activities. The direct emissions are those which are a direct result of the production processes in a specific branch (in cement production, for example, the air emissions by a cement burning facility). In contrast to direct emissions, indirect emissions are estimated by adding up the emissions resulting from the production of the intermediate consumption. This calculation is made (up to this moment) by using the inverted (monetary) input–output table. Two crucial assumptions must be used to allow this calculation:

(a) sectoral homogeneity: the production of each sector is homogeneous with regard to the emissions (constant emission factor for all goods, for example: 20 tonnes of CO_2 per million of turnover);
(b) price homogeneity: each product is sold at the same price to any consumer.

As these assumptions are very restrictive, the calculation of the indirect emissions on this basis must be understood as an approximation. This approximation has already been improved by the integration of the price structure of the electricity producing branch. The following examples demonstrate the potential evaluations which can be deduced from the emission structure.[5]

Combined presentation of economic and environmental figures

Table 3 shows economic and environmental characteristics concerning 11 economic branches, and for each characteristic (gross value, employment, primary

How No.	Selected institutional sectors	Gross value total	Gross value percentage	Workers total	Workers percentage	Uses of energy total	Uses of energy percentage	Carbon dioxide direct emissions	Carbon dioxide percentage	Carbon monoxide direct emissions	Carbon monoxide percentage	Sulphur dioxide direct emissions	Sulphur dioxide percentage	Nitrogen dioxide direct emissions	Nitrogen dioxide percentage	Expenditures for air pollution control	Expenditures for percentage
		bn. DM	%	1000	%	Petajoule	%	mln. t	%	1000 t	%	1000 t	%	1000 t	%	bn. DM	%
1	Energy, water and coal mining products	65	6	450	3	7.585	48	263	59	82	5	371	48	356	41	4.280	51
2	Chemical products (incl. nuclear fuel)	63	5	552	4	1.365	9	22	5	22	1	88	12	47	5	1.230	15
3	Manufacturing of refined petroleum products	25	2	18	0	3.979	25	13	3	6	0	87	11	24	3	360	4
4	Manufacturing of plastic products extraction of stones and clays	60	5	758	5	421	3	25	6	178	10	47	6	130	15	450	5
5	Manufacturing and production of metal	55	5	685	4	984	6	55	12	1.105	64	71	9	44	5	960	11
6	Manufacturing of steel, machines, vehicles and autom. data processing equipment	208	18	2.569	17	251	2	9	2	55	3	9	1	30	3	520	6
7	Electrical machinery, equipment and appliance tools and finished metal products	139	12	1.848	12	139	1	6	1	36	2	6	1	18	2	130	2
8	Wood, paper, leather and textiles, manufacturing of wearing apparel	84	7	1.428	9	313	2	14	3	47	3	39	5	40	5	260	3
9	Food products, beverages, tobacco products	74	6	801	5	253	2	13	3	38	2	26	3	56	6	170	2
10	Building and civil engineering works, installation and building completion works	128	11	1.937	13	209	1	/	2	116	/	/	1	//	9	20	0
11	General government	253	22	4.302	28	398	3	18	4	53	3	15	2	45	5	30	0
12	Total	1.155	100	15.348	100	15.897	100	446	100	1.739	100	767	100	866	100	8.410	100

*) Former territory of the Federal Republic of Germany
Source: own computations, rounded.

Table 3. Economic figures and emission structure, 1990 (former territory of the Federal Republic of Germany).

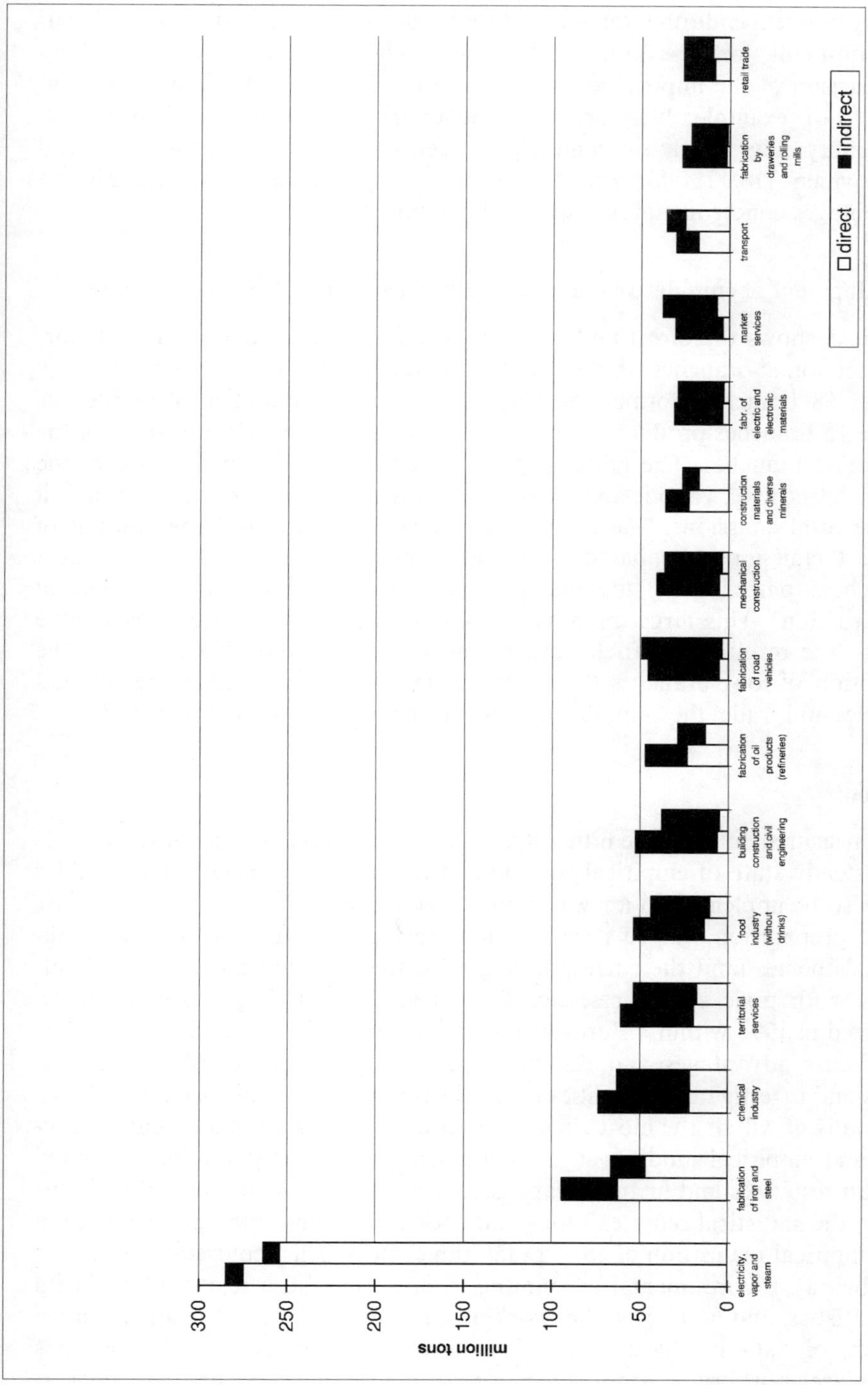

Figure 5. Cumulative CO_2 emissions by branch, 1980 and 1990 (former territory of the Federal Republic of Germany)

energy use, expenditures for air pollution control facilities and the main air pollution emission types) the absolute values and the percentages. This allows comparison of the importance of each sector with regard to different dimensions. For example, branches 6 (manufacturing of steel) and 7 (electrical machinery) have nearly the same high percentages concerning gross value and employment (18/17% for branch 6 and 12% for branch 7) but only modest percentages concerning air emissions (maximum 3%).

Development of cumulative emissions between 1980 and 1990 in Germany

Figure 5 shows the direct and indirect CO_2 emissions of the 15 most important economic branches (those with the highest cumulative emissions in 1980) out of 58 branches (former territory of the Federal Republic of Germany). These 15 branches produce more than 80% of the whole CO_2 emissions of the economic branches. The graph illustrates particularly the importance of the sector 'electricity, vapour and steam', which is responsible for more than half of the total emissions. The graph also shows the relatively large amount of indirect emissions (compared with the direct emissions) produced by most branches, particularly 'manufacture of road vehicles' and 'mechanical construction'. This large quantity of indirect emissions is, to a considerable extent, the result of electricity utilization. It is also interesting that with the exception of four branches (transport, market services, manufacture of road vehicles and trade) the cumulative emissions have decreased in this decade.

Outlook

The measurement of the emitter structure for air emissions has already come to a steady state of empirical routine and maturity: complementary modules will also be implemented for water pollution and waste within the near future. More preparation will be necessary for the incorporation of other environmental themes (and the corresponding pressures), including biodiversity, pollution with toxic substances, etc. The physical input–output tables will be finalized in 1996 within a Eurostat research project. This research project aims to discuss advantages and disadvantages of PIOT for describing material flows and investigate possibilities for analysis in physical units using PIOT.

Finally, it will be the most challenging task to integrate the different conceptual and empirical modules in a comprehensive information system. Limited human resources and financial capacities available for environmental accounting in the statistical office call for good coordination and high performance in the empirical realization of conceptual frameworks. This requirement seems to be obvious. Environmental accounting is, however, still a relatively new field of statistics and it affects the working areas/interests of a relatively large variety of data producers/users. The development of conceptual solutions which are appropriate from the perspective of all engaged parties and which at the same time respect the financial constraints is far from simple. It can, however, be expected that an iterative approach and the efficient use of information technology can help to soften the dilemma between (too) high expectations and (too) low budgets.

Notes

1. See e.g. German Bundestag commission of enquiry into the protection of mankind and the environment (ed.): Die Industriegesellschaft gestalten – Perspektiven für einen nachhaltigen Umgang mit Stoff- und Materialströmen. Bonn 1994. – German Bundestag commission of enquiry into the protection of the earth's atmosphere (ed.): Schutz der grünen Erde – Klimaschutz durch umweltgerechte Landwirtschaft und Erhalt der Wälder. Bonn, 1994. – Looß, A./ Katz, C. (Office for the Assessment of the Consequences of Technology at the German Bundestag, TAB): Abfallvermeidung: Strategien, Instrumente und Bewertungskriterien. TAB-Arbeitsbericht Nr. 16. Bonn, 1993. – Publication of the findings of the 'Materials Management for Regional Sustainability' workshop held by P H Brunner and P Baccini in Vienna, April 1994 (in preparation). – Proposals in the United Nations 'System for Integrated Environmental and Economic Accounting' (SEEA), Chapter III.
2. See e.g. the work of Faucheux *et al.*, Centre d'Economie et d'Ethique pour l'Environnement et le Developpement, Université de Versailles Saint-Quentin-en-Yvelines, France.
3. An example at national economy level is Adriaanse's comparison with the Netherlands' National Environmental Policy Plan; a similar thing occurs in local units with eco-audits.
4. The SEEA Handbook gives an overview of both types of approach.
5. For more details see Thomas, J: Luftemissionsentwicklung der Produktionsbereiche. In: Wirtschaft und Statistik No. 1, 1996.

13
Material flow accounts indicating environmental pressure from economic sectors

STEFAN BRINGEZU, RALF BEHRENSMEIER AND HELMUT SCHÜTZ

The environmental performance of human activities is largely determined by the quantity and quality of the associated material flows. The extraction of raw materials, on the one hand, and the emissions of waste materials on the other, exert pressures on the environment. The material input to the economy (from nature) and the material output to the environment (from the economy) can be accounted for in a balanced manner. Based on recent data on the overall material flow account of Germany, the policy relevant information that can be derived from such kinds of physical satellite accounts is described. In order to avoid a shifting of environmental problems to other regions, the linkage of material flows with the production of imports and exports (their 'ecological rucksacks') have to be considered.

Based on data on the domestic and transnational material input of the German economy an input–output methodology can be used to allocate the different material flows to intermediate and final demand. In this manner, key sectors for dematerialization can be addressed. For instance, the direct and indirect material inputs of the industrial sectors that deliver goods to final demand can be quantified. The dependence of the industrial sectors on material-intensive supply can also be determined and material inputs can be related to the main categories of private and public final consumption (e.g. housing, nutrition, health, education). Preliminary results are presented, based on monetary tables for Western Germany and on physical data of domestic and transnational material flows.

Introduction

The environmental performance of human activities is largely determined by the quantity and quality of the associated material flows. Huge amounts of primary materials are taken from nature year by year, only a part of which is used for economic purposes. Large quantities, such as the non-saleable production of mining spoils, are released to the environment as waste. In addition, many of those materials that have an economic use are short-lived within the

K. Uno and P. Bartelmus (eds.), Environmental Accounting in Theory and Practice, 213–227.

technosphere. Their turnover between raw material extraction and final waste management represents the current status of industrialized countries: the 'throughput economy' (Daly, 1992; Ayres and Simonis, 1994).

In order to proceed towards a sustainable development, the industrialized world must progressively reduce its material and energy throughput (Schmidt-Bleek, 1994; see also Meadows *et al.*, 1992; Weterings and Opschoor, 1992). The Factor-10-Club (1994) states that there is no fixed relationship or magic between the total value of our economic activity and material throughput. We should work towards halving our present global non-renewable material flows, including minerals, freshwater and non-renewable energy carriers. To achieve this, a political commitment to a ten-fold increase in the average resource productivity of the presently industrialized countries would be a prerequisite for meeting the goal of long-term global sustainability. As this strategy is based on present conditions, increases in world population and further economic expansion in the industrialized world would obviously require a factor higher than 10.

Irrespective of the specific degree of dematerialization that will have to be agreed upon by political institutions, there is a need for accounting instruments to monitor the actual throughput of materials and energy on a national level and to relate the major flows to the different actors on the market. This applies especially to the sectors of industry, in order to indicate priorities for measures towards dematerialization.

Statistical reporting increasingly accounts for material flows on a national scale. The Wuppertal Institute has contributed to the development of an overall material flow account of Germany, recently published by the Federal Statistical Office as part of an integrated environmental and economic accounting project (Bringezu, 1993; Schütz and Bringezu, 1993; Federal Statistics Office, 1995).[1] One major objective of material flow accounting is the indication of progress or regress with respect to environmental sustainability (Bringezu *et al.*, 1995). Interlinking economic activities with their environmental pressures, material flow accounts are expected to contribute significantly to international programmes of integrated environmental and economic accounting (United Nations, 1993a).

The latest approach to physical accounting by the German Federal Statistical Office is described by Radermacher and Stahmer in Chapter 12. Complementary to their presentation, this contribution focuses on the further analysis and interpretation of material (including energy) flow accounts. The presented results have also been part of a study on 'Sustainable Germany' which was recently issued in German (BuND, 1996).

Overall material flow account

Focusing on those material flows that are linked to economic activities, a domestic overall material flow account of Germany has been established. It comprises the physical mass balance of the domestic extraction from the environment, domestic deposition and release of residuals to the environment imports and exports. The aims are to provide an overview of the physical basis of the economy, combine information from different statistics (e.g. production statistics and environmental statistics) into a coherent framework, establish a structured information base that can be used to derive indicators for progress towards sustainability and develop a physical satellite account that can be used for integrated economic and environment reporting.

Table 1 shows that first, the throughput of water dominates the account. This category should be treated separately, because the sum of all inputs and outputs would only be meaningful in terms of water use. A distinction is made between used and unused water input: the latter comprises drainable water, e.g. groundwater from mining, or rainwater drainage, and amounts to nearly one-third of the total input. Second, the domestic input of abiotic (non-renewable) raw materials exceeds the input of biotic (renewable) inputs by a factor of about 50 (based on dry weight of the plant biomass from cultivation). Third, most of the abiotic raw material input remains unused. This is mainly due to the non-saleable products of coal mining which are dumped as they have no economic value. Landfill and mine dumping (on the output side) exceed the mass of all other wastes that are disposed of at controlled sites by over 10 times.[2] The relation of the non-used input to the input of used raw material indicates the 'eco-efficiency' of the corresponding extraction process (Bringezu and Schütz, 1995). This relationship can be monitored over time. For example, from 1960 to 1990 the eco-efficiency of the domestic extraction of lignite decreased. Finally, the input of biotic raw materials from cultivation is associated with an amount of erosion that exceeds even the dry weight of the raw materials. The relationship between the biotic input from cultivation and the associated erosion may also be monitored over time. Bringezu and Schütz (1995) showed that this relation decreased from 1980 to 1989 in Western Germany. Obviously, renewable inputs cannot be regarded as 'free' with respect to environmental pressure.

On the output side, it is interesting to see that the CO_2 emissions into air amount to 1 billion tonnes. This is more than one-third of all waste disposal (excluding incineration) and corresponds to about 13 tonnes per capita. Pesticides have been included because of their special importance with respect to their biocidal and metabolic disruption potential. Pesticides, nitrogen and phosphorus emissions to surface waters are not weighted by quality aspects. The amounts of release or dissipative use are accounted in order to lay the basis for a comparison of orders of magnitude over time.

The balance of inputs and outputs (without water) is 0.8 billion tonnes. This amount results mainly from the material that is added to the stock of infrastructures, buildings etc., but also from material losses not yet considered (e.g. other releases to surface waters); it also includes statistical errors. The order of magnitude is in accordance with the approximation of Schütz and Bringezu

Table 1. Domestic overall material flow account for Germany, 1991.

Input (million t)		Output (million t)	
Abiotic raw materials	3993	Waste disposal (excluding incineration)	2891
used: Materials*	829	Controlled waste deposition(*)	222
Ores*	0.4	Landfill and mine dumping*	2669
Energy carriers*	366		
unused: Non-saleable production*	2532	Soil	166
Excavation*	266	Erosion	129
		Dissipative use of products*	
Biotic raw materials	82	Fertilizers*	35
Plant biomass from cultivation (dry weight)*	82	Mineral fertilizers*	8
Fishing/hunting*	0.4	Organic fertilizers (dry weight)*	27
		Sewage sludge (dry weight)*	0.8
Soil		Compost (fresh weight)*	0.8
Erosion (anthropogenic)	129	Pesticides*	0.03
Air	1070	Emissions into air	1599
O_2 for combustion*	1014	CO_2^*	1032
Production of O_2 and N_2	38	NO_x^*, SO_2^*, CO*	20
O_2 for steel production	18	Others*	17
		Water	530
		Emissions into water	34
		Dredge excavation into North Sea	34
		N and P from sewage released to surface waters	0.4
Total	5274	Total	4690
Imports*	433	Exports*	211
Total	5707	Total	4901
		Balance: material added to technosphere	806
Waste for utilization (stored)	64	Waste for utilization (stored)	64
Water	69,290	Waste water	69,290
used*	46,874	treated*	13,857
unused*	22,416	untreated*	32,573
Water imports	0.2	Water losses and evaporation	5395
		Water diversion	17,459
		Water exports	8

* Categories documented by the Federal Statistical Office (German) (1995). Whereas official statistics record waste generation, in this table the actual deposition is documented. Official statistics report fresh weight of plant biomass. These differences are marked by an asterisk.

(1993) of the additional stock of Western Germany in 1989 of 0.7 billion tonnes which was derived from the production of durable materials.

Input and output (without water) are of the same order of magnitude. The statistical difference (additional stock) represents only about 20% of the inputs (from the environment and by imports). Input and output are mainly determined by throughput flows, which are released to the environment after a short-term use. This applies for energy carriers, non-saleable production, excavation, most of the biotic raw materials, erosion, air, and water. Storage flows, which are used for long-term products and will be released on the output side with a certain time lag, represent a minor quantity. Building minerals, a certain amount of the ores, and a part of the biotic input (e.g. wood) are examples for such long-term products.

The major general information that can be derived from the overall material flow account is the interlinkage of material inputs and material outputs of the economy. Every material extracted from the environment will sooner or later burden the environment also on the output side. Any pressure related to the outputs (releases to the environment, wastes etc.) can only be diminished successfully if the input of primary materials to the economy is reduced.

The national overall material flow account may be established at different degrees of aggregation. A detailed picture is necessary to set priorities for improvement measures. A medium aggregation (like that shown in Table 1) may be used to provide an overview, monitor the major material flows with respect to their interlinkages, and to derive the sub-indicators mentioned above. An aggregation to less than five categories (abiotic raw materials, biotic raw materials, soil, air, water) does not seem to provide meaningful information in general. However, if the detailed picture is provided, it may also be meaningful to provide higher aggregates, e.g. for the purpose of international comparison. For instance, in Germany the domestic extraction of materials from the environment amounts to 53 tonnes per capita (without water and air).

The value of information will rise tremendously if the overall account is reported regularly for trend comparisons. Even more important than the absolute amount (and the relation to other categories) is information on the change over time. Such information is necessary for the planning and control of effective measures to improve the environmental performance of the economy. For instance, it is important to know the trend of the inputs of non-renewables and renewables, the pressures on cultivated soil (e.g. by erosion), the consumption of air and water, as well as the trend of waste disposal, dissipation of products, mineral extraction and emissions to air and water.

Transnational material input

Figure 1 shows the domestic overall balance as a flow diagram (right side of the broken line). In addition, those material flows are shown that are interlinked with the production of the imports on a cradle-to-border basis, i.e. including their ecological rucksack (left side).[3] These data have been accounted by conservative calculations based on available statistics. They represent minimum values mainly including the non-saleable production and extraction wastes of mining and the soil erosion by agriculture that burden the environment in the countries of origin (Bringezu and Schütz, 1995).

The transnational extension of the domestic flow account is a necessary prerequisite for the evaluation of progress towards sustainability. If the possible impacts by material flows, induced by the German economy, are to be evaluated in a global context (and this seems indispensable with respect to

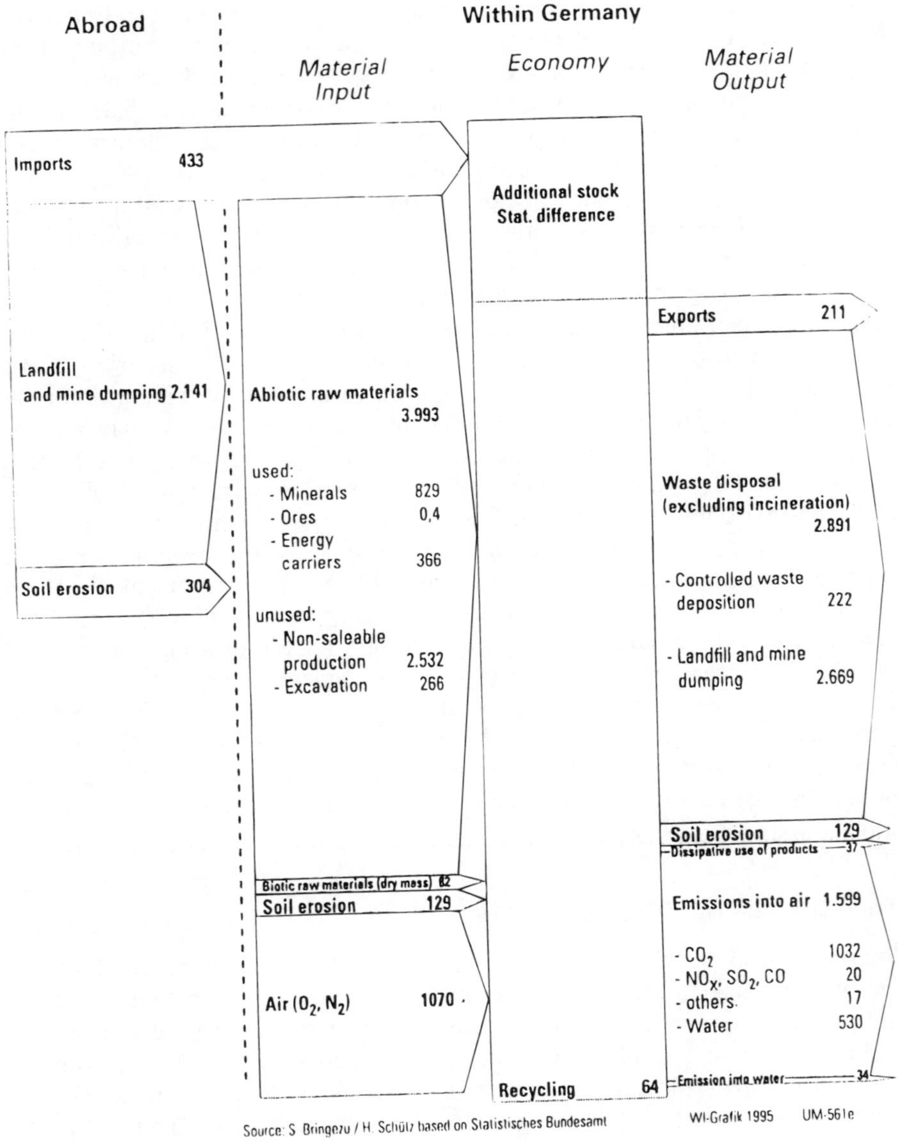

Figure 1. Extended overall material flow account for the German economy, 1991 (million tonnes).

sustainability), then the global material flows interlinked with German production and consumption must also be accounted for. One result of the preliminary account is that the transnational material flow has nearly the same order of magnitude as the domestic extraction from the environment (without water and air). Thus, the transnational material flows, interlinked with the German economy and burdening the global environment, cannot be neglected when monitoring the global impact of the national economy. This may pose problems to national statistics which are usually confined to the national territory. However, in a world with global problems, national statistics need to monitor the global pressure on the environment in terms of the global material consumption of the domestic economy (as in the case of energy consumption). Total material input as shown here may be included in a stepwise extension of integrated economic and environmental accounting. The additional effort for a sufficient approximation of these material flows is estimated to amount to one person-year per country.

Total material input (TMI) comprises the national and transnational (i.e. global) material extraction from the environment. It may be regarded as a highly aggregated indicator that relates to the global environmental pressure associated with the physical basis of an economy (Bringezu *et al.*, 1995) and may be related to other indicators, providing for example an indicator of the material productivity of GDP (Bringezu *et al.*, 1994). TMI does not reflect the fact that the economy also exports products which are again connected with certain material flows. Thus, one may be interested in the global total material consumption (TMC) by considering also those cradle-to-border flows that are associated with exports (Bringezu, 1993). Figure 2 shows the calculation of the material input of the exports based on the assumption that there is no basic difference in the ecological burden of products consumed within Germany and exported products (latest results of Behrensmeier and Bringezu indicate that the material productivity of the production of exports is 1.2 times less than the material productivity for the domestic consumption based on relations to GDP or labour).

Material input of industrial sectors

In order to support improvement measures, the overall picture of material flows has to be related to the various actors. The most important actors are those on the market. This section describes possibilities of relating the direct and indirect material inputs of the German economy to the different sectors of industry.

In order to determine priorities for dematerialization measures, the material intensity of the German industry was analysed on the basis of preliminary data (Behrensmeier and Bringezu, 1995a). Four questions were answered:

(a) Which branches extract material from the domestic environment directly?

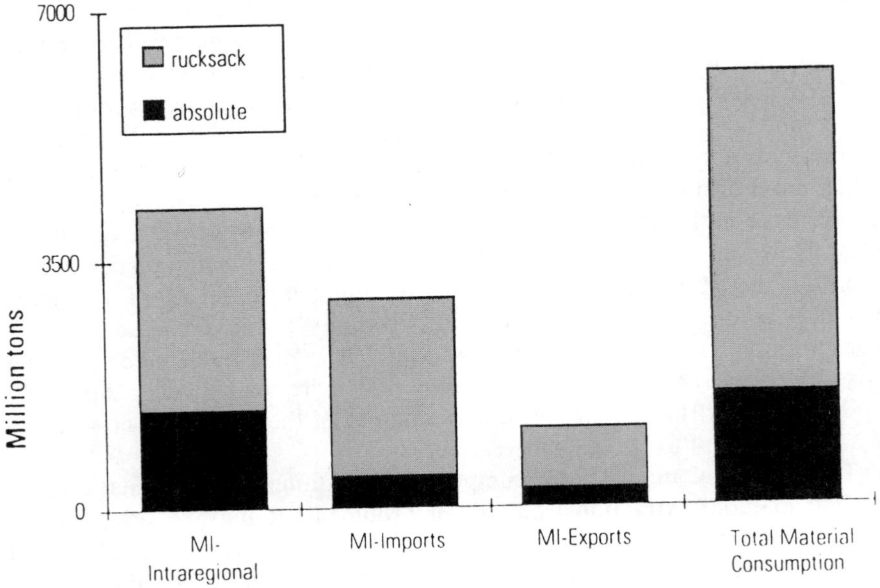

Figure 2. Domestic material input (MI) and imported and exported MI (Germany, 1991). Absolute amounts of products plus estimation of interlinked upstream flows.

Total material consumption results from MI-intraregional plus MI import minus MI export. Preliminary data representing all materials other than water and air.

(b) Which branches import material-intensive products (including the ecological rucksacks)?
(c) What is the TMI associated with the goods that are delivered to the final demand?
(d) Which branches are most dependent on a material-intensive supply (based on current technology)?

Each of these questions leads to a different answer, providing a multi-faceted picture of the material intensity of industries. The following refers only to material inputs without water and air. In order to answer questions (c) and (d), 58 branches were analysed by input–output analyses. The results are documented on the basis of 23 aggregated sectors of the West German industries in 1900.

Domestic extraction of primary materials

Only five sectors are involved in direct extraction or harvest of domestic primary materials (Figure 3, right side). The mining of fossil energy carriers is associated with the highest volume of material movement, most of which is represented by the non-saleable run of lignite mining. The input of the mineral

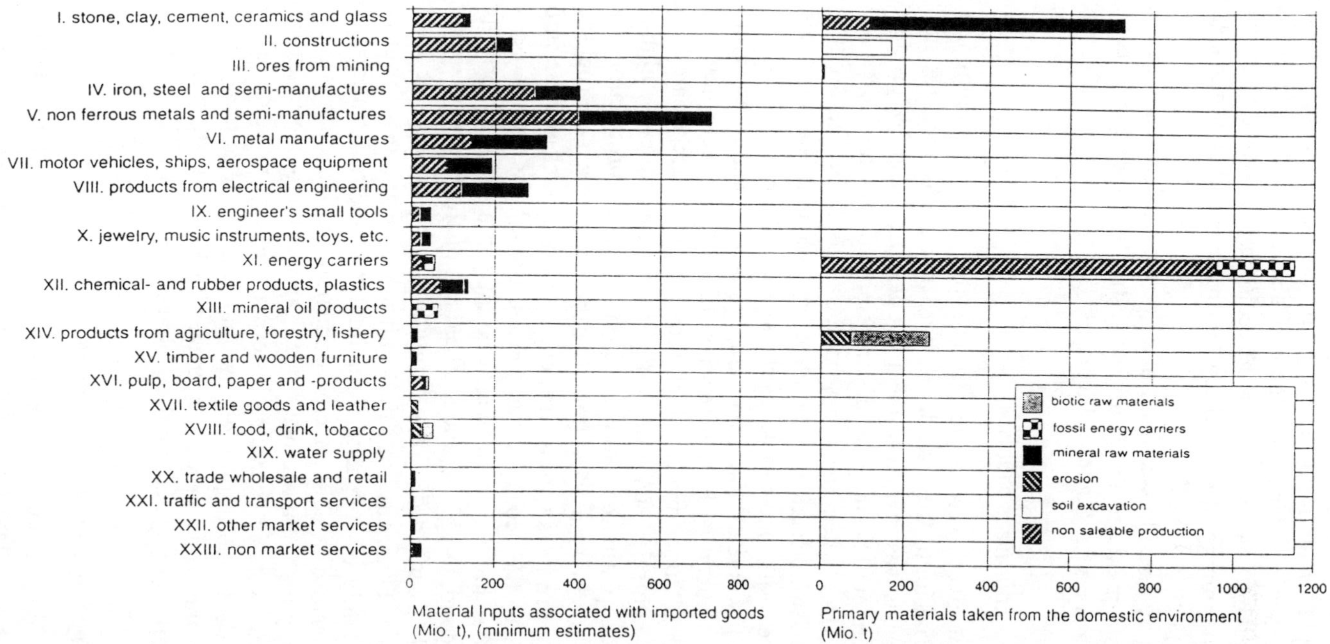

Figure 3. Material input and input of primary materials from the domestic environment, West Germany, 1990. Preliminary data: source Behrensmeier and Bringezu (1995a).

mining sector is dominated by sand and gravel and other minerals for building purposes. The input of biotic materials is far less than the input of abiotic (non-renewable) resources. However, there is also a considerable amount of erosion which is taken as a stand-in for the ecological rucksack of agricultural biomass input. The mining of ores within Germany is rather negligible. Most of the ores are imported (see below).

These sectors may be addressed as primary targets of dematerialization. The measures and instruments to foster an eco-restructuring of these sectors are discussed elsewhere (see e.g. Schmidt-Bleek, 1994).

Import of material-intensive products

The imports of products were related to data that allow a minimum estimation of the material input from cradle to border. For instance, the import of abiotic raw materials was related to the amount of non-saleable production in the country of origin. The imports of semi-finished and final products were related to the TMI of their main base material used for the production in Germany (e.g. cars were related to the life-cycle-wide material input per tonne of steel). Thus, the results have to be taken as preliminary.

The non-ferrous metal industry imports raw materials and semi-manufactured products with the relative highest indirect material input (Figure 3, left side). This is due to a high input of raw materials and its associated rucksack of non-saleable production. The latter is extremely high in some cases, e.g. copper, platinum and gold. Iron and steel production and processing also imports raw materials (mainly iron ore) with a large rucksack. The sectors importing products with a large ecological rucksack play an important role with respect to the needs for dematerialization. First, from an ecological perspective it does not matter in which country the same material is taken from nature. Second, as shown above, the TMI of the German economy via imports is of the same order of magnitude as the domestic extraction of primary materials. Thus, any statistical indication of progress towards dematerialization will only be sufficient, if the import of indirect material inputs is considered. Otherwise, a shift of environmental burden may remain undiscovered. This may be of importance if, for example, the domestic extraction of primary materials were impeded by lowering existing subsidies. If the result were a shift towards the import of raw materials or semi-products that are associated with the same or even a bigger rucksack, then such a measure would have a detrimental effect on the global environment.

Total material input of goods for final demand

In order to reflect inter-industry deliveries, an input–output calculation was performed, using output coefficients based on monetary tables and data on the physical inputs via domestic production and imports. In this first approach, the monetary relations of deliveries are assumed to reflect physical relations. The method itself is designed to work on physical input–output tables which have just been prepared by the German Federal Statistical Office together with the Wuppertal Institute (see Radermacher and Stahmer, Ch. 12). Thus, the data presented here are preliminary.

The estimation of the TMI of the German economy was related to the goods that are delivered to final demand (Figure 4). Final demand, as defined by input–output statistics, comprises private consumption, public consumption, investment, exports and changes in inventories. The result of the calculation is dependent on whether the final demand is supplied directly. For instance, ore mining does not deliver goods to the final demand; therefore, its material inputs are related to other intermediate sectors that are supplied by ore mining.

The construction industry delivers most material-intensive goods to final demand. This is due to material-intensive modes of construction and the use of base materials such as concrete with a big rucksack. The manufacturing of metals, and the construction of vehicles, vessels and aeroplanes are also associated with a high material input of global origin. The energy sector is important, although it serves primarily intermediate demand.

Looking for strategies towards dematerialization at the interface between production and consumption, these sectors may be expected to be of special importance. On the one hand, these industries are challenged to meet the final demand in a dematerialized manner. On the other hand, the consumers of those goods have a special opportunity to contribute to sustainability by demanding dematerialized products and services.

Dependence on material-intensive supply

Intermediate sectors, which depend on the supply of semi-products and do not directly deliver goods to final demand, may be affected by a dematerialization of the economy. In order to identify the relative dependence of material-intensive supplies, the above mentioned input–output calculation was extended. The iterative procedure by which the domestic and the imported material input was projected to the final demand was used to integrate the calculated input of each sector on each step. High values are the combined result of a high direct or indirect material input and a high consumption within the sector ('*Eigenverbrauch*').

Based on current technology, the relative dependence on a material-intensive supply is greatest in the energy supply sector, the iron and steel industry, and the construction sector. The energy sector has already been mentioned to be mainly responsible for high direct extraction of domestic primary materials. The construction sector was found to be important, delivering the most material-intensive goods to the final demand. The third grid of analyses indicates the iron and steel industries to be important actors, too. These

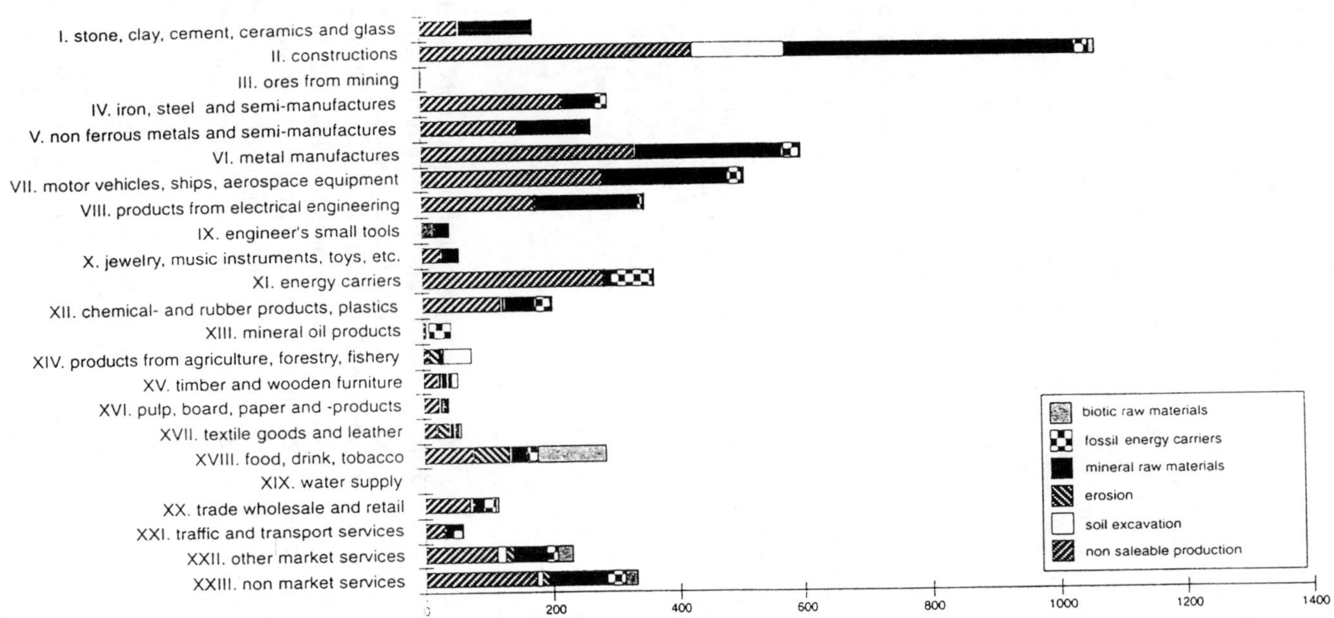

Figure 4. Direct and indirect material input of goods delivered to final demand by industries. The allocation reflects the inter-industry deliveries by an iterative input–output calculation. Preliminary data: source, Behrensmeier and Bringezu (1995a).

sectors are especially challenged to decrease the material intensity of their supply.

The possibilities for eco-restructuring of those industries are currently being studied at the Wuppertal Institute (see e.g. Liedtke *et al.*, 1994; Hinterberger *et al.*, 1994). Guidelines and examples to improve eco-efficiency in business are given by the Business Council for Sustainable Development (1995) and Weizsäcker *et al.* 1995). Thus, corresponding to the flow of materials, and with respect to the options of the actors to influence them, key sectors of the industries can be indicated where materials are extracted from the domestic environment, products are imported, products are delivered to final demand, and industries using current technology are dependent on a material-intensive supply.

Material input of private and public demand

In order to support decision-making towards sustainability on the consumer side, overall information on the environmental burden associated with the main activities of private and public demand can be provided. The above-mentioned input–output methodology was further developed by projecting the TMI of the West German economy, used for domestic consumption, to the final private and public demand. Because investments and the material flows necessary to realize them may be regarded as a technical necessity to deliver goods to the private or public sector, investments were treated like intermediate demand using monetary investment tables for Western Germany 1990. The goods delivered to the private and public demand were related to main fields of demand (housing, nutrition, clothing etc.), using the statistical classification of private use, as well as own classifications (Behrensmeier and Bringezu, 1995b).

Housing is the most material-intensive demand (Figure 5). The building of new houses and the repair and maintenance of old ones (including heating) were all considered. The huge material input is mainly due to construction materials (concrete, bricks, steel etc.) and the energy carriers consumed together with flows of non-saleable production.

Nutrition is second in material intensity. This is mainly a consequence of the material- and energy-intensive processing of feed and food. The material and energy consumption of agriculture, the food industry and households, used for processing and transport, is considerable. About one-third of the material input for nutrition is associated with the consumption of meat and milk and their products which are characterized by high specific material inputs per unit of product. Some plant products like sugar also exhibit high values of specific material inputs.

Leisure time is increasingly spent on material intensive activities. Most important is the high mobility for travelling, especially during vacations. Sporting and audio-video equipment is not only a high-expenditure item, but

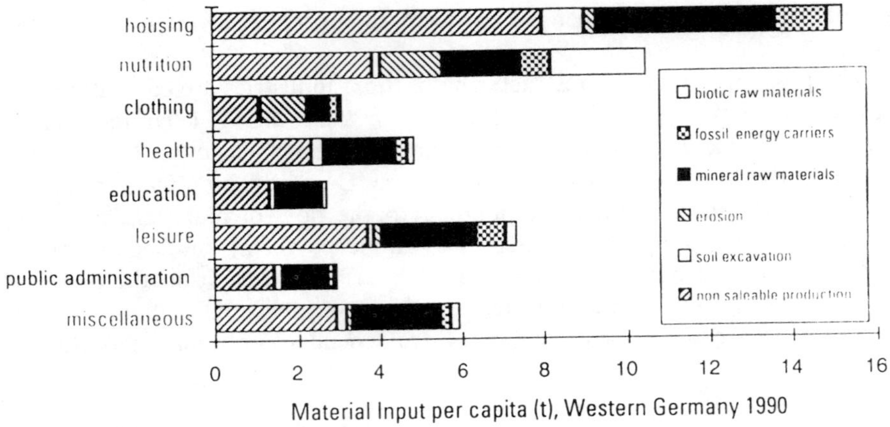

Figure 5. Material input for main demands (without water and air).
Source: Behrensmeier and Bringezu, 1995b.

is also associated with high volumes of material flows. The same applies for cars.

The results should not be interpreted as a call for less housing, nutrition, education etc. Rather, consumers are given the information, enabling them to look for other possibilities to meet their needs. Indeed, there are ways to satisfy different demands by less material and energy input. However, currently there is still a lack of information on the life-cycle-wide material and energy input of products and services. It is a challenging task for statistical services to provide consumers and producers with this kind of information (e.g. by an information exchange on Low-MIPS service delivery).[4]

Conclusions

Environmental pressure associated with economic activities can be indicated by a comprehensive overall material flow account of a national economy; such an account is already part of the integrated environmental and economic accounting programme of the official statistics in Germany. This account supports decision-making towards eco-restructuring and can be used to monitor regularly the environmental performance of the economy. However, the ecological rucksack of the international trade flows, interlinked with national production and consumption, should not be neglected when indicating the environmental pressure of national activities with respect to global sustainability. Accounting for the physical basis of the economy is possible by considering the domestic extraction of primary materials, as well as a conservative minimum calculation of transnational extraction, leading to an approximation of the TMI of the economy.

In order to reflect the interlinkages of environmental pressure and production activities, the different material flows (domestic and foreign) can be related to the various sectors of industry in order to indicate priorities for the implementation of measures to improve sustainability (in terms of dematerialization). Using input–output analysis, different questions may be answered to provide a multidimensional picture. In order to provide analogous information for consumers (and their associations), the different material inputs of the economy can be related to private and public demand. Thus, priority areas can be addressed where consumers can foster sustainable development by satisfying their needs in a dematerialized manner.

Future work is intended to progressively use and analyse physical input–output tables, thus substituting economic valuation of physical data in order to better reflect interactions with the environment.

Notes

1. A first overview of the physical inputs and outputs of an economy was presented by Steuer (1992) for Austria on the basis of available data.
2. There is not sufficient knowledge to assess the environmental harm potential as generally less than that of all other wastes. For instance, the leaching of SO_2 due to the oxidization of sulphides extracted from the underground by worldwide mining is estimated to equal the same order of magnitude as the world-wide emission of SO_2 to the air due to the burning of fossil fuels.
3. The 'ecological rucksack' of an imported product is defined as the 'cradle-to-border' material input that is not incorporated in this or other products, thus representing wastes or emissions to water and air.
4. MIPS is the Material Input Per Service Unit indicator developed by the Wuppertal Institute.

14
Land use accounting – pressure indicators for economic activities

WALTER RADERMACHER

Introduction

Environmental accounting in physical terms has been proposed as a means of adequately treating interlinkages between the economy and nature. The concrete meaning of this approach is, however, not very precise and the applications are widespread. In particular it has to be decided which part of a driving forces–pressures–state–response chain, can and should be represented by an accounting framework. Based on that decision the accounting methodology has to be specified. Last, but not least, the necessary basic data have to be compiled.

A Task Force of the Conference of European Statisticians tested the feasibility of physical environmental accounting in relation to a small number of environmental phenomena, one of which was the change of land use and land cover and the corresponding impacts on environmental quality. The Federal Statistical Office of Germany participated in the land use accounting group. Within the German Environmental–Economic Accounting programme, methods of remote sensing, GIS and geo-statistics play a fundamental role. This has to be seen in close relation to the work in the Task Force: the development of an accounting framework for land use changes has to be directly linked to the improvement of basic data. Starting from the CORINE Land Cover database and a cluster analysis of abiotic parameters for the classification of land types, an ecological area sample is drawn that detects the distribution of biotope types and allows a description of quality aspects of landscapes and biotopes. The results of the sample survey are aggregated to indicators that represent the structural changes in the state of environment. These data are put into an accounting framework that deals with the interlinkages of economic activities, environmental pressures and changes in the ecosystems. The highest priority in the near future will be given to the accounting of driving forces and the related pressures.

What is the problem of land use?

Historical trend in western Germany

A regression analysis of the growth of the economy and of settlement and

K. Uno and P. Bartelmus (eds.), Environmental Accounting in Theory and Practice, 229–243.

traffic areas within a period of 33 years (1960–1993) shows a stable linear correlation of those two developments (Figure 1). This does not mean that the relative growth rates were identical (average growth rate of GDP in constant prices: 3.0% per annum; average growth rate of settlement areas: 1.6% per annum), but it indicates that in years of high economic growth settlement growth rates were also high, and vice versa. This is not surprising, since it corresponds to what we are used to hearing about the importance of construction industries and mobility (traffic) for the economy. The question is whether this development is a threat to the environment. To obtain a first idea about the time horizon in which such a kind of trend could produce substantially higher scarcities of the available land area, one can extrapolate the regression function:

$$S_t = S_{1993} + b \times GDP_{1993}(i^{t-1993} - 1)$$

where S_t = settlement area in year t, b = regression coefficient (0.823 ha/million DM between 1960 and 1993), GDP_t = gross domestic product of year t in constant prices and i = annual economic growth rate (1.030 between 1960 and 1993). By setting S_t equal to the total area the residual time can be calculated which hypothetically remains until the entire country is covered by settlement and traffic areas.

Table 1 shows the result of this extrapolation. The scenarios are evidently dominated by the dynamics of the exponentiality of economic growth rates. A 'business as usual' scenario (i = 1.03; b = 0.82) shows that the limit of available space in the former territory of the Federal Republic of Germany would be reached in 81 years. To tackle this scarcity problem two strategies can be

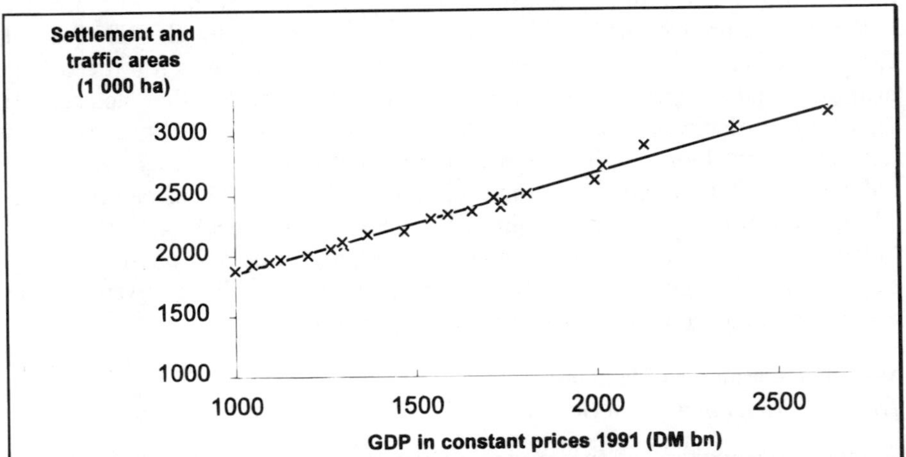

Figure 1. Regression analysis of economic and land use development 1960–1993, former territory of the Federal Republic of Germany.

Table 1. Scenarios for settlement growth potentials in the former territory of the Federal Republic of Germany

Assumed economic growth rate p.a.	Residual time for growth of settlement / traffic areas (years) Assumed regression coefficient (ha / DM mn of economic growth)			
	0.82	0.40	0.20	0.08
3%	81	104	127	157
2%	121	155	189	234
1%	241	308	376	466

followed: the 'more efficiency' strategy, which would loosen the link between economic and settlement growth rates (b) and the 'sufficiency' strategy, which would reduce economic growth rates (i). The calculations show that a strategy which increases only the efficiency but keeps the level of economic dynamics cannot improve the situation considerably. Even a reduction of area 'consumption' by factor 10 would only double the remaining time period. One can argue, however, that this is the normal situation for any exhaustible resource: the market price will dramatically increase and regulate the efficient allocation of the scarce areas. Consequently, the fictive total loss of open space (and the corresponding ecological problems) will never occur. Following this argument, however, leads to the drawbacks for further economic development and the social conflicts concerning the distribution of space. With reference to Ehrlich's, driving forces environmental pressures can be subdivided into population dynamics, activities per capita (production, consumption) and environmental pressures per activity unit. The first pressure factor can be excluded in Germany because the level of domestic inhabitants was stable over the relevant time interval. The economic activities, however, grew considerably (see examples in Table 2).

These relatively clear trends illustrate the driving forces that stand behind the dynamics of settlement growth: a fundamental change in social structures (small households, higher requirements on dwelling conditions) and a rapid increase of mobility. If these factors are directly linked to economic growth (and particularly if they are preconditions for further growth) a spatial growth limit imposes a significant burden on future development in industrialized countries: distributional problems of the available dwellings can no longer be solved by construction of additional houses, which, in addition to the social

Table 2. Selected trends for settlement and traffic, former territory of the Federal Republic of Germany.

	Unit	1960	1990	
				1960 = 100
Settlement / traffic areas	m2 / inhabitant	335	481	*144*
Dwelling area	m2 / inhabitant	23	36	*157*
Inland goods transport	tkm / inhabitant	2 562	4 744	*185*
Passenger traffic	km / inhabitant	4 576	11 445	*250*

problems, implies automatically a deficiency in driving factors for economic growth. Increasing mobility, at least in the space-consuming form of traffic, will produce its own limits. Consequently, there is a need for change of the way and the amount of further economic development.

Statistical reflection on the situation

An integration of these aspects into the frameworks of statistics and economic accounting is necessary. The objective of statistical work in this field is threefold:

(a) changes of land use and cover have to be detected in a suitable way concerning data quality (spatial distribution, periodicity, degree of detail, reliability etc.);
(b) interlinkages with economic activities have to be quantified (sectoral breakdown of trends);
(c) indicators of the environmental burden have to be constructed, which represent quantitative and qualitative changes of the state of environment.

These three aspects are elements of a general scheme of physical environmental accounting. An application of this scheme to the special case of land use accounting will be demonstrated below. It is a part of the final report which was elaborated by a task force of the UN Economic Commission for Europe (ECE) and submitted to the Conference of European Statisticians at the plenary conference in June 1995 (Conference of European Statisticians, 1995).

The characteristic situation for land use is, however, that the data requirements of the accounting framework cannot be fulfilled with the available basic statistics. In particular data which represent widespread geographical differences and their changes over time do not exist, at least for the time being. Consequently, the design of a land use accounting framework must go hand in hand with a statistical programme which tackles the problem of data gaps. With regard to the very limited and decreasing financial capacities of statistical offices this seems to be the bottleneck for further improvements, because the collection of high-quality land use data is extremely expensive. Two questions arise immediately in this context: is it possible to use modern techniques (remote sensing, GIS)?, and can statistics cooperate with external institutions to lower the additional costs?

One specific subject field of the German Environmental–Economic Accounting programme is exclusively oriented towards the solution of these problems (Radermacher and Stahmer, 1994). Recent activities are the construction of the geo-information system known as STABIS (Statistical Information System on Land Use) and the national realization of CORINE Land Cover (CLC). The European project CORINE Land Cover is a part of the information programme and network of the European Environmental Agency in Copenhagen. CLC is defined on the European scale of spatial and substantial resolution of data. Consequently, it does not provide sufficient information on the national or regional scale. It can, nevertheless, be used as an area census and as a basis for the design and implementation of complementary sample surveys which are able to fill the data gaps in a cost-efficient manner. The conceptual elaboration and test of an ecological sample frame for the collection of indicators of the quality of the environment is on the way (Schäfer, 1995a).

In summary, although industrial societies in western Europe are faced with a serious problem of spatial scarcities, the statistical frameworks do not reflect the problem and cannot provide sufficient information for decision processes. A first concept for a physical accounting framework has been developed and tested in a pilot study. The next step will be the realization of the required basic statistics. This, again, will create a need for a revision of the accounting concept. The methodological questions of the accounting framework are considered below. The corresponding elaboration of a statistical land reporting system is discussed by Radermacher (in press).

The UNECE project

The concept and preliminary remarks concerning the quality of data

The general methodology of the UNECE project consists of core accounts and supplementary accounts. Each account is composed of three levels: the stocks, the total flow or change in stocks during one period, and the disaggregation of flows. Core accounts concentrate on the changes in land cover/land use, supplementary accounts are issue-oriented. Because the discussion of issues (representative examples) showed a wide range of potential applications, the accounting framework has to assure a minimum of flexibility. The German contribution to the UNECE group focused on the link between the development of driving forces, the corresponding pressures and the calculation of indicators for groups of environmental impacts by selection of the following subset of issues (Figure 2):

- land use structure of economic sectors,
- partitioning of land (by transport networks),
- sealing of soils (by settlement areas),
- impacts on the artificiality/naturalness of landscape.

The pilot study conducted in the administrative district of Main-Taunus-Kreis (close to Frankfurt) provided the opportunity to test the accounting framework and to obtain a realistic impression of the availability of basic statistical data. In general, the data situation is poor, and additional remote sensing interpretations were required in order to fill the core accounts. Within the group of supplementary accounts it was expected that a lot of information

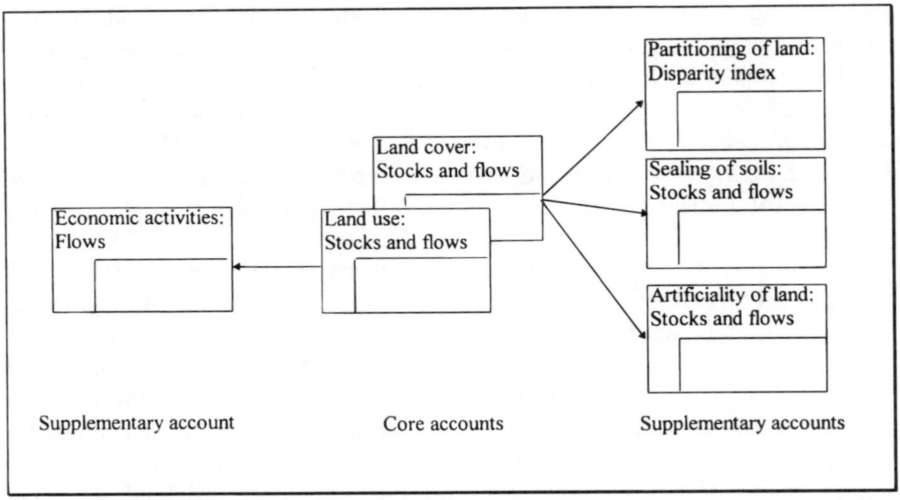

Figure 2. The general model of land cover/land use accounting.

would be available from the local level: this was not the case, mainly due to administrative obstacles. For example, although the artificiality accounts needed information on biotopes which was supposed to be available at the communal administration of that region, it was not possible to obtain the data during the working time of the pilot study. Even if it had worked it would not have been a valid test for a national application, because of the lack of standardization of data at the local/regional level. As already mentioned, an accounting framework for land use depends on the creation of new primary data to a greater extent than other accounting systems. On the other hand the systematic structure of an accounting framework can be used for the development of cost-effective reporting systems.

The following examples are a small sample from the results of the pilot study. They give only a weak impression of the entire information package which is available for the study area. Primary output from GIS calculations are maps. They are basic for statistical evaluations and the filling in of all matrices and accounts which represent an important tool for the presentation of complex information that is complementary to tables/accounts which can sometimes be better understood by decision makers and the public.

Core accounts: land use/land cover

Available data

The accounts for land use and land cover describe the stock and the flow in a selected district (Main-Taunus-Kreis) for the period 1952–1992. Two different

Table 3. Main characteristics of the methods used for data collection

	STABIS 25	STABIS 100 (CORINE Land Cover)
Source of information	Aerial photographs topographic maps	satellite images topographic maps
Method	visual interpretation	Computer-aided visual interpretation
Mapping scale	1:25 000	1:100 000
Size of smallest unit	1 hectare	25 hectares
Number of land use / land cover categories	70	44
Feature type	areas linear features point features	areas (linear features wider than 100 m as areas)

methods of data collection were used. Table 3 shows the characteristics of both methods. STABIS 25 data relating to land use were collected for the years 1952, 1972 and 1992. Using the STABIS 100 method data were collected for 1984 and 1991. Since STABIS 25 contains more categories of land use, has a larger mapping scale and a smaller minimum size per unit, the quality of information of the STABIS 25 data is better than that of STABIS 100 data. The CORINE Land Cover data set is more aggregated then the STABIS data set. The mapping scale is smaller, the minimum size per unit is larger, and the number of different categories of land use/land cover is smaller in the CORINE Land Cover data set.

The linkage of STABIS 25 data with economic data is easier because of the greater number of items in the nomenclature compared with CORINE Land Cover. The presence of items of mixed use in the nomenclature of CORINE Land Cover is also a disadvantage for linking the accounts. The data model of CORINE Land Cover does not include linear features such as traffic networks or small rivers. Therefore it is not possible to create accounts on partitioning using CORINE Land Cover data. For these reasons it would be desirable to use STABIS 25 data for land cover/land use accounting. For the time being, however, only CORINE Land Cover data are available for the total area of Germany due to budgetary restraints. A substantial improvement would be to include linear features into a GIS system and to reduce the size of the minimum mapping unit in CORINE Land Cover to 15 ha to cover smaller features. Using the same source of information (satellite images) in CORINE Land Cover would even permit enlarging the mapping scale to 1 : 50,000.

In the German statistical office the CORINE Land Cover data are used as a sampling framework for collecting geo-referenced data on biotopes. With reference to the work in the UK (Barr *et al.*, 1993) these data can be integrated into an environmental accounting system to represent the state of environment.

The separation of land use and land cover

Both STABIS 25 and STABIS 100 are mixtures of land use and land cover, i.e. artificial areas defined by their use, whereas natural and semi-natural areas are primarily defined by their type of land cover. Since the distinction between use and cover is one of the general questions that should be clarified by the Task Force, a test was made within the pilot area: the STABIS 25 and STABIS 100 data were separated into land use (for linkage with socio-economic data) and land cover (for linkage with data on the state of environment). Classifications for land cover and land use were, therefore, developed in a way that both classifications can be used for the separation of both STABIS 25 and STABIS 100 data (Table 4). In practice, however, such separation did not appear to be useful because this involved a loss of information. It was, therefore, decided to use the original STABIS 25 and STABIS 100 nomenclature to establish a link between land use/land cover and the supplementary accounts. As a consequence, the distinction between land use and land cover is dropped unless the corresponding data were collected independently of each other. For international comparisons the ECE standard classification for land use (which is also a mixed concept) should be applied.

Results for stocks and flows; causes of land use change

Two different flow matrices showing the changes in land cover and land use were derived from the data based on STABIS 25 (1953–1972 and 1972–1992) and one flow matrix for the data collected with STABIS 100 (1984–1991). Using these matrices it is possible to describe the flows of land use/land cover and analyse the type and the dynamics of the changes. Results for land use change derived from STABIS 100 were not sufficient. It was not possible to judge whether the period 1984–1991 is generally too short or whether the test area was too small to detect minor changes in settlement areas.

Based on the flow matrices for land use/land cover an attempt was made to disaggregate the changes according to their potential cause – human impact or natural development. Since there is no specific knowledge about the reason for the change of individual surface units, standardized rules for the determination of the causes have to be established (e.g. if a unit appears to be artificial at the end of the period the change will be classified as human impact). The results are integrated in the flow matrix for land use/land cover (Tables 5 and 6). The majority of changes in land cover/land use in the test area is caused by human impact. Changes in land use/land cover due to accidents were not integrated because this requires additional knowledge about the individual surface units which is not available in most cases. The disaggregation of causes based on STABIS 100 data seemed to be more difficult than of those based on STABIS 25 data because the level of aggregation concerning the nomenclature is greater.

Supplementary accounts

In order to analyse the structure and the extent of human impact on nature in more detail supplementary accounts were combined with the core accounts on

Table 4. Core matrix – land use/land cover stocks (Main-Taunus-Kreis) (hectares).

	Land Cover 1953							
Land Use 1953	1 Artificial cover	2 Woody vegetation	3 Grass and other non-woody vegetation	4 Woody and non-woody vegetation	5 Land without or with sparse vegetation	6 Water surfaces	**Land Use total [ha]**	Land Use [%]
01 Dwelling	874	-	-	-	-	-	**874**	*3.5*
02 Industrial and commercial use	79	-	-	-	-	-	**79**	*0.3*
03 Housekeeping / commercial use	747	-	-	-	-	-	**747**	*3.0*
04 Private / public administration, culture, education, health service	57	-	-	-	-	-	**57**	*0.2*
05 Supply services	5	-	-	-	-	-	**5**	*0.0*
06 Disposal services	-	-	-	-	-	-	**-**	-
07 Mineral extraction	-	-	-	-	84	-	**84**	*0.3*
08 Transportation	312	-	-	0	-	-	**312**	*1.3*
09 Recreation areas	106	-	-	65	-	-	**171**	*0.7*
10 Agriculture	4	3421	11716	-	-	-	**15141**	*61.2*
11 Forestry	-	7010	-	-	-	-	**7010**	*28.3*
12 Mixed use of water bodies	-	-	-	-	-	163	**163**	*0.7*
13 Mixed use of natural and semi-natural areas	-	8	-	51	-	-	**59**	*0.2*
14 Transition land and construction sites	48	-	-	-	-	-	**48**	*0.2*
Land Cover total [ha]	**2233**	**10439**	**11716**	**116**	**84**	**163**	**24750**	
Land Cover [%]	*9.0*	*42.2*	*47.3*	*0.5*	*0.3*	*0.7*		*100.0*

Table 5. Core matrix – land use flows based on STABIS 25 (Main-Taunus-Kreis) (hectares).

	Land use (STABIS) 1972											
Land use (STABIS) 1953	1 Built-up areas	2 Pit heaps, open pits, disposal areas	3 Traffic areas	4 Leisure and recreation areas	5 Agriculture areas	6 Forest	7 Bodies of water	8 Wetland, heathland, rocks, dunes	9 Fallow areas	**Land use ´53 total**	Land use ´53 [%]	Decrease (total)
1 Built-up areas	1746	2	3	4	7	0	0	-	0	**1763**	*7.1*	17
2 Pit heaps, open pits, disposal areas	10	30	1	11	9	7	0	-	16	**84**	*0.3*	54
3 Traffic areas	5	0	298	0	8	1	0	-	0	**312**	*1.3*	14
4 Leisure and recreation areas	56	-	2	103	6	2	0	-	2	**171**	*0.7*	67
5 Agriculture areas	2000	266	129	162	12263	112	9	-	200	**15141**	*61.2*	2878
6 Forest	76	8	8	10	36	6920	0	-	10	**7069**	*28.6*	149
7 Bodies of water	0	0	0	0	1	0	161	-	0	**163**	*0.7*	2
8 Wetland, heathland, rocks, dunes	-	-	-	-	-	-	-	-	-	**-**	-	-
9 Fallow areas	1	-	16	-	32	-	0	-	-	**48**	*0.2*	48
Land use ´72 total	**3894**	**306**	**457**	**290**	**12363**	**7042**	**171**	**-**	**228**	**24750**		
Land use ´72 [%]	*15.7*	*1.2*	*1.8*	*1.2*	*50.0*	*28.5*	*0.7*	-	*0.9*		*100.0*	
Increase (total)	2148	276	159	186	100	122	10	-	228			
Net change	2131	222	144	119	-2778	-27	8	-	180			
Flow coeff. [%]	*38.3*	*84.7*	*22.5*	*55.1*	*10.8*	*1.9*	*3.7*	-	-	*13.0*		

	Area [ha]	**Area [%]**
Total Area	24750	*100.0*
Change due to human impact	24420	*98.7*
Change due to natural process	331	*1.3*

Table 6. Core account – land use stocks 1972/92 – flows (Main-Taunus-Kreis) (hectares).

Land Use (STABIS classification)	Stock initial state 1972 (ha)	Net flow due to human activity (ha)	Net flow due to natural processes (ha)	Stock final state 1992 (ha)
1 Built-up land	3 893.8	1 128.7	-4.7	5 017.8
2 Quarries, pits, mines and land under disposal facilities incl. dumps	305.9	-81.1	-107.3	116.7
3 Land used for traffic purposes	456.7	-80.2	0.0	376.7
4 Recreational land	289.6	204.5	-2.8	491.2
5 Agricultural land	12 363.0	-1 025.0	-80.4	11 257.7
6 Forests and other land under trees and shrubs	7 041.8	31.2	33.2	7 106.3
7 Waters	171.3	32.9	0.0	203.9
8 Wetlands and arid land. Land with sparse vegetation	0.0	0.0	89.0	89.0
9 Fallow land	228.3	-210.2	73.1	91.2
Total Area (abs.)	24 750.5	4 136.1	221.1	24 750.5
Total (%)	*100.0*	*16.7*	*0.9*	*100.0*

	Area (ha)	Area (%)
Total change 1972 - 1992	4 357.2	*100.0*
Change due to human activity	4 136.1	*94.9*
Change due to natural process	221.1	*5.1*

land use/land cover. The supplementary accounts can be divided into two groups, one describing the demand for natural resources (land) by different economic activities and the second describing the environmental impact of this demand.

Economic activities

A matrix that crosses economic activities with land use was created which shows the spatial demands of the different economic sectors (including private households). It was created using a coefficient matrix that estimates the share of each economic sector in land use. The coefficients are still rough and rather subjective estimates to show the feasibility of the linkage between land use and economic activities; but they have to be verified with statistical data at a national level in a next step. This leads to the problem of multipurpose use of land, e.g. a forest is used both for wood production and recreation. Methods have to be developed to calculate the shares of the forestry industry and private households in the land cover of 'forest'. An additional class (not classified) was created to contain land use items that even by subjective estimations could not be assigned to specific economic sectors during the pilot study because of a lack of additional information, e.g. transportation by economic branch.

Because of the problems of linking economic production and consumption activities directly to land use, socio-economic indicators were selected to describe the background of changes in land use/land cover. An attempt was made to find statistical data for selected indicators such as population, human settlements, economic development, employment, housing, transportation, leisure and recreation. Linkage between socio-economic data and data on land use is achieved by spatial aggregation of the data on land use/land cover. However, the availability of statistical data is usually better at higher spatial aggregation levels (national, Länder levels), whereas data on land use/land cover are referenced to a geodetic system and have to be aggregated. Therefore methods have to be developed that, on the one hand, aggregate geo-referenced data in order to combine them with statistical data and, on the other hand, disaggregate the combined data in order to show spatial distributions.

Flow matrices for land use of economic activities for the periods 1953/72 and 1972/92 were calculated (on the assumption of constant land use coefficients per economic branch). The matrices show the changes in land use of economic activities derived from the core accounts on land use. As different economic activities can take place on one surface unit the flow matrices could only be calculated making assumptions on the surface distribution of the referenced units.

Partitioning of land

The traffic network in Germany is very dense and the impact of traffic on the environment is manifold. One effect is to divide habitats, creating small islands in which species are isolated. This effect was described by calculating a disparity index for the potential effects of partitioning by the traffic network. This was done by comparing two size distributions: the first shows the distribution of sizes of the surface units before a specific traffic network was built, while the second shows the distribution with the traffic lines existing. The index measures the disparity between the two distributions (Table 7). The methodological backgrounds of the approach are the Lorenz curve and the Gini Index, both of which are well-known in economic statistics as measures of relative concentration (disparities). Only the surface units divided by roads and railways were considered. The surface units were derived from the STABIS 25 data on land use by combining all non-built-up areas (open land) with the exception of water surfaces and classifying the units according to their size. By comparing the distribution of the areas with the traffic networks for different times it was possible to calculate a flow coefficient for this index (Table 7). The work on this theme showed that the index is highly dependent on the sizes

Table 7. Supplementary matrix – size distribution of open areas with traffic networks for different years (1953/1972; Main-Taunus-Kreis).

With traffic network 1953					**With traffic network 1972**				
class [ha]	Freq.	area [ha]	area [%]	area cum.[%]	class [ha]	Freq.	area [ha]	area [%]	area cum.[%]
0	0	0.00	*0.00*	*0.00*	0	0	0.00	*0,00*	*0,00*
< 10	24	82.81	*0.37*	*0.37*	< 10	131	289.10	*1.43*	*1.43*
10-<20	8	114.94	*0.51*	*0.88*	10-<20	13	198.50	*0.98*	*2.41*
20-<50	16	531.38	*2.36*	*3.23*	20-<50	24	816.40	*4.03*	*6.44*
50-<100	19	1439.58	*6.39*	*9.62*	50-<100	18	1319.90	*6.52*	*12.96*
100-<200	13	2015.22	*8.94*	*18.56*	100-<200	16	2236.70	*11.04*	*24.00*
200-<500	22	7508.29	*33.31*	*51.87*	200-<500	23	7534.60	*37.20*	*61.20*
500-<1000	8	5498.89	*24.39*	*76.26*	500-<1000	6	4063.20	*20.06*	*81.26*
1000-<2000	4	5351.76	*23.74*	*100.00*	1000-<2000	3	3794.60	*18.74*	*100.00*
2000-<5000	0	0.00	*0.00*	*100.00*	2000-<5000	0	0.00	*0.00*	*100.00*
5000-10000	0	0.00	*0.00*	*100.00*	5000-10000	0	0.00	*0.00*	*100.00*
>10000	0	0.00	*0.00*	*100.00*	>10000	0	0.00	*0.00*	*100.00*
	114	22542.87				234	20253.00		
									Index: *10.9 %*

used for the classification. Since only relative distributions are compared the absolute sizes of the surface units have also to be considered for interpretation of the changes.

Soil sealing, vegetation index

With the growth of built-up and traffic areas in recent decades the size of sealed surfaces (isolation of the soil from the atmosphere, hydrosphere and biosphere by human impact) has also increased. The isolation of the soil inactivates its natural functions as a filter and buffer and has negative effects on the water balance, local climate and flora and fauna. The climatic effects of soil sealing are higher average temperatures, a decrease of air humidity and a decreased exchange of air in the cities.

To describe the issue of soil sealing a vegetation index was derived from the satellite images that were used for the STABIS 100 interpretation. Analysis of the data showed that the index was useful for describing characteristics of the typical patterns of urban land use. The index was, however, too variable to describe temporal changes. This was mainly due to seasonal changes between the times the satellite images were taken. Another reason could be that the changes in land cover are too small to derive significant figures.

Artificiality of land cover

Artificiality of land cover is a measure for the extent of human impact on land cover and for changes in the composition of the vegetation caused by human impact. Human impact means all direct activities and indirect effects that can be referred to as 'civilization' or 'civilization effects'; artificiality of land cover, therefore, includes all direct and indirect human effects on ecosystems. Civilization has effects on the habitat, the species and the biotopes, the intensity of which can be classified into seven categories. The methods to assess the intensity by human impact can be divided into three groups:

(a) the individual assessment of biotopes or species;
(b) the description of groups of species according to their special history of immigration or their way of living; this procedure however is controversial;
(c) the typical combination of species in a specific habitat which can be related to categories of artificiality of land cover.

The last method appears to be suitable as a basis for a statistical approach, although the results are restricted by several objections: first, the method does not consider the historical development of landscape; second, it is difficult to date the beginning of human impact; third, diffuse inputs are not recorded; and, finally, the categories of artificiality do not allow any statements on the occurrence of local species. Nevertheless, artificiality of land cover is a practical method to assess the extent of environmental pressures caused by changes in land use and land cover. During the pilot study the application of this method was prevented due to administrative problems. For this reason the concrete realization of the accounting framework for this issue has to wait until the ecological data from the area sample surveys become available.

Conclusions

The aim of the pilot study was to assess the feasibility of an accounting scheme on environmental statistics with physical units. This was achieved by establishing an accounting system on land use and land cover. Introducing a Geographic Information System (GIS), a new method for calculating the balances of land use and the analysis of the changes was used. With a GIS it was possible to fill in the core accounts.

Looking at the results of the pilot study it seems to be possible to establish an accounting system with physical parameters using georeferenced data on land use/land cover. There are several advantages of an accounting system. The accounting scheme allows highly aggregated information to be produced on different environmental issues. Moreover it allows quantitative analysis and determination of relationships and dependencies between different environmental issues. In the pilot study a framework of core and supplementary accounts and matrices was set up from which it was possible to derive aggregated indicators for the related core and supplementary accounts.

What lessons could been learned in the UNECE group for the setting of priorities and the planning of subsequent steps? In general, the advantages and shortcomings of an accounting approach are clearer after the pilot study: accounting is not a magic wand that brings light into the darkness of cause–effect chains. Its system and elements can, however, help to aggregate and structure data. Furthermore, it can be used to link accounts following business accounting principles (which is quite different from cause–effect relations in the sense of natural sciences). Finally, one of the advantages of an accounting approach is the opportunity to check data consistency. An essential result of the German experience is that following the same approach for land use/cover as for material/energy flows seems to be suitable in the sense that a sectoral breakdown of land use changes can be linked to economic activities and can give a first approximation for specific pressure indicators.

After carrying out the project it is obvious, however, that the statistical linkage of the economic activities with accounts on the state of environment through the core accounts of land use/land cover is still unsatisfactory at the conceptual and at the data level and requires further investigation. Hence, this is the next step to be taken in the German work plan. Also on the list for future work is the linkage of issue indicators with economic activities (for example, the contribution of economic sectors to soil sealing). Finally, it will be determined whether indicators of landscape changes caused by agriculture changes (like in the UK contribution to the group) can be integrated. In connection with these tasks the main objective for the next years will be the improvement of the underlying database.

15

Linking land cover, intensity of use and botanical diversity in an accounting framework in the UK

ANDREW STOTT AND ROY HAINES-YOUNG

Environmental accounts have been constructed for rural land cover change in Great Britain. The accounting framework was developed with the UN Economic Commission for Europe (ECE) Task Force on Physical Environmental Accounting. The accounts show the opening and closing stocks of rural land cover and the flows between land cover types in units of area (hectares) over a six year period, 1984–1990. A typology of the flows of land cover is developed which relates to the intensity of agricultural land use, forestry and built development. In addition to these major changes from one cover type to another, the more gradual changes in botanical diversity within a cover type over a longer time period, 1978–1990, are presented. Uses of the accounting framework within the context of policies for sustainable development and conservation of biodiversity are discussed.

Policy context and background

The UK Department of the Environment (DOE) sponsors research to inform current and future Government decisions, to guide the execution of policy, to monitor the achievement of policy goals, and to stimulate and promote innovation and spread best practice. The 1990 Environment White Paper (Department of the Environment, UK, 1990) and the 1995 Rural White Paper (Department of the Environment, UK, 1995) both underlined the UK Government's commitment to basing policies on the best available information, to environmental monitoring and to making presentations of environmental data more coherent and comprehensive. At the Earth Summit in Rio de Janeiro in July 1992 the UK Government became a party to the Convention on Biological Diversity. The Convention has important implications for national research because it requires contracting parties, among other things: to identify, and monitor through sampling and other techniques, the components of biological diversity; to identify activities which have a significant adverse impact on biodiversity and monitor their effects; and to maintain and organize data about these activities. In January 1994, the Government published specific commitments to draw up a set of costed targets for biodiversity, to

K. Uno and P. Bartelmus (eds.), Environmental Accounting in Theory and Practice, 245–260.

improve the accessibility and co-ordination of existing biological data sets and to develop a set of key indicators of sustainable development (Department of the Environment, UK, 1994a; 1994b).

The DOE aims to improve the information available to policy makers. Better integrated and more efficient frameworks for data collection and presentation are required, including the use of environmental indicators, performance measures, targets and environmental accounts. The longer term aim is to develop fully integrated national systems of economic and environmental accounting which clearly express the environmental impacts of economic activity (Department of the Environment, UK, 1994b). A first step in this direction is to explore existing data sources and work with international bodies to resolve outstanding methodological issues.

Rural land cover change is a key environmental parameter in the context of policies relating to agriculture, forestry, control of development, biodiversity and quality of life. Changes in land cover reflect the shifting balance of competing demands on the finite land resources of the nation. Whilst these resources (measured by area) are not generally 'used up' by human activities, they may be changed irreversibly, changed in such a way as to limit the choices of future generations, or changed in value. The degree of reversibility of a land cover change can be related to the type and intensity of use. The more intensive the use the less easy it is to reverse. The stock and change of land cover is thus a fundamental indicator of sustainable development and biodiversity.

In 1993 the UK joined the UNECE Task Force on Environmental Accounting and commissioned the University of Nottingham to develop and demonstrate the applications of physical accounts for land cover and biodiversity. The accounts were based on the data obtained from the series of national 'countryside surveys'. The results of this work have been published in the report to the Conference of European Statisticians (1995) and in the final contract report to DOE (Haines-Young *et al.*, 1996).

Data resources

The accounts are based on the results of Countryside Survey 1990 (CS1990) (Barr *et al.*, 1993). CS1990 includes a complete census of land cover in Great Britain, derived from the analysis of remotely sensed satellite data, and detailed land cover and associated ecological information collected by a sample-based field survey. The sample data can be related to two earlier surveys in 1978 and 1984. These data allow the extent of change in land cover and botanical diversity to be described. The survey data also permit the changes in plant species to be examined and related to changing management intensity.

The methods of data collection and analysis are significant for the accounting framework. Four aspects are of particular importance:

(a) The field survey component of CS1990 uses a sample of the 1 × 1 km square cells of the Ordnance Survey National Grid. The squares were selected using a stratified random sampling design derived from the ITE Land Classification. Each 1 km square cell of the national grid is assigned to one of a set of 32 'land classes' determined by a multivariate classification (TWINSPAN) of environmental parameters. Altogether 508 squares were sampled in the field in 1990. Of these squares, 381 had been surveyed previously in 1984 and 256 in 1978. Detailed ecological information was obtained from five permanent vegetation plots (200 m^2 quadrats) located randomly in each square. A total of 2500 such plots were surveyed in 1990, of which 1200 had been recorded previously in 1978. Mean area estimates for land cover are calculated for each land class strata and the totals for Great Britain are weighted by the extent of each land class. The data presented are estimates and have associated sampling errors.

(b) A further limitation of the CS1990 field survey data is that they only relate to rural areas of Great Britain (GB). Squares with more than 75% developed land (urban areas) were excluded from the sample. Although Northern Ireland has been covered by a similar survey (Cooper and Murray, 1992), the results have not been integrated to produce a complete UK data set.

(c) The random sampling approach adopted provides a good representation of land cover types and habitats which occur commonly in rural areas. Other rare or locally concentrated features (such as coastal sand dunes) are not well represented. Thus, the data for botanical diversity reflect the wider, predominantly agricultural countryside rather than sites of particular importance for nature conservation.

(d) Although the analysis of the remotely sensed data provides a complete census type survey without sampling error other errors are associated with image classification. The degree of correspondence between the satellite and field survey data varies from 80%, for the more distinctive cover types, to 30% for those which are less easy to characterize (Wyatt *et al*., 1994). The remotely sensed data are only classified into 17 broad land cover categories and are currently only available for 1990; they have not been used in the accounts.

Land cover and land use comparisons

For the purpose of international comparisons it is important that the UK data can be presented in standard classifications of land cover and land use such as CORINE and UNECE. Some preliminary work on the comparison of definitions used in the UK has been completed (Wyatt *et al*., 1994) and, using this approach, the UK data on land cover stock have been converted to the international frameworks of CORINE and UNECE (see Table 1 for the latter). However, a comparison of definitions is only a starting point, and quantitative inter-calibrations between survey data are required to convert data between systems with some degree of confidence. For example, a pilot study to convert the remotely sensed data for Great Britain to the standard CORINE land cover map has shown that the different spatial resolution of the classifications resulted in a loss of information, as land cover parcels below the minimum mappable unit were aggregated in mosaics with other cover types (Fuller and Groom, 1994).

Major practical problems include the different survey approaches required for collecting data on land cover and land use, the common occurrence of multiple land uses, and the widespread use of hybrid classifications (Dunn and Harrison, 1995). In this study changes in the type and intensity of use were not directly measured but were inferred from the characteristics of the land cover transformations and botanical changes.

Land cover accounts and intensity of use

The basic land cover account for GB has limited applications. It does not show the transformation of one cover type to another, it gives no indication of the process of change, it presents only net change within categories and provides no information on relative values. Following the approach agreed by the UNECE Task Force (Conference of European Statisticians, 1995) aggregated flow accounts were constructed by assigning the various types of land cover transformation to different categories of change of use. The categories of change of use were:

(a) **intensification** – a flow which represents the transition of a land cover type associated with low intensity use to a higher intensity use (e.g. semi-natural cover type to arable cover);
(b) **extensification** – a flow which represents the transition of a land cover type associated with high intensity use to a lower intensity use (e.g. improved grassland cover type to semi-natural cover type);
(c) **afforestation** – a flow which represents the planting or natural regeneration of trees;
(d) **deforestation** – a flow involving the clearance of trees;
(e) **development** – a flow involving the transformation of open land to urban, industrial or transport uses; and
(f) **reclamation** – a flow involving the creation of open land from areas previously developed (e.g. the reclamation of mineral workings).

In addition, there are transformations between cover types which remain within the same major categories, which do not involve a change of use. These flows include crop rotations, grassland rotations, semi-natural changes (e.g. succession), woodland rotation (e.g. conversion of broadleaf to conifer plantation), and redevelopment (e.g. conversion of industrial to residential cover type). An example of the account is presented for woodland land cover types in Table 2. The account shows the opening and closing balance, in terms of the 1984 and 1990 stock, of four woodland cover types: coniferous woodland, mixed woodland, broadleaved woodland, and scrub. The units of accounting are hectares. In the case of coniferous woodland (in the first row) the estimated stock increased from 1,250,000 ha to 1,316,000 ha, a net increase of 66,000 ha or 5.4%. The columns between the opening and closing balances show the flows to and from other cover types. Thus, the 'afforestation' column shows the estimated amount of tilled land, grassland or semi-natural land cover which was converted to coniferous woodland in this period, in this case 107,000 ha. This gain in woodland was counter-balanced by the felling of 41,000 ha. A small amount of coniferous woodland was gained by conversion from other woodland types, and a small amount was lost to development.

The two right-hand columns show the gross loss of the cover type in hectares and as a percentage of the 1984 stock. Thus, for broadleaved woodland, an estimated 67,000 ha or 7.5% of the original stock was lost, despite a small overall net increase of 27,000 ha of broadleaved woodland. The presentation of this account has some important policy implications. Current forestry policy in the UK aims to increase the extent of broadleaved woodland and conserve existing areas of ancient woodland. These results show that in the period 1984 to 1990 as much broadleaved woodland was 'lost' (i.e. converted to other cover types) as was gained. It is generally argued that, in terms of biodiversity, losses of older established broadleaved woodland, especially ancient semi-natural woodland, will not be immediately compensated for by gains as a result of new planting. However, we do not know from the land cover data whether this was in fact the case.

Table 2 also presents information about the standard error of the estimate of the net change. This can be used to give an indication of the variability of the change in the sample. However, because the data are not normally distributed it does not provide a good measure of accuracy (Barr *et al.*, 1993). Changes in land cover, for example afforestation, are highly variable within the sample; a few sample squares recorded substantial coniferous afforestation and many squares recorded none.

The account for selected semi-natural land cover types is shown in Table 3. For many categories there were only small net changes and, as reported in Barr *et al.* (1993), this gives the impression that historical rates of loss of these habitats had been curbed. However, the accounts show large transfers into and out of, and between, the semi-natural categories such that generally between a tenth and quarter of the area in each type was lost in the period 1984–90. This may have important implications for biodiversity because habitats which are ancient or long established often have a greater value for wildlife than recently created habitats. Thus, whilst the overall extent of a cover type or habitat may be stable, its quality may be diminished.

Table 1. Rural land cover stock in Great Britain, 1984 and 1990. Data transformed to UNECE Statistical Classification of Land Use (Source: Haines-Young *et al.*, in press; Wyatt *et al.*, 1994).

CS1990 CODE	UNECE CODE	LAND USE CATEGORY	1984 STOCK ('00 sq km)	1990 STOCK ('00 sq km)
1-17	1.1	Arable land	508	487
19	1.2	Land under permanent grass	8	6
21-26	1.3	Land under permanent meadows and pastures	602	588
53	1.4	Other agricultural land	15	15
18	1.5	Fallow agricultural land	8	26
37,41	2.1	Land under coniferous forest	127	136
39	2.2	Land under non-coniferous forest	89	91
38	2.3	Land under mixed forest	20	22
40	2.4	Other wooded land	10	9
54	3.1	Residential land	65	71
55,57	3.2	Industrial land	31	34
58	3.3	Quarries, mines etc	1	3
51,52	3.7	Land used for transport	49	49
56,20	3.9	Recreational land	36	36
35,36	4.1	Mires	204	202
45,47	4.3	Other wet open land	32	34
32-34	5.1	Heathland	122	125
27,29,30	5.3	Montane grassland	189	181
28,31,50	5.4	Other dry open land	66	68
42,49	6.1	Bare rocks	7	7
48	6.2	Sand beaches, dunes etc	2	2
46	6.3	Other unvegetated land	13	13
43,44	7.1	Inland waters	36	36
		TOTAL	2240	2241

The classification of changes in use assists the interpretation of the processes of land cover change. For example, the 7% net loss of 'drier northern bog' was nearly all attributable to afforestation and the 11% net loss of 'dense bracken' was mostly associated with changes to other semi-natural cover types, such as heathland. Similar accounts have been constructed for tilled land, managed grassland and built-up land (Haines-Young, 1996) but these are not presented here for reasons of space.

To summarize, three key points arise from this presentation of the land cover accounts:

Table 2. Land cover account for woodland, 1984 to 1990 (Units: '00 sq km and per cent, as indicated)

COVER TYPE	**1984 STOCK**	ROTATION	AFFORESTATION	DEFORESTATION	DEVELOPMENT	RECLAMATION	**1990 STOCK**	NET CHANGE	STANDARD ERROR	**% CHANGE 1984-90**	1984 STOCK LOST	% 1984 STOCK LOST
Coniferous Woodland	**125.0**	0.1	10.7	-4.1	-0.1	0.0	**131.6**	6.6	4.8	**5.4**	5.1	4.1
Mixed woodland	**20.3**	0.7	0.8	-0.3	-0.1	0.2	**21.5**	1.2	0.8	**6.1**	1.5	7.3
Broadleaved woodland	**88.8**	-0.2	6.8	-3.3	-0.8	0.1	**91.2**	2.7	1.5	**3.1**	6.7	7.5
Shrub	**9.7**	-0.7	1.8	-2.0	-0.1	0.0	**8.7**	-1.0	0.6	**-10.1**	3.5	35.9
Perennial crops (orchards)	**8.5**	0.0	1.4	3.3	-0.1	0.1	**6.5**	-2.0	2.1	**-23.6**	3.5	40.7
ALL WOODED LAND	**252.0**	**0.0**	**21.5**	**-13.1**	**-1.3**	**0.4**	**259.6**	**7.6**	**2.0**	**3.0**	**20.2**	**8.0**

(Source: Haines-Young *et al.*, 1996).

Table 3. Land cover account for selected semi-natural cover types, 1984 to 1990 (Units: '00 sq km and per cent, as indicated)

COVER TYPE	**1984 STOCK**	SEMI-NATURAL CHANGE	INTENSIFICATION	EXTENSIFICATION	AFFORESTATION	DEFORESTATION	DEVELOPMENT	RECLAMATION	**1990 STOCK**	NET CHANGE 1984-90	STANDARD ERROR	**% NET CHANGE 1984-90**	1984 STOCK LOST	% 1984 STOCK LOST
Unimproved grass	**15.1**	0.9	-3.1	5.3	-0.3	0.1	0.0	0.1	**18.1**	3.0	2.4	**19.6**	4.0	26.5
Calcareous grass	**5.8**	0.0	-0.2	0.7	-0.2	0.2	0.0	0.0	**6.3**	0.5	0.5	**8.8**	0.4	6.8
Dense bracken	**41.6**	-5.2	-1.0	1.2	0.7	0.9	0.0	0.1	**36.9**	-4.7	4.1	**-11.4**	12.2	29.4
Moorland grass	**86.1**	-1.4	-0.7	0.2	-1.0	0.0	-0.3	0.0	**82.9**	-3.2	2.0	**-3.7**	11.7	13.6
Open canopy heath	**63.7**	3.9	-0.1	0.3	-2.0	0.2	0.0	0.1	**66.1**	2.4	3.1	**3.8**	7.6	11.9
Drier northern bog	**49.5**	-0.3	-0.2	0.0	-3.2	0.0	0.0	0.0	**45.9**	-3.6	3.2	**-7.3**	5.3	10.6
Wet heaths/bog	**154.7**	0.9	0.0	0.0	0.0	0.1	0.0	0.0	**155.7**	1.0	0.5	**0.7**	3.0	2.0
Felled woodland	**1.7**	0.2	-0.2	0.0	-0.4	2.8	0.0	0.0	**4.1**	2.4	1.5	**135.5**	0.9	52.9
Unmanaged grass/tall	**22.3**	1.0	-4.7	11.9	-1.1	0.9	-2.4	0.8	**28.7**	6.5	2.6	**29.0**	9.6	42.9
Other semi-natural	**253.2**	-0.7	-7.9	9.5	-5.7	1.0	-0.5	0.2	**250.4**	-3.0	n/a	**-1.2**	35.5	14.0
TOTAL	**693.7**	**-0.7**	**-18.1**	**29.1**	**-13.2**	**6.2**	**-3.2**	**1.3**	**695.1**	**1.3**	**1.3**	**0.2**	**90.2**	**13.0**

(Source: Haines-Young *et al.*, 1996).

(a) Changes in land cover can, at a generalized level, be used to infer changes in land use. The identification of types of transformation is informative in a policy context.
(b) The reporting of net change on its own will tend to underestimate the amount of land cover change over a survey period. This highlights the issues of compensation – how far the gain of new stock compensates for the loss of established stock in terms of biodiversity, and reversibility – how easy (or expensive) it is to reverse a given change.
(c) Measurement of change in extent of land cover types only gives a partial picture of the full impact of human activity on the environment. From a policy perspective it is not only the change in area which is important but the change in quality or value (however measured). It will also be essential to know whether the parcels which remained stable in terms of land cover are as ecologically valuable at the end of the survey period as at the beginning.

Biotope accounts

To address the issue of change in quality the accounts have been extended to include botanical data from vegetation quadrats surveyed in 1978 and 1990. The data provide a measurement of botanical diversity in a stratified random sample of rural land in Great Britain. These are called the 'biotope accounts'. The term 'biotope' is used to describe the smallest land unit which is 'still a holistic unit' defined in terms of a characteristic assemblage of species (Haines-Young, 1996). In CS1990 biotopes were formed by grouping vegetation plot data using a multivariate classification technique, TWINSPAN. Originally, 29 vegetation classes were derived (Barr *et al.*, 1993) but for ease of presentation these have been aggregated into six major biotope groups: arable fields; lowland improved grassland; lowland semi-improved grassland; woodland; upland grass mosaic; and moorland. The approximate area of these biotopes in Great Britain is shown in Figure 1. The statistical algorithm used to classify the biotope units ordered them on an approximate gradient which can be interpreted as a gradient of management intensity or nutrient status (see below). At one end of the gradient the vegetation is associated with intensively managed, high nutrient status arable land, and at the other end the vegetation is typical of extensively managed, low nutrient status moorland and bog.

The biotope groups recorded in 1990 are distributed amongst the 58 land cover types also recorded in 1990. The units in these tabulations are numbers of plots which, because they are obtained from a random sample, are approximately proportional to the area of rural land. There is a strong relationship between the biotope groups and the cover types in which they occur, and these data show the botanical composition of the different cover types. For example, about half the plots in the coniferous woodland cover type have botanical characteristics of woodland biotopes, a quarter have affinities with the upland

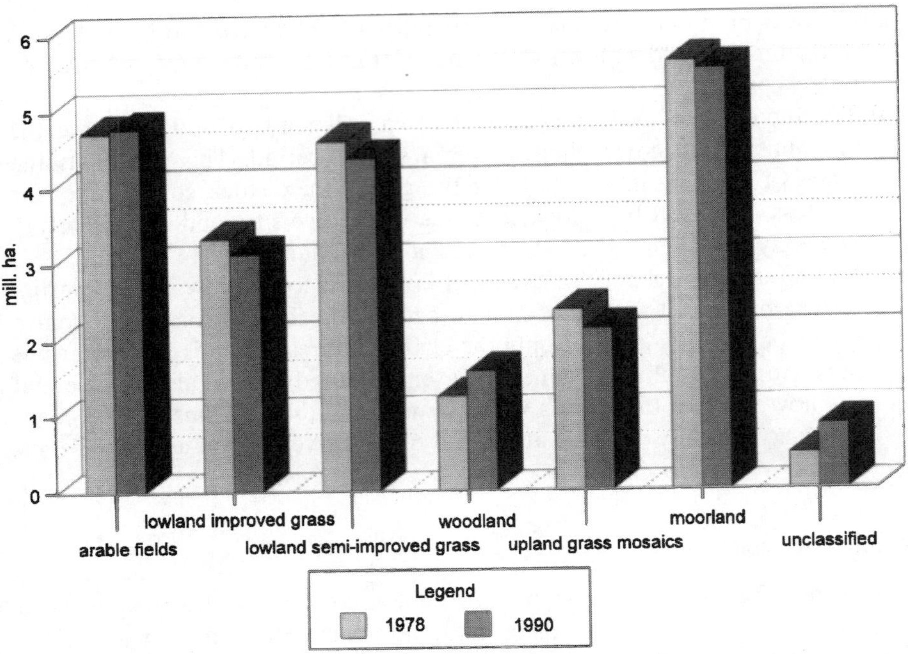

Figure 1. Estimated area of Great Britian with vegetation in each of the main biotope groups in 1978 and 1990 (million hectares).

grass mosaic biotopes and another quarter have affinities with moorland biotopes. These tabulations can be used to assist in the interpretation of the ecological impacts of land use change. Thus, intensification of agricultural use, characterized by the replacement of 'non-agriculturally improved grass' cover with 'pure rye-grass', would be associated with a change in the biotope composition from lowland semi-improved grassland to lowland improved grassland, with a loss of species diversity and a shift up the intensity gradient. These data also show that, because of the random sampling procedures used, some of the less common cover types (e.g. sand dunes) are not adequately represented in the vegetation data set.

The next step in the development of the biotope accounts was to construct a change account using the data from 1978 and 1990. Table 4 has a similar structure to the land cover accounts (Table 2) and shows the net change in the numbers of plots in each biotope group and the flows between the groups. For example, the number of woodland biotope plots increased by one-third. Most of these plots were gained from the upland grass and lowland semi-improved grass biotope groups. Despite this substantial net increase in woodland plots, 10% of the original woodland plots in 1978 had changed to grassland biotopes. Overall 72% of all plots did not change between the major biotope

Table 4. Biotope change account 1978–90: net change in the number of plots in each biotope group and flows between groups (number of plots)

Biotope group	1978	loss to arable	loss to lowland improved grass	loss to lowland semi-improved grass	loss to woodland	loss to upland grass	loss to moorland	gain from arable	gain from lowland improved grass	gain from lowland semi-improved grass	gain from woodland	gain from upland grass	gain from moorland	1990	% 1990 stock of plots carried over
arable fields	204		-37	-7					40	16				216	74.1
lowland improved grassland	164	-40		-35	-3			37		39	3		1	166	51.8
lowland semi-improved grassland	241	-16	-39		-10	-8	-2	7	35		4	9	4	225	73.8
woodland	69		-3	-4		-4	-1		3	10		16	6	92	62.0
upland grass mosaics	103			-9	-16		-20			8	4		25	95	61.1
moorland	249		-1	-4	-6	-25				2	1	20		236	90.3
total	1030	-56	-80	-59	-35	-37	-23	44	78	75	12	45	36	1030	71.8

(Source: Haines-Young *et al.*, 1996).

Table 5. Changes in the average decorana scores for paired plots recorded in 1978 and 1990 (decorana score)

Biotope Groups	Average Decorana score of all plots in 1978	Average Decorana score of core plots in 1978	Average Decorana score of core plots in 1990	Change in Decorana score of core plots 1978-90	Averge Decorana score of plots lost from 1978 stock	Average Decorana score of plots gained by 1990 stock	Average Decorana score of all plots in 1990	Change in Decorana score 1978-90 (all plots)
arable fields	**681**	687	720	33	661	699	**714**	33.0
lowland improved grassland	**568**	572	567	-4	564	556	**565**	-2.8
lowland semi-improved grassland	**477**	475	474	0	483	487	**478**	0.5
woodland	**358**	354	352	-2	378	335	**345**	-12.9
upland grass mosaics	**297**	293	292	-2	302	283	**288**	-8.8
moorland	**156**	147	156	9	210	211	**161**	5.2
All plots	**430**	**408**	**416**	**8**	**n/a**	**n/a**	**442**	**11.8**

(Source: Haines-Young *et al.*, 1996).

groups between 1978 and 1990 and could be considered relatively stable in botanical characteristics at this crude level of analysis.

Another, more sensitive, way of measuring the change in botanical quality involves the calculation of an index of intensity of management. This index is produced by the technique of detrended correspondence analysis (Barr *et al.*, 1993). The analysis produces an ordination (ordering) of the vegetation plots according to their 'decorana' scores. This ordination, or gradient, is associated with management intensity and nutrient status. Thus, a vegetation plot representing an intensively farmed arable field has a high decorana score and a plot from a moorland bog has a low decorana score. Plots within each of the biotope groups can also be ordered on this gradient. The decorana score for each plot has been calculated for 1978 and 1990.

Changes in the average decorana scores for each of the biotope groups are shown in Table 5. For all plots in Great Britain, the average decorana score increased from 430 to 442 between 1978 and 1990; this represents a slight movement up the intensity gradient. The largest increase in decorana score occurred in the arable biotope.

There are two components to the changes in decorana score recorded: changes which occurred in plots which remained within the same biotope group ('core plots'), probably as a result of changes in intensity of management, and those which occurred in plots which moved between biotope groups, probably as a result of a major land use change.

In arable fields, for example, the decorana score in the core plots increased from 687 to 720, suggesting increased intensification within the biotope. This was associated with a reduction in the number and variety of arable fields. New arable field plots which were recruited from other biotopes had by 1990 a decorana score of 699, lower (i.e. less intensive) than the core plots. At the opposite extreme in moorland, the decorana score in the core plots increased slightly from 147 to 156. New moorland plots which were recruited had by 1990 a decorana score of 211, higher (i.e. more intensive) than the core plots. The woodland biotope recorded the largest overall decrease in decorana score. The core plots were virtually unchanged, and the overall decrease was a result of the loss of plots with a higher score (i.e. conversion of more productive woodland to another use) and recruitment of plots with a lower score (i.e. afforestation of less productive upland grass or moorland).

Although the decorana score is a useful way of organizing the vegetation data in relation to intensity of management it cannot be simply related to the botanical diversity or conservation value of the biotope. Each biotope group will have examples of habitats of particular importance for biodiversity. An alternative and more easily understood measure of biodiversity is the mean number of plant species per plot. For the six biotope groups, mean species number per plot in 1990 varied from 3.8 in arable fields to 22.4 for upland grass mosaics (Figure 2). Changes in mean species number in paired plots between 1978 and 1990 have also been calculated (Barr *et al.*, 1993; Haines-

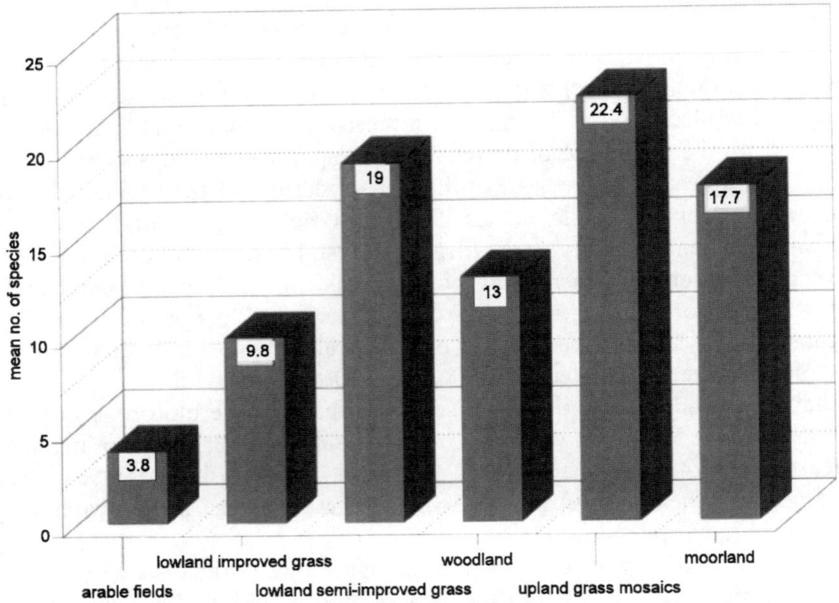

Figure 2. A simple measure of botanical diversity, the mean number of species in each of the main biotope groups for all plots surveyed in 1990, weighted by area.

Young, 1996). These show significant losses of botanical diversity within some biotopes. For example, mean species number in plots which remained within the woodland biotope fell from 15.5 to 12.4 between 1978 and 1990, which represents a 20% loss of botanical diversity in 12 years. Further work is in progress to derive improved measures of changes in the biodiversity of vegetation.

The biotope accounts show that important ecological changes occurred in those land parcels which had not apparently experienced a change in land cover. There was, for example, evidence of increased intensification and reduced diversity within the arable fields and lowland semi-improved grassland biotopes. The results show the importance of detailed ecological information to support the interpretation of land cover accounts for policy applications.

Zonal accounts

The accounts and data so far presented relate to all rural land in Great Britain. The sampling frame used in CS1990 allows the disaggregation of the accounts into four major environmental zones or 'landscape types': arable lowlands; pastural lowlands; marginal uplands and true uplands. Each of these zones has distinctive environmental characteristics, land cover composition and associated land use (Barr *et al.*, 1993). Presentation of the zonal land cover accounts demonstrate different rates and trends in land cover change. Nationally, for example, there was a slight net increase in broadleaved woodland between 1984 and 1990. The zonal accounts show that this increase occurred mostly in the arable lowland zone, where woodland cover increased by 10%. In contrast, there was a small net loss of woodland in the pastural lowlands. Government policies aim to increase the area of broadleaved woodland nationally and to protect existing woodland of special nature conservation value. These policies appear to have been more effective in arable than in pastural lowland areas.

In general, the zonal accounts demonstrate that at different scales of resolution the accounts may lead to different interpretations. National accounts may obscure important aspects of ecological and environmental change which are occurring at a sub-national level.

Summary and discussion

Information about land cover and the way it changes over time is a vital ingredient in environmental decision making. This presentation has examined how environmental accounting can be used in the analysis and presentation of such data, in order to help with the tasks of policy development and appraisal in relation to issues of sustainable development and the conservation of biodiversity. It draws upon the data resources of CS1990 and refers to work undertaken as part of the UK contribution to the UNECE Task Force on Environmental Accounting.

This chapter has shown that a set of land cover accounts can be presented which classify, in general terms, the types of changes in intensity of use observed over the period 1984–90. The accounts showed that the analysis of net changes in land cover was often misleading. While many ecologically valued cover types, such as broadleaved woodland, showed an increase in area, elements of the original, and perhaps more diverse, stock were lost. Thus, it is recommended that in future reports on countryside change both net and gross change in land cover should be presented. The classification of transformations of land cover into change of use categories provides a simplified comprehensive summary of the dynamics of land.

Quantitative analysis of land cover change only gives a partial picture of impacts on environmental quality. It is recommended that quantitative data are set alongside the analysis of qualitative changes in ecological characteristics. A statistical classification technique was used to define a set of objective 'biotope' groups, which represent the major patterns of botanical variation found in the wider countryside. Two indices of biodiversity, the intensity gradient and the mean species number, were used to present qualitative changes in vegetation. These indices have shown that at the national level some 'biotopes' experienced a decline in species diversity and a trend towards vegetation more characteristic of intensive management in the period 1978–90.

Disaggregation of the data into zonal accounts demonstrated that different zones or 'landscape types' showed quite different trends which are not evident in the national accounts. Disaggregation into other policy relevant units, such as designated areas, would be useful but cannot be achieved with existing data. In particular, the CS1990 data do not provide adequate information about changes in rare or localized habitats which are of national and international importance for the conservation of biodiversity. The geographical context of land cover change is an essential characteristic of change which is absent in national accounts.

The land cover and biotope accounts can be used to define a rigorous and defensible framework from which a range of land-based indicators of sustainable development may be derived. Such indicators which express the proportion of the initial stock of a natural asset which is 'carried over' during the study period may be used to address issues of compensatory changes in land cover. Further work is required to improve techniques for the qualitative assessment of land cover and biotopes using statistical procedures. Qualitative assessment of land cover change should include measurements of biodiversity, reversibility and compensation.

Acknowledgements

The Department of the Environment commissioned this work within its Countryside Research Programme. Dr Andrew Stott was the Department's nominated officer for the contract. The work was undertaken by Dr Roy Haines-Young and Dr Charles Watkins at the Department of Geography, University of Nottingham. Access to and advice on the interpretation of data was provided by Dr Bob Bunce and Mrs Caroline Hallam at the Institute of Terrestrial Ecology, Merlewood. Members of the UNECE Task Force on Environmental Accounts contributed to the development of the conceptual approach adopted.

Part IV

Concepts, Methods and Analysis

16
The value of nature: valuation and evaluation in environmental accounting

PETER BARTELMUS

Sustainable development calls for the integration of economic and environmental concerns in planning and policy making. To achieve integration economists and environmental scientists apply their tools and values to the other field. The result is a dichotomy of monetary valuation in environmental economics and accounting and the development of non-monetary indicators and indicator frameworks.

The System of integrated Environmental and Economic Accounting (SEEA), developed by the United Nations Statistics Division, applies three categories of valuation in different versions (modules) of the SEEA: market valuation of natural resource stocks and use, maintenance costing of natural asset depletion and degradation, and contingent and other valuations of welfare effects of environmental degradation. The three valuations are examined as to their consistency with economic theory and national accounts conventions. Sustainability criteria provide the rationale for extending the (produced) capital consumption concept of national accounting to non-produced natural assets and their maintenance. For reasons of data availability and compatibility with conventional accounting rules, maintenance costing and a simplified approach to market valuation, the net price or Hotelling-rent method, are considered to become the standard valuation in environmental accounting. In the case of non-renewable resources this valuation could be supplemented by a user-cost allowance, reflecting a weaker (income) sustainability concept. Damage or welfare valuation should not be introduced in production-orientated environmental (national) accounting but could be applied in more limited studies of particular environmental concerns.

Measuring the sustainability of development (beyond economic growth) requires supplementary indicators of non-economic amenities such as health and recreation, biodiversity, equity, political freedom etc. The policy relevance of indicator sets can be improved by relating indicators to social standards and targets or by combining them into overall indices. Valuation (by the invisible hand of the market) is replaced by evaluation (by the visible hands of standard setters or index builders).

The author is a staff member of the United Nations. The views expressed here are his own and not necessarily those of the United Nations.

K. Uno and P. Bartelmus (eds.), Environmental Accounting in Theory and Practice, 263–307.

Consensus building and standardization of concepts and methods in environmental accounting and indicator development are urgently needed. The proliferation of overall indices, indicators and valuations has created confusion in national projects of environmental statistics, indicators and accounting. Standardization in capacity building should not discourage pluralism in research and experimentation.

Introduction: value concepts

Environment and economy interact. Policy failures in both environment and development can be traced back to the neglect of this interaction in 'compartmentalized' line ministries and agencies (World Commission on Environment and Development, 1987). Discussions and publications before and in the wake of the Earth Summit, the United Nations Conference on Environment and Development, provide ample evidence of environment–economy interrelations (World Commission on Environment Development, 1987; United Nations Environment Programme, 1992; Bartelmus, 1994a; Brown L. R. *et al.*, various years). The obvious answer is policy integration. Reorienting planning and policies towards integrative sustainable development has been the generic theme of the Earth Summit and follow-up programmes and action plans.

Both economists and environmentalists[1] looked into their analytical tool kits for possibilities of applying them to the other field. In doing so, they imposed their own values with the result of distinctly different views of the merged field of environmentally sound and sustainable development. The question of values and valuation is at the heart of integrated environmental and economic accounting. It has created considerable dissent about the nature, use and usefulness of such accounting. This chapter shows that some of this dissent might evaporate when its theoretical foundations are examined and when it is confronted with real-world situations and data availability. Hopefully, this will pave the way towards harmonization and standardization of integrated accounting and related data development.

Economic values

Economists derive the value of a resource from its scarcity and usefulness in meeting human needs and aspirations. Resources that are available in unlimited supply, however essential they may be to survival, do not have economic value. They are 'free'. Valuation of scarce economic goods is to facilitate rational (optimal) choices about their use by economic agents, including policy makers. Economic theory tells us that, under conditions of perfect competition, there is no need to search for appropriate values. In this situation, existing market prices reflect accurately relative scarcities and the most efficient use of goods and services. They also reflect a utility maximizing choice of goods and services by individual (final) consumers at a particular distribution of income.

However, those conditions hardly exist in reality due to oligopolistic and monopolistic influences, externalities (e.g. from environmental impacts) and ignorance about the comparative qualities of goods and services. In addition, the existing distribution of income may be far from desirable. Observed market values may thus reflect a situation of allocation inefficiency and inequity. Observed prices are no more indicators of relative scarcity (see Box 1).

Box 1: Perfect markets – myth or reality?
Perhaps the most eloquent attack on the myth of perfect competition in atomistic markets was launched by Galbraith (1986). He argued that large parts of the economy are governed by oligopolies of national and international corporations and state monopolies. Thus, the applicability of microeconomic optimization strategies and optimality criteria in general equilibrium situations can be doubted. It is quite curious that this attack on the fundamentals of widely taught and applied economics has brought about so little change in economic analysis and policy. Galbraith himself points out that this is due to a combination of denial from and alliance of corporations and State; both, government and corporations, hide their common goals of growth and power behind the screen of allegiance to (powerless) perfect competition. The 1994 Nobel prize award in economics to pioneers in game theory might reflect a need to explain economic behaviour more and more in strategic terms rather than by the invisible hand of atomistic markets.

There appear to be two distinct responses to this controversy. One is to deny or neglect market imperfections and live with the 'semi-fiction that market prices accurately reflect relative scarcities' (Solow, 1992, p. 10). The widespread use of market-price based national accounts aggregates in economic analysis and policy advice attests to this attitude. The other response is to forgo reality in favour of idealized models of general equilibrium or optimal development, producing convenient 'shadow' prices.[2] Shadow prices are used in cost–benefit analysis for project and programme selection and in more generic analyses, e.g. of computable general equilibrium (CGE) models, for policy formulation and evaluation.[3]

Justifications provided for either approach are not convincing. It is argued, for instance, that action based on an ideal situation might contribute to achieving this situation, possibly by a 'sequence of policy reforms' (Dasgupta, 1994, p. 42). In a similar vein, it has been pointed out that 'economics in a vacuum' might still throw light on complicated problems; in addition, hypotheses made in an oversimplified model of perfect competition might become valid in the long run: for instance, monopolistic market imperfections might thus become 'eroded by competing technologies' (Samuelson and Nordhaus, 1992, p. 295). The surge in mergers of large US corporations in the 1990s does not lend support to this argument.

Externalities of environmental destruction and degradation provide further ammunition to those who doubt the relevance of conventional economics for long-term policies of sustainable growth and development. Environmental economists therefore seek to incorporate scarce environmental goods and services into their value system. Their premise is that the environment is perceived and treated as an economic commodity. Consumers can thus be prompted to reveal their (dis)preferences for environmental services or service losses in money terms (Jacobs, 1994). In the absence of markets for this classical public good, supply and demand curves can at least be imagined, if not simulated.

To this end, the concept of economic value is broadened to include environmental concerns. The economic values of environmental concerns can be divided into direct (for current production and consumption) and indirect (for ecological support) use values of environmental functions, as well as non-use (option or existence) values derived from the knowledge about preserved ecosystems or species (Munasinghe, 1993, p. 22). A variety of valuation techniques to capture these values have been developed for cost–benefit analyses of the environmental impacts of projects and programmes. Box 2 lists commonly applied valuation methods. Objective valuations of physical impacts and damage are distinguished from subjective valuations, revealed in surrogate or hypothetical markets of environmental quality or health services.

Box 2: Valuation techniques in environmental cost–benefit analysis

1. Objective valuations use damage functions which assess the physical damage caused by offending activities. They include the measurement of effects of environmental impacts on

(a) changes in productivity;
(b) cost of illness (morbidity);
(c) change in human capital (mortality);
(d) replacement and restoration costs of produced and non-produced natural assets.

2. Subjective valuations measure possible environmental damage as expressed or revealed in real or hypothetical markets. They are related to individual utility functions and include the estimation of

(a) preventive/mitigative expenditures for health effects and changes in productivity of produced and natural assets;
(b) hedonic approaches of assessing property/land value changes due to changes in productivity and environmental quality;

(c) travel cost (increase) due to environmental deterioration;
(d) contingent valuation (sample surveys of preferences for environmental goods and services).
Source: Dixon *et al.*, 1994, p. 30.

Environmental economists have also advocated the introduction of shadow-priced environmental welfare effects into national income or net national product. They argue that, under the ideal conditions described above, both indicators can then be interpreted as the result of an optimization process of maximum welfare generation with limited resources (Mäler, 1991; Pearce, 1994).

Environmental values

Environmentalists, notably those of the 'deep' kind,[4] strongly contest the notion that the environment can be treated as a commodity. People see or should see the environment from a cultural or ethical point of view as an indivisible social good. Toward such a good people express attitudes or moral convictions rather than preferences in terms of economic costs and benefits. To an environmentalist it is quite repugnant to see the value of a public good or national heritage subjected to the 'willingness to pay', which in reality would probably reflect an 'ability to pay' (Jacobs, 1994). The question is not how to generate economic value for an underpriced commodity; the question is how to prevent being 'colonized by the economy' through misleading economic valuation (van Dieren, 1995, p. 7).

Since, according to this reasoning, the real value of nature cannot be expressed in monetary terms, new indicators or indices of the quality of life (OECD, 1973), 'human development' (United Nations Development Programme, 1991) or sustainable development (Bartelmus, 1994b) need to be developed. Aggregation of the different facets of these concepts would have to be done by appropriate weighting of the quantifiable aspects of priority concerns and related activities and results. Such indicator selection and weighting can be seen as an attempt to incorporate social and economic values and parameters into environmental mindsets and data frameworks – colonizing the economy? Lacking the powerful aggregative capacity of economic valuation, full integration appears to be an elusive goal for indicator construction. Most integrative environmental data systems content themselves, therefore, with loose data 'linkage' in frameworks that cover human activities and their environmental impacts and repercussions. There have been attempts, though, to use monetary valuation to incorporate a broad spectrum of social and environmental values into overall indices of economic welfare. The use and usefulness of indicator and index construction, outside the national accounts system, is briefly discussed later.

Accounting values

National accountants try to stay away from this truly value-laden discussion by focusing on observable exchange values of market prices. From an accounting point of view, market prices provide a common and measurable numéraire for aggregating the results of economic activity. 'The power of the SNA as an analytical tool stems largely from its ability to link numerous, very varied economic phenomena by expressing them in a single accounting unit' (Commission of the European Communities *et al.*, 1993, para. 3.70). Indeed, the purpose of national accounting is not to measure economic welfare, or the quality of life, but economic performance as the sum of market transactions. The question is whether performance has any meaning, beyond showing short-term changes in the amount of economic activity.

The types and patterns of performance do provide insight in the workings or frictions (disequilibria, e.g. in domestic markets, foreign trade, employment or price changes) of the economy. In particular, data on the structure of the economy permit the modelling of the effects of economic policy measures on the different sectors and institutions, e.g. by means of input–output analyses. However, if performance cannot be linked unequivocally to the net benefits derived therefrom, it cannot be evaluated and used for steering the economy in desirable directions. In the absence of any other standard welfare measures, national accounts aggregates continue, therefore, to be used as proxies for economic welfare, despite disclaimers by the producers of these aggregates (van Dieren, 1995, p. 77; Commission of the European Communities *et al.*, 1993, paras 1.68-82, 2.178-80).

Environmental (national) accountants seek to modify this focus on market transactions by incorporating alternative or adjusted valuations of non-marketed environmental processes and impacts into the national accounts. The purpose is to improve the assessment of truly 'net' benefits of economic activities, taking their environmental effects into account. Monetary values of environmental depletion and degradation are thus introduced by using or adapting the valuation techniques developed by environmental economists.

The proposed System of integrated Environmental and Economic Accounting (SEEA) (United Nations, 1993a) proposes three main categories of adjusted or imputed valuations in different versions or modules of the system. They are

(a) market valuation of economic natural assets and their changes;
(b) maintenance costing of environmental depletion and degradation; and
(c) contingent valuation of environmental damage resulting from depletion and degradation.

The different techniques of assessing these values and their relative merits and drawbacks are examined in the following.

Market valuation of natural resources

Valuation and system boundaries

Observable market prices provide the common numéraire for the universally adopted System of National Accounts (SNA) (Commission of the European Communities *et al.*, 1993). The use of this numéraire in aggregation and accounting balances also delimits the scope and coverage of the SNA. This is evident in the system's focus on market transactions which produce the market price as the agreed exchange value between transactors. However, a rigorous restriction of the SNA to market transactions would seriously limit its analytical power. A number of exceptions are therefore made to include goods and services, and assets and liabilities that are not exchanged in markets, but that could or should become a market transaction.

The 'could' criterion refers to goods that are produced for own-account purposes as well as for market exchange; thus, markets for these goods already exist. When markets are absent, due to indivisibilities of public goods or political priorities for the non-commercial supply of social services, the incorporation of these goods and services is more questionable. The argument of market failure (Commission of the European Communities *et al.*, 1993, paras 9.84 and 9.92), in this case, reflects the 'should' criterion, i.e. a normative judgement about the coverage of public goods consumption. Such judgement has its roots in political processes where decisions about the desirability and amounts of public goods are made, irrespective of individual preferences, though possibly checked by processes of democratic decision making. National accountants thus deviate from the neutral criterion of measuring real or simulated market activity: they include social-value based goods and services such as security, research, public health and environmental protection in the system's production and consumption boundaries.

It is important to note these exceptions as they provide a justification for incorporating further desirabilities, notably of an environmental nature, by expanding the system boundaries. The SEEA thus expands the SNA's asset boundary for incorporating environmental stocks and flows. Also, as in the case of social goods and services, environmental effects are valued at cost (see below). Clearly, changes in the system boundaries affect valuation. In turn, the introduction of non-market or non-economic values affects the scope and coverage of the accounting system.

Coverage of natural assets

The scope and contents of the SNA are delimited by three main inter-related boundaries of production, consumption and asset categories. They all focus on market transactions and related income generation (with the above-described exceptions). The economic asset definition of the SNA already includes certain natural assets 'over which ownership rights are enforced and from which economic benefits may be derived' (Commission of the European Communities *et al.*, 1993, para. 10.2). These natural assets can be produced, such as agricultural products of fruit trees or livestock, or non-produced, such as land, mineral deposits or forests in the wilderness.

Produced natural assets still represent a part of the (more or less) natural environment, e.g. agricultural landscapes. They make their appearance through economic (agricultural) production, recorded as fixed capital formation or inventory increase in the SNA. Non-produced, economic, natural assets, on the other hand, are not generated by human beings. They are the result of geological processes (land/soil, subsoil assets), natural growth (biota) or cyclical replenishment (water bodies). In the SNA they are therefore considered to be outside the production boundary but inside the economic asset boundary. They make their economic appearance in the asset accounts as transfers (Commission of the European Communities *et al.*, 1993, paras 12.18-19) from the environment. Of course, no counterpart changes of environmental assets are shown in the economic asset accounts. Those changes would have to be captured in environmental asset accounts (see below).

Changes in the availability of economic, non-produced natural assets, resulting from depletion, degradation or transfers from the environment, are accounted for in the SNA as 'other volume changes'. Contrary to capital consumption and formation, other volume changes do not affect the production and income accounts. This somewhat schizophrenic approach, recognizing the economic relevance in one type of accounts but denying it in another, is rectified by the SEEA to some extent. It is achieved by shifting the market value losses from depletion and degradation as environmental cost into the production accounts and as environmental asset use into net capital formation of the asset accounts. As a result, certain 'green' indicators are generated such as environmentally adjusted value added (EVA), environmentally adjusted net domestic product (EDP) and environmentally adjusted capital formation (ECF). To the extent that environmental (depletion) costs are already accounted for in microeconomic budgets such adjustement of conventional aggregates is simply a correction of inaccurate national accounts aggregates. However, when microeconomic cost internalization has not taken place, the deduction of imputed environmental costs from macroaggregates is more controversial.[5]

The SEEA also applies this approach of depletion costing to non-economic (in the SNA sense) assets over which ownership is not enforced but which may provide economic benefits to actual or potential exploiters. These assets include significant natural resources such as fish in the ocean, timber in tropical forests or (traded/hunted) biota in the wilderness. Since these assets yield economic benefits they should indeed be considered as economic assets. They are, therefore, accounted as natural economic capital in the stock and flow accounts of the SEEA, irrespective of institutional arrangements of ownership. In this manner, a broader concept of economic capital is introduced while remaining fully consistent with the principle of market valuation of conventional accounting. This is one of the major advantages of the SEEA over other natural resource accounting methodologies and index calculations that are delinked from the framework of national accounts.

To maintain, as much as possible, consistency and comparability with the economic concepts of the SNA, only the asset boundary is expanded in the SEEA,[6] but the production boundary is maintained in its more conservative versions. As a consequence, the economic appearance of natural assets, through discovery of new reserves or revision of estimates about existing reserves, is treated, as in the SNA, as a 'transfer' (rather than capital formation) from the environment to the economic system. While some disagree,[7] it can be argued that these (non-produced) assets were created by nature without human intervention, i.e. not through an economic production process. Both SNA and SEEA therefore record these transfers as other changes in volume in asset accounts, similar to holding gains, outside the production and income accounts. These transfers should not be confused with expenditures for exploration of mineral resources or improvement of land which are treated by both SNA and SEEA as capital formation of intangible (discovery through exploration) and tangible (land improved) capital.

Depletion and degradation of renewable natural resources (e.g. forests or fish in the ocean) are included in the SNA as other volume changes. However, at least some of these assets do not seem to appear in the balance sheets as opening or closing stocks. The reason is that they are not considered to be economic unless large-scale exploitation takes place (Commission of the European Communities *et al.*, 1993, para. 12.19). A somewhat artificial construct is suggested in this case for environmental accounting, namely that 'transfer to economic uses and depletion take place at the same time' (Commission of the European Communities *et al.*, 1993, para. 21.163). In principle, however, the SEEA should fully account for the potential economic value of natural resources, embedded in the environment. This value would thus include the exploitable timber content of forests in the wilderness or fish stocks in the ocean, irrespective of the scale of actual exploitation.

The nature of natural resource stocks

There has been some debate whether natural resources are more in the nature of single-use commodities stored in inventories, or of fixed assets that can be used repeatedly over more than one accounting period. This distinction has direct consequences for the treatment of asset depletion as intermediate consumption (from inventories), affecting gross value added (GVA) and its sum total GDP, or as capital consumption (of fixed assets), generating (by deduction from GDP/GVA) the corresponding net values NVA and NDP. These considerations would also affect the valuation of depletion, taking into account, or not, future income streams from resource use (see below).

From a conventional national accounting point of view, natural resources are neither fixed assets nor inventories, since they are not 'produced as outputs from production' (Commission of the European Communities *et al.*, 1993, para. 6.185). They did not enter the economic system by means of a transaction, and their use can thus not be accounted for, in the SNA, as a re-allocation (over time) of previous capital formation to intermediate or capital consumption, i.e. as cost in current production. It is not plausible, therefore, to argue that SNA asset accounting principles require that these depletion values be shifted from other volume changes in asset accounts as depletion cost to the production accounts (Hill and Harrison, 1995, p. 5; El Serafy, 1989, p. 17). However, if SNA principles are modified to include non-produced natural resources in expanded asset boundaries, as in the SEEA satellite accounts, the case can be made for imputing depletion costs in addition to capital consumption in the production accounts. Deducting these costs from gross or net value added obtains a more sustainable (Bartelmus, 1994a, p. 70) measure of net production and income generated than the SNA measures of economic performance, GDP or NDP. In this sense, an altogether different objective of sustainability of production and income generation is explicitly introduced into neutral, i.e. transaction-oriented, duplication-free conventional accounting.[8]

When accounting for (improved) sustainability in the broader framework of the SEEA, SNA analogies do not really have to be conjured up to justify depletion cost deduction, either from gross or net values. Depletion and degradation represent values of capital consumption, which should be deducted in any measure of (more) sustainable economic performance or growth. On the other hand, it is a fact that natural resource stocks such as subsoil resources, timber in forests and fish in the ocean are not readily available for single use like an inventory of produced goods. Rather, they form an entity (deposit of ores or ecological system) from which the raw material has to be extracted in costly and lengthy production processes. As a consequence, the returns from using these entities occur with delays and over long periods of time. They need therefore to be discounted when assessing the value and changes in value of the entity. In this sense, natural resources exhibit characteristics of fixed assets that are worn down by use over more than one accounting period.

Market valuation in asset and production accounts

For many natural assets, there are no markets for trading the assets due to their common property nature. This presents, of course, the problem of finding a market value for these assets (and their use), consistent with SNA valuation principles.[9] There appears indeed to be market failure since the assets themselves are not traded in the market but their outputs, resulting from resource exploitation, are. A number of methods for estimating market values have been advanced for this case. Invariably, they are based on observed and predicted market prices and costs of the resource outputs. The more common methods are presented in the Appendix. The following discusses the underlying assumptions and practicability of these valuation methods.

Given the long-term exploitation characteristics of economic natural assets, there seems to be no dispute about the applicability of the opportunity–cost principle of valuing economic assets. Economic assets will be acquired and put to a particular use in production, assuming rational economic behaviour and perfectly competitive markets, if their discounted expected net returns exceed the discounted returns from any other investment opportunity available. The agreed market price for the sale/purchase of the asset will reflect these expectations of future returns. In the case of using non-marketed natural resources, no such transaction takes place, but a market value can be simulated by applying the same rationale for exploiting the resource. The discounted flow of expected net returns from resource exploitation can thus be assumed to reflect the expenditure investors would be willing to incur if the resource could be purchased in the market. Obviously, this valuation can be applied to both, the resource value at a particular point in time (in balance sheets) and the change in value between two points in time (accounted for in asset accounts).

The present-value method (see equation 1 of the Appendix) of estimating the natural asset value is thus the generally accepted method for (non-marketed) asset valuation in the SNA (Commission of the European Communities *et al.*, 1993, para. 13.28) and the SEEA (United Nations, 1993a, paras 163–167). However, there are several problems in applying this method. First, there is the question of what are the components of income and value added from asset use and what constitutes capital consumption, i.e. depletion cost. A second question is to what extent changes in the prices of resource outputs should be reflected in depletion cost or should be accounted as holding gains or losses in separate revaluation accounts. A third question is how to deal with negative returns from asset use, i.e. potentially negative asset values.

Sustainability and capital maintenance

The demarcation between income and capital (consumption) has been the preoccupation of economists for the last 200 years. The debate seemed to have found a resting place in national accounts conventions, notably the recently revised SNA and its elaborate principles of asset accounting (Commission of the European Communities *et al.*, 1993, Chs X–XIII). However, newly discovered scarcities of non-produced natural assets have raised new questions about the fine lines between income from (natural) capital use, discovery of natural resources and the costs of capital consumption and maintenance.

As pointed out above, the consumption of non-produced assets cannot be costed as a shift from a previously produced inventory of capital goods and other commodities into production and consumption. Rather, one has to resort to the more normative criterion of sustainability of production and income generation which, at least theoretically, could disrupt the consistency with SNA measurement and valuation principles. Sustainability measurement aims to locate that part of actual revenue generated during an accounting period which needs to be retained (saved) to sustain long-term per-capita (final) consumption through reinvestment in production capacity. The residual income could then be used for current consumption – a significant aspect of the standard of living or economic well-being.

There are major difficulties of assessing sustainability, considering in particular the role of technological progress, changes in lifestyles (consumption patterns) and disasters in lengthening or shortening the life of production capacities. For purposes of actual measurement, the sustainability question is therefore usually reduced to the principle of 'keeping capital intact' (United Nations, 1993a, para. 257) as 'a reasonable indicator of baseline sustainability' (Hartwick, 1994, p. 32). In this manner, the sustainability criterion is shifted from (welfare-oriented) income/consumption maintenance to (more technology/cost-oriented) production maintenance.[10] This approach is indeed quite consistent with SNA's capital maintenance or replacement philosophy.

Further problems arise if natural assets are exhaustible, i.e. non-renewable. In this case, the sustainability criterion, at least for capital maintenance, has to be weakened. Several sustainability concepts have been advanced, reflecting different degrees of capital conservation (Daly, 1991b; Serageldin and Steer, 1994). For purposes of environmental accounting, three categories can be usefully distinguished. Strong sustainability would require full preservation of the natural resource. Conditionally strong sustainability would take possibilities of substitution of natural resource inputs in the same or similar production process(es) into account. Weak sustainability, on the other hand, would focus on the preservation of income flows rather than production processes without regard to capital composition. Such income maintenance could be achieved by reinvesting revenues into capital formation in other production processes, i.e. through diversification, or through tapping financial markets. Strong sustainability in capital maintenance seems to be reasonable in the case of renewable resources which could either be exploited up to the point of natural regeneration or beyond if appropriate restoration (e.g. reforestation) is undertaken. The depletion cost would, in the latter case, be the actual or potential (annualized) restoration expenses.[11]

The more problematic case is when natural resources cannot be replaced because they are non-renewable or, even if renewable, because they have been depleted close to extinction or the original quality levels can no longer be attained. The latter may be the case when tropical forests are destroyed for agriculture or timber production, or land is severely eroded, e.g. by agricultural malpractices. Determining substitution possibilities and costing them, i.e. applying the conditional sustainability criterion, might introduce inconsistencies in choosing more or less arbitrarily technological alternatives. On the other hand, the strong sustainability criterion, i.e. the full preservation of the resource, is not a viable solution for non-renewable resources; non-use of an exploitable natural resource can, from an economic point of view, be considered as squandering part of a country's development potential. This leaves us with the weak sustainability criterion of maintaining the overall capital base for production or income generation. The question then is how much of the income generated from resource extraction is true income, available for consumption, saving or net investment, how much is depletion cost, required for reinvestment and to be deducted from gross value added, and how much is just gain or loss from holding assets when prices change.

Depletion, income and value added

Considering natural resources to be similar in nature to fixed assets permits use to be made of SNA criteria when defining depletion or degradation as capital consumption. These criteria describe replacement costing and measuring current value change of capital assets as two sides of the same coin (Commission of the European Communities *et al.*, 1993, paras 1.63 and 6.179).[12] Thus, when replacement costing is not possible, due to the non-renewability of exhaustible resources, the only valuation consistent with SNA's capital consumption valuation is to measure depletion cost as part of the change in the economic value of the natural assets, due to their use in the production of resource outputs;[13] this value was defined above as the present value of future net returns to capital.

The change in value of a non-renewable natural asset, due to extraction or harvest,[14] ΔV, amounts, at the end of the accounting period (see Appendix), to

$$\Delta V = V_1 - V_0 = -R_0 + \frac{V_1}{1+r} \times r \tag{1}$$

ΔV is thus composed of the actual net return R_0 from sales of the extracted resource during the accounting period, offset partially by the (discounted) interest gained by moving up one period in time (Roy, 1977, as cited in Hartwick and Hageman, 1993, p. 220) on the remaining capital V_1. This is the income that would be earned if V_1 were sold and its value invested elsewhere (in the next best available investment opportunity). This opportunity cost has been referred to as true income available for consumption; the value remaining after the deduction of true income from net returns (profit), i.e. ΔV, is termed depreciation (Hartwick and Hageman, 1993) of the asset and could be reinvested to maintain the capital value.

The net return R_0, realized from the sales of resource outputs, is defined as the revenue received minus all costs of exploitation, exploration and development, including a normal return to fixed capital employed in these activities.

Box 3: Rents, rentals and royalties

Different definitions of rent are used in economic analysis and accounting, creating a good deal of confusion. The Hotelling rent is defined as the net return from the sale of a resource output in a situation where the so-called Hotelling rule of net-price change applies (see below). In this case, the Hotelling rent amounts to the product of the physical decrease of an exhaustible resource times the net price, i.e. the market price minus all marginal costs of exploitation. The Hotelling rent concept thus differs from the 'pure economic rent' concept which refers to the return to perfectly inelastic (regarding its supply) production factors (Samuelson and Nordhaus, 1992, p. 158). Such rent can be siphoned off by private and public owners of the asset by means of royalties, fees, bonuses or taxation without changing supply and demand patterns and thus resource allocation (efficiency) in the economy. These payments, also called rent by the SNA, are made to the owner of the asset and are accounted for as property income; they do not affect value added generated (Commission of the European Communities *et al.*, 1993, paras 7.87 and 7.91). The use of royalty is an ambivalent term since it may refer to payments for asset services, also called rentals in the SNA (Commission of the European Communities *et al.*, 1993, para. 6.181), e.g. provided by patent authors, but also to property income when paid for the exploitation rights, notably of subsoil assets.

This value is frequently referred to as rent, or somewhat inexactly as 'Hotelling rent' (see Box 3 and below). Rearranging equation (1), realized net return can be expressed as

$$R_0 = -\Delta V + \frac{V_1}{1+r} \times r \tag{2}$$

i.e. as a reduction in the value of the asset from extraction during the accounting period and a positive element of potential interest gained on capital. The interest factor can be interpreted as receivable interest on the remaining asset and is thus more in the nature of property income. As such it should not affect operating surplus or value added, i.e. income generated in production (Commission of the European Communities *et al.*, 1993, para. 7.91). This makes for the total net return component R_0 to account for the cost of natural resource depletion (from extraction), i.e. capital consumption (rather than depreciation as defined above), in the calculation of environmentally adjusted value added of the depleting sector.

It has been argued, however, that only part of the net return or rent needs to be re-invested to keep capital intact, while the remainder should be retained as operating surplus and part of value added generated by resource use (see below). The determination of this amount is a matter of considerable controversy in the case of non-renewable resources where replacement costs cannot be determined. When introducing the consumption of natural capital into production and income generation accounting, the basic principle should probably be to stay as close as possible to capital stock maintenance and actual physical change in capital used in production, rather than to introduce, at this stage, income maintenance and depreciation criteria.

Criteria of (weak) income sustainability would have to consider the possibility of drawing interest on cash returns. As property income, interest is identified as part of the primary distribution of income, rather than the generation of income in production. Thus preference should be given at the initial stages of net production (net value added, NDP) accounting to the full deduction of the net return from resource extraction. Further analysis could then be conducted on the basis of income, i.e. national income or disposable income maintenance and distribution, for consumption and welfare generation. This would apply in particular to the proposed user-cost allowance, discussed below.[15]

Valuation and revaluation

The change in value of an asset ΔV, defined in equation (1), may include elements of holding gains or losses due to actual (during the past accounting period) and anticipated price increases. It appears quite impossible to extricate anticipated price changes in net returns in order to obtain a value of income forgone free from potential holding gains/losses and resulting from extraction only. However, for price changes during the accounting period some adjustment could be made 'with reference to a constant set of prices' (Hill and Harrison, 1994, p. 411; see below, for the special case of Hotelling rents).

Holding gains and losses of asset stocks during the accounting period have to be determined as a residual, after deducting from the current value of closing stocks (1) the value of opening stocks, calculated as the present value at the beginning of the accounting period, (2) the value of resource extraction and (3) other volume changes (from growth, discovery, natural disasters, revisions of stock estimates, or change in use). Flows (2) and (3) would have to be valued in principle at the unit present value when extraction or other value change occurs. For practical reasons, the average unit value (at the beginning and end of the accounting period) might have to be applied (Hill and Harrison, 1994, p. 411).

Equations (1), (2) and (3) in the Appendix show that the present value of an asset varies with the amount of net returns obtained and anticipated, and the discount rate applied. The latter is assumed to be constant, for simplicity reasons. If the rate varies over time it reflects changing expectations about the returns of alternative investments, which affects the preferences about the speed of extraction (now or later) and thus about gains or losses from holding the assets or running them down. In this sense, value changes due to changes in the discount rate can indeed be interpreted as revaluation elements (Hill and Harrison, 1995, p. 2).

Negative net returns

The case of net returns turning negative due to real losses from non-profitable assets or from doctored ones, e.g. through transfer pricing, has been raised as a problem in asset valuation (Born, 1995, pp. 6/7). Two cases can be distinguished. On one hand, assets may continue to be exploited, despite actual losses, in the hope of better times (price increases or higher quality discoveries); on the other hand, if expectations of future revenues are pessimistic, operations may be terminated, involving closing costs such as the clean-up or rehabilitation of the site of operations. In both cases, actual losses reflecting negative net returns are fully accounted for in production and income as negative operating surplus. The change of the asset from one that has generated positive net returns to one which, because of price decrease or cost increase, produces negative net returns should be recorded as economic disappearance of a non-produced asset, in line with SNA conventions (Commission of the European Communities *et al.*, 1993, para. 12.28). Contrary to economic disappearance due to depletion, accounted for as a production cost in the SEEA, this type of disappearance is not due to a production process and should be retained as other change in volume in the SEEA. Price changes or new technologies could of course bring about the (re)appearance of the (possibly physically diminished) asset as a proven reserve which per definition entails a positive potential net return from exploitation. The economic appearance of the asset would also be accounted for as other volume change in both SNA and SEEA.

Simplified approaches: net rent and user-cost allowance

Present value calculations require information about future net returns, based on estimates of the lifetime of the resource, revenues from anticipated extraction and future costs of extraction, exploration and development of the resource. In addition, an appropriate discount rate needs to be determined. The choice of the discount rate is highly controversial as it is used for justifying social time preferences such as the preservation of natural capital for future generations, as well as economic preferences for saving and increased earning through investment. As a result, proposals for discount rates have ranged between 0 and 22% (Born, 1992). Of course, the discount rate may also change in the future, though constancy of the rate is usually assumed in present value calculations. A number of suggestions have been advanced to deal with uncertainties regarding the discount rate and the prediction of future net returns.

A priori one could attempt predicting costs, prices and discount rates by the usual forecasting methods based on extrapolation of past trends, regression analysis of relevant determinants of revenues and costs, and more sophisticated modelling of future market conditions. However, such estimations are fraught with questionable assumptions about future long-term trends in volatile markets, possible discoveries, resource qualities that may affect relative prices, and technological developments (Landefeld and Hines, 1985, pp. 11–12; Born, 1992, pp. 53 *et seq.*). At least two proposals have been made, therefore, for circumventing uncertain predictions by simplifying the original present-value method. One does away with future returns by applying the net price or net return from the sale of the resource output to stocks and output. The other method calculates a user-cost allowance, assuming constant net returns over the lifetime of the resource. The validity of these simplified approaches is discussed in the following.

Hotelling rent assumptions in net price valuation

The net price method reduces the present value calculation to simply taking the net return from the last unit of resource output extracted during the accounting period, as the value with which to multiply stocks and stock changes. Equation (3) defines the net price N_t as the difference between the market price per unit of the resource p_t and the unit marginal cost of extraction, development and exploration c_t at the time of stock taking or transaction:

$$N_t = p_t - c_t \qquad (3)$$

If net price costing of the depletion of natural resources is already common practice in corporate accounting (Landefeld and Hines, 1985, p. 14; Commission of the European Communities *et al.*, 1993, para. 21.157), these costs are fully reflected in market prices. In this case, national, environmental accounting would simply retrieve depletion costs from the residual value of operating surplus for the adjustment of environmentally inflated value added. Unless comprehensive surveys of microeconomic practice are carried out, we cannot be sure whether market pricing of resource outputs is generally reflecting depletion cost.

On the other hand, the net return from an exhaustible resource has been considered in economic theory as the amount which needs to be invested to ensure non-declining (sustainable) consumption in the long term, and thus intergenerational equity in economic production and growth (Solow, 1974; Hartwick, 1977). The success of such substitution of natural resource capital by produced capital is based, however, on a number of quite unrealistic assumptions. These include again the vacuum model of perfect competition and foresight, optimal resource allocation and income maximization (see above). A further restriction is the use of an aggregate (Cobb–Douglas) production function, which permits perfect (full) substitution of all production factors with further restrictions regarding their productivity (Hartwick, 1977).

Only under these conditions, can the convenient Hotelling rule be applied. This rule states that in long-term equilibrium markets, the net price of the marginal unit extracted from a scarce resource rises with the discount rate, i.e. the rate of return on next-best alternative investment. As shown in the Appendix, the increasing revenue from the sale of one unit of the resource neutralizes in this case, more or less, the effect of the discount factor. 'More or less' refers to the further assumption that the net price increase (at the discount rate) during the accounting period be considered as revaluation rather than part of the depletion cost. Another way to look at this assumption is to assume constancy of prices during the accounting period with a sudden rise at the end of the period (Landefeld and Hines, 1985, footnote 30). Using average prices as suggested above thus approximates only the required value shift to the revaluation accounts, due to the Hotelling price increase $N_0(1 + r)$.

The net price method has been widely applied, because of its simplicity, in adjusting GDP (Repetto *et al.*, 1989; Solórzano *et al.*, 1991), NDP and net value added (in several projects of integrated environmental and economic accounting),[16] and in valuations of stocks and stock changes of particular resource accounts (Landefeld and Hines, 1985; Born, 1992). However, the validity of this method can be questioned due to the above-mentioned assumptions that do not seem to hold in reality. Apart from market imperfections (see Box 1), perfect substitutability of natural resource capital by produced or human capital is unrealistic (Daly, 1991a). Also, where a natural resource reflects different qualities, marginal exploitation costs increase with lower quality resources extracted, and the rents on marginal tons would increase at a rate lower than the discount rate. As a consequence, the Hotelling rent would overstate natural resource depletion. This effect is compounded if average costs are used instead of marginal costs in estimating the net price, since marginal costs generally exceed average costs. The net price method can thus be considered as an 'upper limit on economic depreciation' of the resource (Hartwick and Lindsey, 1989, as cited in Born, 1992).

The question is whether it is necessary to separate out this overstatement of the Hotelling rent for purposes of estimating true depletion costs which are free from income elements resulting from market power (market imperfections), differing resource qualities (Ricardian rents) or other factors (such as locational advantages). Given the general problematique of market price accounting in a far from perfect economy, such separation would be inconsistent with national accounting practice and its practical value, given obvious data gaps at the national level, remains doubtful. All in all, theoretical arguments (of Hotelling rent and sustainability), national accounting practice (of distinguishing between capital consumption/maintenance and revaluation), microeconomic accounting practices and data availability seem to argue for using the net price in assessing the loss of natural capital from its use in production.

Constant returns in the user-cost allowance

The net price method has been criticized as overstating the cost of resource extraction by obliterating the income generated in this activity. This is despite 'the fact that countries with marketable natural resources are evidently better off than those without such resources' (El Serafy, 1989, p. 13). However, this argument seems to confuse microeconomic income concepts such as Hicksian income with the broader macroeconomic objective of national income generation from a sustainable production base. From this point of view, there is of course considerable advantage in being able to produce value added from resource extraction in terms of returns to labour and employment and produced capital, irrespective of the particular returns to natural resource owners. This appears also to be the view taken in national accounting when measuring economic performance before accounting for disposable income (see above).

More importantly, the author then proceeds to suggest a user-cost approach to depletion costing. This approach can be shown (see Appendix) to be a simplification of the generic present-value method by assuming constancy of net returns from resource extraction. The basic idea is to convert a time-bound stream of (net) receipts R from the sales of an exhaustible resource into a permanent income stream X by investing a part of the receipts, i.e. the user-cost allowance $R - X$ over the lifetime of the resource. Only the remaining amount X of the receipts should be considered true income. Given a particular amount of net receipts R_0 for an accounting period, the calculation of the user-cost allowance $R_0 - X$ is straightforward, requiring only two additional parameters, the discount rate (r) and the lifespan (T) of the resource (El Serafy, 1989; see also Appendix):

$$R_0 - X = \frac{R_0}{(1+r)^{T+1}} \tag{4}$$

At first sight, this method seems to address the narrow assumptions of the net price method which tends to overstate the depletion cost allowance. Total net returns or Hotelling rents[17] are split up into a capital cost and income component similar to the original depreciation formula (see equation 2). However, this is achieved by introducing another assumption of constant net returns R which implies constant extraction rates and net prices. This assumption is of course as questionable as presuming net price increase at the discount rate. The latter, though, is based on a standard model of economic equilibrium, providing at least a theoretical justification for its assumptions. The net price method also has the advantage of doing away with the arbitrariness in selecting a discount rate for assessing time preferences or opportunity costs. In addition, it can be shown that the user-cost allowance is assuming sudden death of the capital asset at the end of its lifetime (Hartwick and Hageman, 1993, p. 221), irrespective of the physical wear and tear of the asset. This allowance is thus even further removed from the national accounting concept of capital consumption.

A further question is whether the microeconomic focus on income maintenance (by reinvestment in financial markets) is appropriate for macroeconomic national accounting and policy. It can be argued that maintaining the national productive capacity, impaired by a run-down of physical capital, is of greater priority in policies of self-reliant domestic economic performance and growth, than providing for a permanent provision of returns from financial investments. Indeed, the user-cost valuation cannot be applied in asset accounting as defined in the SNA. This is because this valuation does not distinguish between actual depletion or extraction cost, other volume changes and revaluation, required for the compilation of these accounts.[18] The user-cost approach might thus be of some relevance in estimating a sustainable income figure but introduces non-comparability (with produced capital) when attempting to measure the maintenance (sustainability) of productive wealth and production.

On the other hand, the determination of a lower, income sustaining, value of depletion cost by the user-cost approach and of an upper, production sustaining, value by the net price method provides a range of estimates of comparative interest. This is the reason why in case studies of the SEEA both values were routinely estimated. Perhaps the ultimate, though theoretically not necessarily satisfying, criterion of deciding for one or the other method will come from corporate accounting and reporting practices. Unfortunately, corporate environmental accounting rules are only in the first stages of discussion and development (Bartelmus, 1994a, p. 51).

The cost of environmental degradation

Beyond markets: pricing the priceless

Accounting only for the depletion of economic assets, as for instance in the case studies of the World Resources Institute (Repetto *et al.*, 1989; Solórzano *et al.*, 1991), can be criticized for neglecting or denying the real social costs of production and consumption. Those costs are the result of external effects of economic activities on the environment which in turn affect the quality of human life, if not life itself. The costs of destroying life support systems through pollution and other forms of environmental degradation, such as destructive land use, can be far greater than the expenditure for economic natural capital maintenance, notably in industrialized countries. Box 4 confirms this argument for a few case studies, even when 'degradation' is measured conservatively as the maintenance cost of environmental services.

Box 4: Depletion and degradation cost in case studies of SEEA application

A number of case studies applying the SEEA, have been or are being conducted in developing countries (Colombia, Ghana, Indonesia, Mexico, Papua New Guinea, Thailand), a newly industrialized country (Republic of Korea) and an industrialized country (Japan). Preliminary and not always fully comparable (due to coverage, data availability and experimentation with SEEA methodologies) results show a widely differing significance of depletion of natural resources vs. environmental quality degradation in developing and industrialized countries. The following ratios of depletion over degradation cost (valued at maintenance cost, see text) have been compiled from the above case studies:

- Japan (1990): 2.4%
- Republic of Korea (1992): 3%
- Mexico (1985): 74%
- Papua New Guinea (1990): 300%
- Ghana (1993): 112%.

The importance of total environmental (depletion and degradation) cost, expressed in per cent of NDP, also varies considerably:

- Japan (1985): 2.8% and (1990): 2.3%
- Republic of Korea (1985–1992): 2–5%
- Mexico (1985): 13%
- Papua New Guinea (1986–1990): 3–10%
- Ghana (1991–1993): 1.2–12%.

Sources: *Ghana*: Ghana Statistical Service (1996); *Japan*: Department of National Accounts (1995); *Mexico*: van Tongeren *et al.* (1991); *Papua New Guinea*: Bartelmus *et al.* (1992); *Republic of Korea*: Korean Environmental Technology Research Institute (in prep.).

Environmental accounting: satellite system or modelling?

The term economic is used in national accounting in reference to market valuation and market transaction boundaries. This narrow definition facilitates measurability and the analysis of allocation efficiency. However, it represents a radical reduction of the scope of economics which is generally perceived as dealing with scarcity of goods and services, irrespective of the existence of markets or other allocation or distribution mechanisms. This is indeed the major reason for environmental economists and environmental accountants to broaden the scope of conventional economics and accounting to include environmental assets and the costs and benefits of their change.

However, conventional economists and national accountants have resisted the incorporation of environmental concerns (beyond natural resource depletion) in national accounts, even in satellite accounts. They emphasize allocation efficiency of markets and the measurability aspect of observable market phenomena. The costing of environmental externalities is considered as an issue of research (Commission of the European Communities *et al.*, 1995), modelling (van Dieren, 1995) or ancillary systems (Hill and Harrison, 1995), rather than a matter of statistical measurement in economic and natural resource accounts. Specifically, reference is made to (1) the absence of commercial (corporate) accounting practices, contrary to natural resource (depletion) accounting (Commission of the European Communities *et al.*, 1993, para. 21.157), (2) the introduction of notional (shadow) prices and effects into a system based on market prices and (3) the 'dislocation of accounting links between flow and stock accounts' (Hill and Harrison, 1995, p. 5).

The first two arguments also seem to apply, though perhaps to a lesser degree, to depletion accounting. As indicated above, there are no general accounting rules for depletion at the corporate level, though some resource exploiting industries might already make an allowance for running down their resource stocks. In fact, balance sheet provisions for pollution have also been made by some chemical concerns in anticipation of potential liability for the clean-up of toxic wastes.[19] The third argument is not convincing either. To date, hardly any country has routinely compiled asset accounts, but all have routinely published production and income accounts that do stand on their own in economic analyses. As to the shadow pricing argument, the widely propagated Hotelling rent calculation for resource depletion has been justified by a quite unrealistic model of optimization and general equilibrium; the corresponding net price costing of depletion is thus indeed a shadow pricing approach. On the other hand, widespread incorporation of depletion costs in corporate accounts would justify a corrective cost allowance in value added and NDP, just as for produced capital consumption.

In the case of environmental externalities, such practice of environmental cost accounting can probably not yet be assumed. As shown below, the deduction of hypothetical degradation cost can thus be justified only at the microeconomic level where cost internalization would hardly affect relative prices and production and consumption patterns of the economy. At the national level, total potential cost could only be compared, in principle, with the sectoral and national aggregates that would result from cost internalization and corresponding structural adaptation of the economy. The determination of these aggregates would indeed require the modelling of an alternative scenario, based on assumptions about production and consumption technologies and market behaviour.

Given the fact that environmental impacts actually occur during the accounting period, the case can be made, however, for applying maintenance or prevention costs as weights for the aggregation of environmental impacts from economic sectors and the economy as a whole. Such costing would at least provide a snapshot of the immediate environmental impacts of economic activity, indicating what costs should have been borne (internalized) by those who caused them. The comparison with actual value added and NDP generated reveals the initial impact of economic production and consumption on the environment without entering the realm of modelling alternative levels and structures of economic performance.

On the other hand, such modelling would indeed be useful for allocating the initial costs generated to those who ultimately have to bear them. Immediate impact costing, at hypothetical avoidance cost, provides the link between ex-post accounting for environmental impacts and ex-ante modelling of the ultimate effects of full (environmental) cost pricing on the economy. The incorporation of hypothetical costs of environmental maintenance would then generate hypothetical (modelled) aggregates such as an 'analytical green GDP' (Vu and van Tongeren, 1995), an 'optimal net domestic product with regard to environmental targets' (see Meyer and Ewerhart, Chapter 22) or a 'maximum NDP generated in an economy in which the burden on the environment is reduced to a sustainable level' (de Boer *et al.*, 1994). The different assumptions behind each particular model and corresponding ranges of results are the price to pay for abandoning reasonably standardized measures from environmental and national accounts for the assessment of hypothetical costs in hypothetical situations.[20]

Much of the resistance of national accountants towards monetary environmental accounting might thus reflect an anxiety of 'official' statisticians to change well established systems on which their reputation of objectivity has been built. They believe that much of this goodwill might be lost if controversial concepts and modelling assumptions are let in through the backdoor of supplementary systems. There is, however, a trade-off when choosing between accurate but less relevant information and less accurate estimates that address the priority concerns of society. This seems to be the case in the area of environment and development where data users continue to make urgent calls for implementing integrated environmental and economic accounting.[21]

The following discusses the role of environmental functions in production theory and welfare analysis. The purpose is to examine to what extent proposed valuations of environmental assets and their quality changes are consistent with economic theory and conventional accounting principles.

Environmental production cost: factor inputs of environmental services

A link between the (non-economic) environmental and economic assets is provided by treating the economic appearance of natural resources as transfers from the environment to the economy (in the above-described SNA sense). These transfers are made visible from both ends in the SEEA by (1) introducing explicitly a further asset, the environment, in the asset accounts, and (2) reflecting certain changes of the assets in the production/income accounts.

The shift of natural resource assets from the environment to the economy does not explain, however, other potentially more damaging changes in the environment. They result from using environmental media of air, water and land as sinks for wastes and pollutants, and from the direct destruction of ecological systems, e.g. by improper agricultural or touristic activities. Commonly, the effects of these changes are referred to as environmental externalities. They are considered to be qualitative environmental degradation, as distinguished from quantitative natural resource depletion. However, both quantitative and qualitative changes may occur in all categories of assets which, in fact, may exhibit simultaneously environmental and economic functions.[22]

There are basically two justifications for introducing the impairment of environmental functions into the national accounts. One considers the environment as a further non-produced factor of production which provides capital services of waste absorption. The other goes a step further in treating the effects of losing environmental services as a reduction in human welfare, closer in nature to income and consumption effects than to production costs.

Economic activity of production and consumption can both be held responsible for discharging residuals into the natural environment. As long as these residuals are safely absorbed by self-regenerating ecosystems, nature provides its disposal service for free. In this case, the environmental service obtains, per definition, a zero economic value. The capacities of the environment to absorb these pollutants have, however, been frequently exhausted, generating social costs of damage to health, recreation, loss of genetic material and other aesthetic and ethical values (e.g. of species loss). Thus, defensive expenditures (Leipert, 1989) against the loss of environmental services had to be incurred. They reflect a displacement of factor services from welfare generating production and consumption into environmental maintenance.[23] However, when these expenditures increase the quality and quantity of the environment, beyond restoration or avoidance of current impacts, they could be considered as environmental accumulation or capital formation, similar to land improvement. If, on the other hand, social damage costs are not fully mitigated by defensive expenditure they should be considered as unaccounted for cost resulting from the loss of (depleted or degraded) environmental assets. Environmental services losses are thus similar to the permanent or long-term loss of natural resource inputs in production. The main difference is that they are intangible service inputs which are generally not traded in markets, except for the rare cases where tradable pollution permits are issued. As in the case of natural resource depletion, the reduced capacity of nature to provide factor inputs to production should thus be costed. Furthermore, the cost should be allocated to those who use the scarce environmental resource as a sink for pollutants or for other economic purposes (e.g. agricultural or residential development). Of course, and as shown below, the absence of markets, not only for the environmental assets but also for their services, makes valuation more difficult, but not impossible.

A further problem is that, contrary to natural resource use which can always be interpreted as an input into a production process, pollution generated through final consumption by households is difficult to interpret as a (production) cost. By definition, final consumption does not generate direct cost in national accounts. Externalities of consumption may either affect the utility functions of other households or generate diseconomies for enterprises. Diseconomies are already accounted for as production cost in conventional accounts. Since they reflect a picture of distorted allocation efficiency, these costs should be allocated to those agents that cause them rather than to those that bear them, in application of the maintenance valuation approach (see below). Impacts on utility functions, on the other hand, affect the well-being of individuals which is not accounted for in conventional accounts.

It has been suggested[24] that final consumption should be increased by the value of pollution, reflecting the additional consumption of newly scarce environmental (waste disposal) services by households. In a second step the same value should be deducted from capital formation as a loss of natural capital. This procedure would (1) change the consumption boundary of the SNA and (2) effectively exclude social (environmental) cost generated by households from EDP and EVA calculations. If this consumption increase is valued at maintenance cost (rather than through a more controversial assessment of individual preferences, e.g. by means of contingent valuation, see below), it would create an aggregate that is inconsistent with demand-side oriented analysis of final consumption.

For the purpose of production/performance analysis, notably in terms of EDP and EVA, an alternative treatment of household pollution is suggested in the SEEA. Pollution costs generated from household consumption are shifted to the production sector as environmental cost of other industries (United Nations, 1993a, para. 275). This procedure ensures that the total social cost of maintaining environmental quality, through a (hypothetical) production process of conservation, is assessed, irrespective of the origins of these costs and of their ultimate effects on consumption, income and well-being.

Environmental damage cost: welfare effects of degradation

The above refers to the need for recording more comprehensively the total environmental cost of production in national accounts. Economic theory provides an alternative justification, from a benefit-lost rather than cost-incurred point of view. The focus is on the further effects of the loss of factor services on human welfare. Valuation proposals thus address the ultimate effects of contamination on human health and well-being or 'utility', i.e. the costs borne by individuals rather than the costs caused by economic activity (United Nations, 1993a, p. 91).

Undeterred by problems of quantification, theoretical economists typically present elegant explanations of welfare generation under optimal conditions. Again, a similarity to resource depletion can be observed. As in the case of Hotelling rent costing, it can be shown that environmental damage cost needs to be deducted from value added generated in production in order to obtain optimal (maximum) NNP (Dasgupta and Mäler, 1991). In turn, such costing ensures the long-term sustainability of consumption (Hamilton and Atkinson, 1995). The relevance of these models, beyond vacuum analysis (see above), is of course subject to the findings of further empirical research.

Maintenance costing and contingent valuation

Costs caused and borne

In line with the distinction between the direct use of environmental productive services and the consequential welfare effects from loss of consumer services, two main valuation approaches are advanced by the SEEA. Maintenance costing focuses on the direct impacts of production in assessing the costs of keeping environmental assets intact. Contingent and related valuations of the demand/benefits/damage side of environmental services, on the other hand, attempt to measure the loss or generation of ecological and environmental consumer services. Ecological services include repercussions (indirect effects) of direct environmental impacts such as the effects of SO_2 emissions on the quality and habitat functions of forests via acid precipitation. Environmental services consist of aesthetic, health or recreational effects borne by human beings.

Both approaches attempt to find, in principle, market values for measuring environmental costs and benefits. Maintenance costing searches for market values of factor inputs, required for the potential restoration or avoidance of environmental impacts caused by economic agents. Contingent and related demand-side or use valuations attempt to measure preferences for (outputs of) environmental services or dispreferences for their impairment, as if markets existed for these services.

Contingent valuation thus reflects the environmental cost borne by ecosystems and individuals. However, maintenance costing appears to exhibit characteristics of both the cost-caused and cost-borne concepts, if restoration costs are included. The concept of restoration cost was advanced by Hueting (1989) as a supplement to national income. It represents the expenditures that society would have to incur to restore environmental quality lost in the past to desirable levels, expressed in sustainability standards. These quality losses can, of course, not be traced back to those who caused them, e.g. through emissions in past accounting periods or from transboundary pollution by the rest of the world. While reflecting a maintenance cost concept of restoring environmental assets, restoration costs thus also represent the environmental costs accumulated and borne by society over time. As indicated in Box 5, the cost accumulation concept does not fit into periodic income and production accounting and is therefore not further pursued here. It may have some analytical value, however, in assessing 'in monetary terms how far a nation has drifted away from a sustainable development' (Hueting, 1989, p. 37).

Box 5: Restoration cost and environmental debt

Environmental cost accumulation, as reflected in the concept of restoration cost, represents an environmental debt (National Institute of Economic Research and Statistics Sweden, 1994, p. 40). Environmental debt, incurred by past generations (or accounting periods), indicates the responsibility of the current generation (for the sins of their own and past generations) for environmental effects that will have to be borne by future generations. In its reference to an accumulated environmental burden, this concept is not strictly comparable with production and income aggregates, as maintained by Hueting and Bosch (1994, p. 237). Even if this burden is distributed over past and the current accounting periods, it will hardly be possible to allocate it to causing sectors (in the domestic and foreign economies), permitting only an overall adjustment of total national income.

Of course, just as in the case of conventional monetary debt, a comparison with income, saving, and financial assets and liabilities could shed some light on the capability of a nation to reimburse the principal of the environmental debt in future accounting periods. Such analysis seems to be behind the notion of a genuine savings (World Bank, 1995, ch. 7) indicator which could be interpreted as national saving net of an annualized hypothetical repayment of environmental debt.

Cost-caused accounting can thus be taken as synonymous with maintenance or avoidance cost accounting, and contingent valuation is assumed to reflect the (environmental damage) cost-borne approach.

Taking care of nature: market simulation of environmental protection

Monetary valuation of the direct environmental impacts of production and consumption permits expanding the scope of environmental accounting while maintaining closeness and consistency with the production philosophy of the

SNA. Costing the maintenance of environmental 'capital' is the anchor which prevents environmental accounts from drifting away into the realm of welfare measurement and analysis. On the other hand, it is welfare analysis which provides a theoretical justification for incorporating environmental concerns into the accounting framework.

Setting out from the premise that environmental services can be commodified, economic theory can be used to show how their value is determined by the interaction of (simulated) market forces. New scarcities of environmental services are reflected in the generation of social environmental damage costs which in turn incite environmental protection expenditures to prevent or mitigate the social costs. The supply of protection services can thus be compared with the demand for environmental quality by final users.

Valuing public goods is common practice in the SNA where collective services are measured at production cost (Commission of the European Communities *et al.*, 1993, para. 9.91). The same valuation principle could be applied to the environment and its collective services of waste absorption. There is a difference, however, since there are no direct expenditures by the supplier, nature, for inputs into waste disposal. This drawback can be overcome, considering that economic agents would have to step in for nature, once its supply is impaired, i.e. once its services have become scarce. Government, in particular, as caretaker of public goods, would have to ensure that an appropriate supply of waste disposal services is maintained; governmental agencies are usually well equipped to preserve nature's absorptive capacities through environmental protection.

In perfectly competitive markets, equality of the marginal benefits obtained from environmental damage reduction with the marginal cost of environmental protection permits to use either marginal protection or marginal damage cost as the unit value of environmental services. This is of course textbook economics of market clearance, illustrated in Figure 1. The figure plots the supply S of environmental quality E (expressed in terms of reduced emissions e), as a function of the marginal environmental protection (emission avoidance or reduction) costs C, against the demand D for environmental quality, as a function of the willingness to pay (WTP) for these costs.

Market clearance is obtained at the optimal emission reduction level e^*. At this level, the marginal costs c^* of emission abatement (reflected by the supply curve S) are equal to the perceived marginal benefits of abatement (reflected by the WTP of the demand side). The marginal costs c^* can thus be used to value the difference between actual emission abatement e', carried out during the accounting period, and optimal abatement e^*. Equation (5) defines thus the correct value (under the assumption of a perfectly competitive market for environmental protection) of the environmental damage cost (EDC), actually generated during the accounting period, as

$$\text{EDC} = (e^* - e')c^* \tag{5}$$

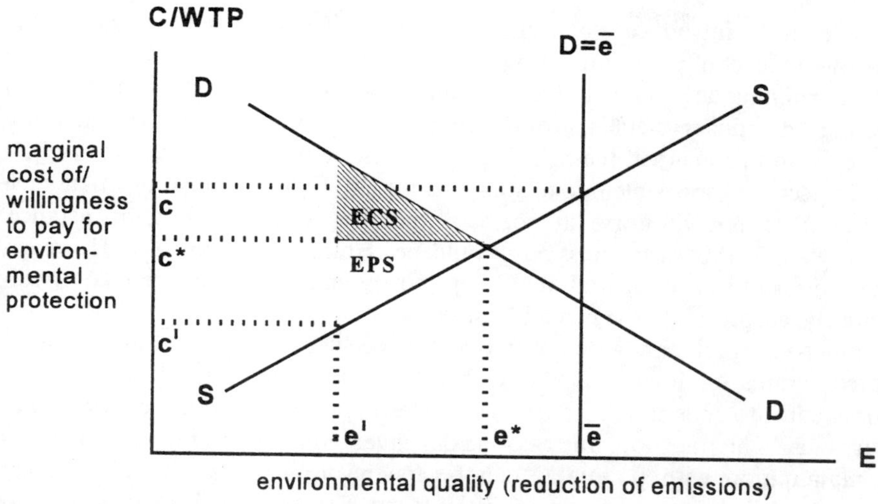

Figure 1. Market simulation of environmental protection.

EDC are the environmental costs caused during the accounting period that have to be imputed (beyond actual protection costs incurred, i.e. *e'c'*) to those who are responsible for their generation. This cost allocation reflects the cost internalization criterion of the polluter pays principle (see below).

It should be noted that this value of EDC is consistent with market valuation in national accounts. It includes an environmental producer surplus, EPS (Møllgard, 1995, p. 8), shown as a white triangle in Figure 1, which would be omitted if actual (marginal) cost incurred *c'* is used for environmental damage costing. A similar omission would occur (depending on the slope of the cost curve) if estimated average cost were applied. Also, in line with market valuation, environmental consumer surplus (beyond actual expenses), ECS (shaded), which is part of contingent valuations of the WTP, is excluded in the value of EDC.

Textbook economics tells us also that EPS is disappearing in the long run through competition. In this case, actual (minimum average) costs of the best available technologies would be a good approximation for (long-term) environmental damage costing. In the short-term, however, marginal costs can be expected to be greater than average costs, and using the latter would understate the value of damage (avoidance) cost. Still, even a horizontal supply curve (see Figure 2) does not solve the problem of determining demand, which would have to resort to assessing individual preferences for environmental quality. Contingent and related valuations have been proposed to this end, without much success at the national (accounting) level (see below).

Maintenance costing of environmental asset use

One way to circumvent the problem of estimating demand for environmental services is to introduce environmental standards that refer to a permissible maximum level of pollution, set by experts or governmental fiat. As in the case of determining environmental debt to future generations (see Box 5), standard setting introduces elements of (normative) evaluation into the valuation process. In this manner, the invisible hand of more or less competitive markets is replaced by the visible one of standard setters. Figure 1 illustrates that such standard setting will normally not be at the optimal level e^* but, for instance, at level $\bar{e}$. In this case, emissions would be reduced at a cost of $\bar{c}$. However, consumers of environmental quality are not prepared to meet these costs voluntarily, according to the demand curve.

Environmental standards are frequently specified as ambient concentration or contamination limits. They need then to be translated for cost-caused measurement into emission standards and corresponding emission reduction standards $\bar{e}$. This is not an easy task, given that concentration of and contamination with pollutants can be the result of emissions from earlier periods and other regions or countries. One way around this problem is to assume, not unrealistically in most situations, that optimal pollution levels have already been exceeded and additional emissions beyond the actual reduction level e' cannot be tolerated. This represents in effect the special case of setting the desirable emission (reduction) standard at $\bar{e} > e'$.

Figure 2 illustrates how the unknown demand curve is replaced by a particular emission reduction standard $\bar{e}$. Demand $D = \bar{e}$ is then confronted with the long-term supply curve $S = c_{\min}$. This produces an EDC estimate as the cost of complying with emission abatement at the level of $\bar{e}$, beyond actual abatement e' (shaded area):

$$\overline{\mathrm{EDC}} = (\bar{e} - e')c_{\min} \tag{6}$$

Figure 2 also shows the case where for technical or economic reasons the cost of further reduction of emissions becomes prohibitively high, i.e. the supply curve becomes vertical ($\hat{S}$), with the marginal cost of abatement approaching infinity. $\hat{e}$ is thus an upper limit of emission reduction (standards). The SEEA proposal for maintenance costing seems to refer to both, the minimum (average) costing for the best available technology (implied in the horizontal supply curve and equation 6) and the special case of optimal emission reduction levels exceeding actual reduction of emissions ($e^* \simeq \bar{e} > e'$). Maintenance costs are thus defined as the costs of using the natural environment which would have been incurred if the environment had been used in such a way that its future use had not been affected (United Nations, 1993a, para. 50). These costs are of course hypothetical because in reality an actual use did take place which affected the environment. It is therefore not surprising that most of the controversy about environmental accounting hinges on the introduction of such a valuation into an information system that focuses on observable transactions (see Vanoli, Chapter 20). On the other hand, it is precisely the omission of equally observable environmental impacts and their ensuing social cost which prompted some of the harshest attacks on the validity of the

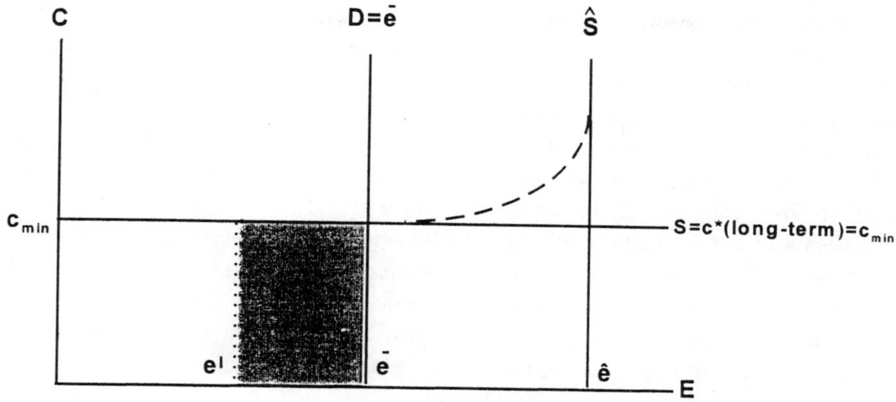

Figure 2. Costing compliance with environmental standards.

national accounts.[25]

The rationale for maintenance costing can thus be based on the following two criteria (Bartelmus and van Tongeren, 1994, p. 16): first, the application of the sustainability concept which has gained a central role in the discussion of integrated (environmentally sound and sustainable) development; and second, the extension of the national accounts concept of replacement cost of the consumption of fixed capital to valuing the use of environmental assets. The sustainability concept reflects a conservationist view of the environment. Uncertainty about possible long-term hazards from disturbing the natural environment and possible irreversibilities of environmental impacts from economic activities call for a high degree of risk aversion, i.e. the maintenance of at least the present level of environmental quality. The notions of sustainability and conservation reflect a macroeconomic view of society's taking care of a public good by means of collective environmental protection services. The link to microeconomic cost concepts and accounting can be found in the fact that the public good is affected directly by a wide range of microeconomic (production and consumption) activities. As a consequence, there is good reason to allocate the macroeconomic (social) costs to those who caused them, according to the polluter/user pays principle. Accounting for accountability is the principle behind the allocation of environmental costs to the different sectors of the economy in the SEEA.[26]

The replacement cost approach for valuing the use of environmental functions is similar to valuing the maintenance of services of man-made capital in the national accounts. The consumption of fixed capital is estimated as the amount necessary to keep the level of the man-made assets intact by means of replacement investments. Such calculation of capital consumption is also hypothetical in nature, because it is not certain whether actual investment expenditures will be incurred at maintenance cost levels.

Another application of the cost-caused and maintenance valuation concepts is for the recording of transboundary flows of residuals from and to the rest of the world. The SEEA brushes over the valuation problem in this case by simply indicating negative exports and import figures (United Nations, 1993a, para. 316). Another approach has been suggested in several country projects of UNSD, considering transboundary pollution flows as negative transfers in kind to/from the rest of the world. In line with SNA accounting rules, such transfers affect (environmentally adjusted) national income rather than domestic product. They should be valued at domestic (for residuals exported) and foreign (for residuals imported) avoidance cost according to the cost-caused principle.[27] Exports of residuals would thus increase the value of national income, while imports would decrease it because of an additional social cost transferred upon the country from abroad.

Maintenance costing covers both, degradation and depletion of natural assets. By applying maintenance valuation care should be taken to avoid possible double-counting in costing the environmental impacts from natural resource use and environmental degradation generated by the same economic activity. This is the case, for instance, when the costs of preserving forests cover both the maintenance of economic functions (timber stocks) and ecological ones (habitat, flood control etc.).[28] In the case of renewable natural resources, replacement costs would reflect maintenance or conservation criteria, applied to exploitation of the resource beyond sustainability levels. However for exhaustible resources a weaker sustainability criterion would have to be applied to economic functions of the resource to avoid squandering the resource.

Alternatively, if full conservation, even of economic functions, for future generations is desired or environmental and economic functions cannot be separated, maintenance costing would value the non-use of the natural resource in terms of value added forgone. Measures of maintaining environmental functions may include, apart from reduction of or abstention from economic activities, the substitution of outputs or inputs in production, preventive environmental protection or the more difficult to allocate restoration of degraded natural systems (Hueting *et al.*, 1991; Bartelmus and van Tongeren, 1994). For all these measures, their immediate environmental impacts, including those of substituted technologies, would have to be accounted for. Of course, as discussed above, the assessment of indirect costs generated by cost internalization requires behavioural assumptions which are the realm of modelling rather than accounting.

Contingent valuation of environmental services

Contingent and other (damage) valuations (see Box 2) are means of assessing the demand for environmental services. The scarcity of these services is reflected in environmental damage resulting from the loss or impairment of environmental functions. The SEEA addresses the measurement of environmental damage in a separate version which introduces the cost-borne valuation into environmental accounting. This valuation combines the market valuation of natural resource depletion, mainly borne by enterprises, with contingent or other valuations of the welfare effects from environmental deterioration, borne by households.

Well-known problems of applying these valuations in cost–benefit analyses at the project level[29] accumulate at the national level. At least for the time being, such valuations do not seem to be applicable in routine national accounting. However, they might be usefully explored in more experimental studies that focus on selected concerns or (subnational) regions.

It is doubtful, however, whether contingent valuations are at all suitable for incorporation into national accounts. National accounts are based on market prices which value transactions and are not capable of assessing welfare due to their exclusion of consumer surplus. As illustrated in Figure 1, contingent valuations do include this surplus when assessing the willingness to pay for environmental services. Mixing these valuations would create aggregates that are neither performance nor welfare measures and thus difficult to interpret.[30]

From valuation to evaluation

Measuring sustainability in growth and development

As pointed out in the Introduction, the measurement of environment–economy interaction is to provide the necessary information for integrative policies of sustainable growth and development. The above discussion of valuation in integrated accounting showed that a high degree of environmental and economic data integration can be achieved within the framework of integrated environmental and economic accounting. However, monetary valuation becomes arbitrary with increasing remoteness of the results of human activity and natural processes from economic output.

Development goals of equity, health, literacy, freedom or political stability are difficult to quantify, even in physical terms, and quite impossible to value in money terms. A comprehensive concept of development would have to cover all these values and amenities. Obviously, additional, non-monetary information is needed for the assessment of the sustainability of development, beyond economic growth. This can be illustrated by categorizing the concept of sustainability in a generic process of providing goods, services and non-economic amenities to society.[31]

Operational concepts of sustainability

Following the supply process of economic goods and non-economic amenities through to the final uses and users permits the distinction of the sustainability of supply/production from the sustainability of use and users. This categorization links the concept of sustainability to well defined economic production and consumption processes and to less well defined qualities of life of human users of goods, services and amenities.

A significant part of the economic sustainability of production and income is already addressed in national accounts by making an allowance for the consumption of capital in production processes. Scarcities of further inputs of natural resources and environmental services make for the extension of the sustainability criterion from produced capital maintenance to the maintenance of natural capital, as described above. In addition, the maintenance of human and institutional capital would also have to be considered in a comprehensive discussion of the sustainability of economic production and growth. Because of conceptual and statistical difficulties in defining and measuring human and institutional capital, integrated environmental and economic accounting, as proposed by the United Nations (1993a), has concentrated on produced and natural capital and its consumption or use.

For the analysis of the final destination of goods, services and amenities, a distinction can be made between final uses and the ultimate users of goods and amenities. Such analysis is more welfare- and people-orientated than the cost/technology-oriented description of production and supply. An attempt to push monetary valuation further into the realm of non-economic (dis)amenities is the application of contingent valuation to environmental effects on human health and welfare. The above-described difficulties of applying this and related (benefit/damage) valuations at the national level disqualify this approach from use in routine accounting.

Despite these difficulties, a number of indices have been proposed for correcting the perceived flaws of national income and GDP as measures of economic welfare. They include in particular an Index of Sustainable Economic Welfare (ISEW; Daly and Cobb, 1989), a measure of Net Economic Welfare (NEW; Samuelson and Nordhaus, 1992) and the ISEW-based Genuine Progress Indicator (Cobb *et al.*, 1995). These indices typically make allowances for the narrow scope of conventional economic aggregates which exclude desirables of household production, leisure and environmental services and for the excessive scope of conventional aggregates, including regrettables or defensive expenditures, e.g. for environmental protection or defence. A number of non-monetary indices, focusing on the ultimate users, i.e. human beings, of goods and amenities have also been proposed. They include an overall human development index (United Nations Development Programme, 1991) and measures of the carrying capacity of land areas for human populations (Food and Agriculture Organization *et al.*, 1982; Vitousek *et al.*, 1986).

To the extent that these indices reflect adequately final use of goods, services and amenities and/or their welfare effects, non-decline of the indices could be interpreted as a measure of sustainable development. However, a number of flaws, regarding scope, coverage, aggregation and compilation, throw doubt on the validity of these measures. This is not surprising since, invariably, they are compiled eclectically outside any consistent framework such as the national accounts with their rigorous accounting rules, definitions, balances and identities.

Valuation and evaluation – a dichotomy in indicator development

The above brief description of operationalizing the concept of sustainability points already to a basic dichotomy in measuring sustainable growth and development, either in physical (non-monetary) or in monetary terms (Bartelmus, 1995). Monetary accounting systems, notably the SNA and its satellite, the SEEA, provide conventional and environmentally adjusted aggregates of economic performance (domestic product, national income, capital and capital formation, consumption, saving etc.). Contrary to such valuation, non-monetary data frameworks, such as those for social and demographic statistics (United Nations, 1975), environment statistics (United Nations, 1984) or for indicators for sustainable development (see Box 6) generate demographic, social and environmental data and indicators in physical terms, with widely differing units of measurement. The SEEA also proposes to link physical and monetary accounts by incorporating physical data on natural assets and natural resource and residual (pollutants) flows as counterparts of the monetary environmental accounts (United Nations, 1993a, ch. III).

Box 6: Framework for Indicators of Sustainable Development (FISD)

One way to capture most of the different approaches and listings of environmental and sustainable development indicators is to combine the concerns of potential data users, as reflected in UNCED's Agenda 21, with the framework for environmental data production, the Framework for the Development of Environment Statistics (FDES; United Nations, 1984), endorsed by the international statistical community. This approach is reflected in a Framework for Indicators of Sustainable Development (FISD) proposed by the UNSD (Bartelmus, 1994b). The framework combines clusters of Agenda 21 chapters with information categories of activities, impacts, responses to impacts and inventories/stocks of the FDES. In this manner, it generates indicator topics and corresponding indicators for and of sustainable development (see below). A similar 'driving force–state–response' framework has been developed from a data use point of view by the United Nations Department for Policy Coordination and Sustainable Development (DPCSD) in collaboration with the UNSD.

Lacking the powerful aggregative capacity of economic valuation, full integration appears to be an elusive goal for physical data frameworks. Aggregation, if at all to be attempted, has to resort to explicit (e.g. expertocratic) or implicit (such as the use of equal weights in simple indicator averages) weighting.[32] A somewhat different attempt was made by Dutch statisticians to bridge the dichotomy of monetary and non-monetary indicators by introducing physical environmental indicators into the national accounts system (de Haan and Keuning, 1995). It remains to be seen if this approach, which has to resort to controversial aggregation methods,[33] will become a useful instrument of decision making in the Netherlands and beyond.

In order to increase the policy relevance of more or less extensive sets of indicators, the setting of targets or standards of basic human needs or the quality of life is typically proposed. It can be shown (Bartelmus, 1994a) that such standard setting turns the analysis of the sustainability of development into one of the feasibility of development programmes; those programmes would have to comply with an exogenously set normative framework of minimum and maximum standards and thresholds. Monetary valuation is replaced in this case by social evaluation of feasible development. The coverage of development concerns is indeed expanded, as compared to monetary valuation in integrated accounting, but at a price. The price is the introduction of normative criteria in lieu of directly measurable costs and prices.

Indicator use: policy analysis, management and score keeping[34]

Environmental accounting and indicators for sustainable development are both relatively new methodologies. Experience with their use and usefulness in decision making is still lacking. This is because, on one hand, there is not yet full consensus on concepts, classifications and methods of compilation and, on the other hand, statisticians are typically quite reluctant to enter the less objective arena of policy analysis. Most projects of environmental statistics and accounting lack therefore an assessment of actual policy use of the data produced. However, some theorizing on potential uses and applications might stimulate further exploration of how integrated data systems can support policies of sustainable development.

Accounting data and aggregates

Physical data of natural resources and residuals provide the necessary counterpart information for monetary accounting. However, physical accounts have limited aggregation capacity, across different natural resources and pollutants. They might be able to combine related resources and pollutants in common equivalency units (such as oil equivalents); but beyond that, the use and usefulness of different natural resources and the impacts of different pollutants cannot be fully reflected in one indicator or compared directly to related economic aggregates. Physical accounts can thus provide valuable information for the management of particular resources and pollutants. They cannot, however, on their own, be the basis for policy decisions that take both environmental and socioeconomic concerns and objectives into account.

Integrated monetary accounts, on the other hand, can be used to assess several key aspects of integrative economic policy (Bartelmus, 1996). They include: the (non)sustainability of economic growth; the structural distortion of the economy by environmentally unsound production and consumption patterns; and the efforts undertaken to address these issues. Conventional indicators of national income or product are typically used in the measurement and analysis of economic growth. EDP or similar aggregates could therefore be introduced into such analyses. Replacing conventional growth indicators, notably GDP or NDP, by an environmentally modified indicator, such as the EDP, provides a handy overall measure of the success or failure of integrative policy. Expanding the scope of key economic variables such as capital, capital formation and consumption in dynamic (growth) models could produce further (early warning) signals about the trends and limits of (more) sustainable economic growth. In addition, these models (within their assumptive limitations) could provide advice about how to reorient economic development toward a sustainable path.

Structural distortions in the economy from under-priced over-use of environmental resources can be addressed by full-cost pricing, i.e. environmental cost internalization into the budgets of households and enterprises. Given the inefficiencies of command and control measures in environmental protection and natural resource conservation, the application of market instruments has been generally advocated. Economic instruments of cost internalization include effluent charges, user taxes, tradable pollution permits, deposit–refund systems etc. They are usually applied to those who can be held responsible for natural resource depletion and environmental degradation, according to the user/polluter pays principle. Integrated accounting can help to define those instruments and measure the appropriate level of fiscal incentives (subsidies) or disincentives (effluent charges etc.).

The deduction of environmental costs from conventional indicators of value added does not necessarily imply that these costs are or are about to be internalized by individual economic agents. These costs are imputations which do not suggest any particular role of environmental costs in actual price formation. Such pricing would have to be modelled according to prevailing elasticities of supply and demand. While such modelling of shadow prices is beyond the object of accounting, imputed environmental cost information can provide the initial input into the simulations of market behaviour of producers and consumers. Even in the absence of such modelling, setting the level of economic instruments at imputed environmental cost level would provide a rational basis for iterative cost internalization policy and the monitoring of its effectiveness.

Environmental protection and the reorientation of macro- and microeconomic policies towards sustainable development are the main efforts to achieve progress in both environment and development. The separate identification of the expenditures for these efforts in environmental accounts permits measuring the relative contributions of different sectors and institutions to the overall national effort on sustainable development. If an unequivocal connection between these expenditures and the change in environmental quality and natural resource use could be made, the efficiency of environmental protection and policies of sustainable development could be assessed in terms of their costs and benefits. Major problems of relating particular measures and programmes to time-lagged and synergistic or antagonistic effects on the environment and human welfare would have to be resolved in such an exercise.

Perhaps more promising would be an effort of tracing current and capital environmental expenditures to their sources, as measured in the financial accounts. Financial accounts are an integral part of national accounts and can thus be directly linked to non-financial transactions in the economy. This link would facilitate the integrative analysis of environment–economy interaction in areas of debt, savings, inflation, fiscal policy and income distribution. However, little work has been done on the financial aspects of environmental accounting which to date has focused on the physical environment and its role in sustaining growth and development.

Indicators for sustainable development

Environmentally adjusted monetary accounting indicators address directly questions of sustainability of economic growth, in terms of maintaining broadly defined capital or national wealth. As discussed above, data support for the broader concept of sustainable development would require the coverage of social goals such as equity, political freedoms, culture and the health of humans and ecosystems. These concerns were therefore incorporated in lists of indicators for sustainable development (see Box 6). The preposition 'for' indicates that this multitude of concerns is not to be addressed by one overall index but by a set of indicators referring to numerous causing activities, processes, effects and policy responses.

On their own, these indicators might very well assist in the management of particular issues but would still lack policy relevance, a recurrent lament in indicator development. It is only when grouped together, for instance in a consistent data framework, that large sets of indicators may gain enough coherence to facilitate integrated policies and analysis of sustainable development. As indicated above, one way to improve on policy relevance is to refer them to a normative set of standards and targets that determine the feasibility of sustainable development. The development of and agreement on such sustainable development standards should be of high priority in the development, implementation and evaluation of integrative policies. Once available, these standards need to be introduced into indicator frameworks for the evaluative measurement of progress towards or violations of those targets and standards.

Another way of circumventing controversial valuation methods is to introduce physical variables into economic models. Mixing physical and monetary variables and parameters in optimization models, with or without restrictions, permits shadow pricing of physical phenomena for purposes of assessing optimal and/or sustainable economic performance.

Indicators of sustainable development

Indicators of sustainable development are usually included (but somewhat hidden) in indicator sets for sustainable development. They are those indicators or indices that aspire to measuring directly sustainable growth, welfare or development. Without further breakdown by sectors or institutions the value of these indices for policy formulation and evaluation is minimal. However, if they are to gain a similar status of acceptance as the routinely compiled aggregates of GDP or national income, they might at least be useful in keeping the score (Department of Environment and Natural Resources, Republic of the Philippines *et al.*, 1994, p. 2; see also Peskin, Ch. 22) of sustainable development, over time and among nations.

This seems to be indeed the ambition of the overall indices described above, notably the GPI. The authors of this index have thus called for the replacement of GDP and even the whole national accounts system by appropriate indicators of social progress (Cobb *et al.*, 1995). Obviously, the socio-environmental (non-monetary) side of the above-described dichotomy of values and (e)valuation is at work here. Given the lack of transparency and the arbitrariness in scope, coverage, weighting and valuation of overall welfare indices, the risk of index manipulation cannot be denied. The capacity of these indices might indeed be more for spreading 'funk by numbers'[35] than for redirecting policy towards sustainability goals.

Outlook: consensus building and standardization

There is no question about the need for harmonization and standardization of indicator and accounting methodologies. The current proliferation of different concepts, methods and classifications, frequently developed outside any organizing framework or system, has generated non-comparable information and confusion in countries attempting to implement these methodologies.

Internationally adopted standards, such as those developed under the aegis of the United Nations Statistical Commission, would facilitate international comparison and a coordinated approach to capacity building. On the other hand, such international adoption can only be achieved if there is consensus, at least on key concepts and methods. Unfortunately this is not yet the case, neither in environmental accounting nor in the development of indicators of sustainable development. There is thus a trade-off between the pragmatic objective of meeting urgent needs for data development and capacity building, and the scientific objective of encouraging pluralism in approaches and methodologies through research and experimentation.

Some consensus building on indicators and integrated accounting is in the making. For environmental accounting, this is facilitated by the existence of an internationally adopted system of national accounts, the 1993 SNA, which is also the basic framework for the SEEA. The Second Special Conference of the International Association for Research in Income and Wealth (IARIW) in Tokyo (5–8 March 1996) revealed commonalities in practical and theoretical approaches, in particular for the measurement and valuation of natural resource depletion. The treatment of environmental externalities remains a controversial issue, though. The Statistical Commission, at its twenty-ninth session (New York, 11–14 February 1997), agreed therefore to build on practical experience for the revision of the SEEA to be undertaken jointly by UNSD and the 'London Group' of national accountants. In the meantime, a group of international organizations and experts, the so-called Nairobi Group, agreed on a joint work programme in preparing an 'operational manual'. The manual is intended to translate the complex methodologies of the SEEA into pragmatic step-by-step guidelines for preparing integrated accounts and using their results in policy analysis.[36]

In the area of indicator development, the Department for Policy Coordination and Sustainable Development of the United Nations, in cooperation with UNSD, has proposed a framework and work programme for indicators of sustainable development, for adoption by the United Nations Commission on Sustainable Development (United Nations, 1995). The Commission, at its third session in April 1995, approved the work programme and encouraged the further development of the framework and list of indicators. An Intergovernmental Working Group on Environment Statistics, convened by UNSD, adopted sets of environmental indicators for international compilation and methodological development by UNSD.

All this augurs well for developing greater consensus on urgently needed practical concepts and methods. At the same time, it should not discourage continuing research on controversial or unresolved questions. The international community of data users and producers, as represented by the United Nations Commission on Sustainable Development and the Statistical Commission, should harness the current diversity of methodological proposals with a view to harmonization and consensus building. UNSD as the Secretariat to the Statistical Commission is prepared to play its part in this process.

Notes

1. To use a crude distinction between advocates of mainstream neoclassical economics and holistic views of the human environment. In a similar vein, for purposes of exposition, reference is

made to environmental economists, national accountants and environmental accountants. Of course, neither the existing schools of thought nor their representatives may fully fit into this simplified categorization.

2. E.g. the 'prices along the optimal trajectory' of maximum intertemporal welfare; those prices are shadow or 'accounting prices' since hardly any economy will usually meet the model assumptions. See for a discussion of such a model, incorporating environmental resources and expenditures for environmental protection, Mäler (1991).
3. For purposes of incorporating environmental impacts in project evaluation, see e.g. Dasgupta (1994). A good example for assessing the interrelationships between macroeconomics and the environment in a CGE model is the Norwegian approach to environmental assessment which appears to skip factual measurement in favour of modelling for policy advice (Alfsen, 1996).
4. Deep ecologists emphasize the need for a 'comprehensive, religious and philosophical world view' and the 'intrinsic equality' of all species (Naess, 1976; Devall and Sessions, 1985).
5. This is because actual internalization would affect the level and structure of economic performance and make the sum of imputed environmental costs non-comparable to observed macroaggregates. See below for a discussion of this issue.
6. The further expansion of the asset boundary for the incorporation of 'non-economic' (environmental) assets is described below.
7. For instance, the Bureau of Economic Analysis (1994) of the USA treats discoveries of mineral resources as capital formation, thus offsetting the depletion of these resources. Similarly, Repetto *et al.* (1989) treated discoveries as national income raising capital investment.
8. Performance measures of the SNA, such as GDP and NDP, are no more than the duplication-free sum of economic production activities. The deduction of intermediate and capital consumption from output avoids double-counting when outputs and capital services enter a particular production process as inputs from other production units. As shown above, any other interpretation of net value added or NDP runs into problems of market valuation which does not reflect maximum efficiency nor maximum welfare (due to market imperfections and exclusion of consumer surplus in market prices). Of course, a certain measure of sustainability, in terms of capital maintenance, is already incorporated into production when costing capital wear and tear, i.e. capital consumption. The sustainability criterion is discussed in some detail below.
9. Of course, when natural assets are marketed, observed prices of market transactions in these assets can be applied; the value of asset services could be estimated by data on rentals or leases paid for the permission to use the assets.
10. Alternative approaches that focus on welfare measurement, applying contingent and non-monetary evaluation or weighting, are discussed in the following sections.
11. For exhaustible resources, such as mineral resources, the cost of discovery could be taken as capital restoration cost. In fact, this approach would treat exhaustible resources as de-facto renewable ones (Born, 1992, p. 27). This so-called 'replacement cost approach' is rather speculative (about expected discoveries). It is also based on the controversial notion that discovery represents a genuine replacement of a depleted resource (see note 7, above). The replacement cost valuation is therefore not further pursued here.
12. This follows directly from the definition of the physical side of capital consumption which refers to the normal (anticipated) obsolescence of produced capital in production and applies current (market) prices in replacement costing and asset (change) valuation. Of course, rational (optimal) re-investment behaviour is assumed, i.e. actual prices reflect accurately the current value of the resource (in terms of discounted net returns).
13. To the extent that the economic value of a natural resource is affected by degradation, notably from contamination with pollutants, such change in the quality of the resource would be reflected in the expected net returns from lower quality resource stocks. Typically such degradation is relatively small by comparison to the quantitative depletion of the resource. 'Depletion' is therefore considered to include such qualitative changes in economic value. The value of impacts of pollution on non-economic assets or non-economic functions of economic ones is the subject of alternative (non-market) valuations, discussed in the next section.
14. Excluding unexpected changes or impacts from activities other than production, e.g. natural disasters or discoveries. Again, this convention is to stay as close as possible to SNA concepts and (production) boundaries, in order to obtain maximum comparability with standard economic aggregates. However, the introduction of the sustainability criterion might open the

door to a different notion of capital (maintenance) and income, taking (positive and negative) 'windfalls' into account when compiling a 'useful measure of real social income' (Hicks, 1946, p. 180).

15. This applies even more to other concepts such as Hicks (1946) income which defines a microeconomic income concept as a residual based on the maintenance of the net worth of the microeconomic agent. Net worth can be affected by holding gains and losses, capital transfers and other events such as natural disasters, it is even more removed from the established – national – income accounting.
16. Carried out with the support of UNSD in Colombia, Ghana, Indonesia, Mexico, Papua New Guinea, Philippines, Republic of Korea and Thailand.
17. The identity between 'receipts net of extraction cost' (El Serafy, 1989, p. 13) and net returns or Hotelling rent is not obvious. The objective of identifying an allowance which through investment would create a permanent income stream suggests that a normal return to (fixed) capital should be included in the user-cost allowance, contrary to the Hotelling rent definition. However, for purposes of national accounts calculation of value added and NDP or GDP, rather than microeconomic income use strategy, a natural capital depletion allowance should not include a normal return to capital which is part of the operating surplus component of value added (point made to the author by J Logarta Jr. of the Environmental and Natural Resource Accounting Project in the Philippines).
18. This is a typical inconsistency of environmental accounting approaches, developed outside the national accounts framework. Other examples are referred to in note 30, below, where environmental costs and benefits were estimated by means of a neoclassical general equilibrium approach to welfare measurement and in Box 5 discussing the concept of environmental debt.
19. According to the *Wall Street Journal* of 23 March 1992, Monsanto, Du Pont, Cyanamid Cos made allowances between US $200 and $400 million at the end of 1991.
20. The author benefited on these questions from suggestions made by S Keuning of the Netherlands Central Bureau of Statistics and C Stahmer of the German Federal Statistical Office.
21. See e.g. the action plans of a recent international conference on 'Taking Nature into Account' (Brussels, 31 May and 1 June 1995), focusing on Europe, and of its follow-up conference (at the global level) on 'Accounting for the Future' (Washington, D.C., 3 October 1995), organized by non-governmental organizations and the World Bank.
22. Environmental 'degradation' may include quantitative losses such as the destruction of an ecosystem or the extinction of species. Similarly, as discussed above, 'depletion' of natural resources may include changes in their quality, e.g. when soil productivity is affected by contamination from pollutants (see also note 13, above). Rather than following the popular distinction between depletion and degradation, the SEEA distinguishes more rigorously impacts on economic (asset) functions from those on environmental ones to avoid double-counting and mix-up of valuations.
23. It is a matter of some dispute if these defensive expenditures should be deducted or not from economic aggregates as suggested by some authors (Daly, 1989; Pearce *et al.*, 1990, p. 108). Apart from problems of determining what outputs of society are defensive or 'regrettable' and of accounting for the indirect costs of antecedent industries, such deduction from GDP or NDP seems to confuse these aggregates with welfare measures (rather than net production or capital formation) (Bartelmus, 1994a, pp. 38/39). See also below regarding the compilation of economic welfare indices.
24. Oral communication by J van Tongeren of UNSD.
25. See e.g. Brown, L. R. (1993, p. 4) who blames the accounting system to 'generate environmentally destructive economic policies' as it fails to reflect the reality of natural capital depreciation and environmental destruction.
26. It should be noted that, contrary to the depletion cost of marketed natural resources, environmental maintenance costs are thus not referring to changes in the availability of natural assets but represent a measure of the total social cost of environmental protection, redistributed to economic agents for possible internalization.
27. Of course, the cost-borne principle cannot be applied to transboundary flows, because it is hardly possible to trace changes in environmental quality back to the countries responsible for the export of environmental degradation.
28. In this case, any depletion cost previously deducted from value added or NDP to obtain environmentally-adjusted value added or net domestic product of type I (EVA I or EDP I)

would have to be added back in to calculate EVA II and EDP II, covering both natural resource depletion and environmental degradation.

29. Contingent valuation, for instance, faces free-rider attitudes, short-sightedness or ignorance of consumers about long-term environmental impacts, and effects of income levels and distribution, when questioning individuals in opinion surveys about their preferences for environmental quality and related social values.
30. An example of such mixture is the 'neoclassical' accounting approach, proposed by Peskin (1991), which applies avoidance costing to measure the 'outputs' of nature's environmental services (and inputs into the economic system) and contingent and other valuations to estimate the damage input into the nature account (i.e. negative output of the economy).
31. See for a detailed discussion of this process and its categories, Bartelmus (1994a).
32. For instance, expertocratically determined weights have been applied in EUROSTAT's pressure index (Jesinghaus, 1994), and equal weights were used in combining 'deprivations' in income, life expectancy and educational attainment in UNDP's (1991, p. 88) Human Development Index.
33. Aggregation is achieved by means of 'equivalent' factors, measuring the contribution of indicators to 'policy themes'. The controversiality lies in both, theme selection and the difficulty of providing across-theme aggregation of environmental indicators.
34. The following review of potential indicator uses is based on Bartelmus (1995).
35. *The Economist* of 30 September 1995, p. 27, in connection with the Genuine Progress Indicator (GPI).
36. See Chapter 2 for a description of this approach.

Appendix: methods of market valuation of natural resources

Present-value method†

The present value V_0 of a natural resource asset, i.e. the value at which the asset would be traded under perfect (market) conditions and optimal behaviour, is the sum of the expected net revenue flows $R_t = N_t Q_t$, discounted at nominal or real interest rates r, assumed to be constant for the life T of the resource:

$$V_0 = \sum_{t=0}^{T} \frac{N_t Q_t}{(1+r)^t} = R_0 + \frac{R_1}{1+r} + \frac{R_2}{(1+r)^2} + \cdots + \frac{R_T}{(1+r)^T} \quad (1)$$

N_t is defined as the total unit value of the asset less the marginal costs of extraction, development and exploration, and Q_t is the quantity exploited over the period t. Equation (1) assesses the value of the resource at the end of the accounting period, after the returns have been fully realized (R_0 in the first period). If the value of the resource is assessed at the beginning of the period, discounting has to start with the first period, changing equation (1) into:

$$V_0^* = \sum_{t=0}^{T} \frac{N_t Q_t}{(1+r)^{t+1}} = \frac{R_0}{1+r} + \frac{R_1}{(1+r)^2} + \frac{R_2}{(1+r)^3} + \cdots + \frac{R_T}{(1+r)^{T+1}} \quad (2)$$

The change in value of a natural asset ΔV during the accounting period can be expressed as the difference in current values of the anticipated returns at the

† Part of the following is based on Bartelmus and van Tongeren (1994), pp. 11/12.

beginning (V_0) and the end (V_1) of the accounting period. The emphasis is on anticipated, effectively excluding any changes in reserves other than planned resource extraction. Given that

$$V_1 = R_1 + \frac{R_2}{1+r} + \cdots + \frac{R_T}{(1+r)^{T-1}} \tag{3}$$

and that V_0 can thus be expressed in terms of V_1 as $R_0 + V_1/1 + r$, ΔV can be formulated as

$$\Delta V = V_1 - V_0 = -R_0 + \frac{V_1}{1+r} \times r \tag{4}$$

when assessing the value change ex post, i.e. at the end of the accounting period. The corresponding ex-ante formula is

$$\Delta V^* = -R_0 + V_0 \times r \tag{5}$$

The change in economic value of the asset (under perfect market conditions) can be expressed as the excess of net returns from exploiting the asset over the maximum return (income, interest) earned $V_0 \times r$ if the asset value were invested elsewhere.

Net price method

This method determines the value V_t of a resource at the beginning of period t as the volume of the proven reserve $Q = \sum Q_t$ (over the lifetime of the resource T) multiplied with the difference N_t between the average market value per unit of the resource p_t and the per-unit (marginal) cost of extraction, development and exploration c_t :

$$V_t = (p_t - c_t)Q = N_t \times \sum_{t=0}^{T} Q_t \tag{6}$$

The net price method is based on the Hotelling rent assumption which states that in a perfectly competitive market the net price of an exhaustible natural resource rises, due to increasing scarcity, at the rate of the interest of alternative investment, offsetting the discount rate.

Equation (1) can therefore be reformulated as[‡]

$$\begin{aligned} V_0 &= Q_0 \times N_0 + \frac{Q_1 \times N_0(1+r)}{1+r} + \cdots + \frac{Q_T \times N_0(1+r)^T}{(1+r)^T} \\ &= (Q_0 + Q_1 + \cdots + Q_T)N_0 = Q \times N_0 \end{aligned} \tag{7}$$

[‡] The following is based on Bartelmus *et al.* (1992), Appendix 4, p. 50.

Under these assumptions, the depletion cost, or change in value of the resource stock between periods 0 and 1, is obtained as

$$\Delta V = V_1 - V_0 = (Q - Q_0) \times N_0(1 + r) - Q \times N_0 \tag{8}$$

Considering the net price increase $N_0(1 + r)$ during the accounting period as a holding gain (to be accounted for as revaluation, assuming marginal exploitation cost as constant), the true depletion cost is deflated by $(1 + r)$ to obtain R_0, the net return during the accounting period, i.e. the so-called Hotelling rent:

$$\Delta V = (Q - Q_0)N_0 - Q \times N_0 = -Q_0 N_0 = -R_0 \tag{9}$$

User-cost calculation

In order to convert the finite (constant) returns R from the use of an exhaustible resource into a perpetual income stream X (even after the exhaustion of the resource), part of these returns, the user-cost allowance $R - X$ should be set aside for investment. It can be shown (El Serafy, 1989) that the user-cost allowance at the interest rate r and the lifetime of the resource of T years amounts to

$$R - X = \frac{R}{(1 + r)^{T+1}} \tag{10}$$

It can also be shown (Hartwick and Hageman, 1993) that the user-cost method is a special case of defining depreciation as the change in the discounted value of a resource, assuming that the periodic net returns are the same for the remaining life of the resource. Calculating ΔV as the difference between V_1 and V_0 (see equations 1, 2 and 3) yields ΔV as

$$\Delta V = \frac{R}{(1 + r)^T} \tag{11}$$

or

$$\Delta V^* = \frac{R}{(1 + r)^{T+1}} \tag{12}$$

and the true income element X as

$$X = R - \Delta V = R - \frac{R}{(1 + r)^T} = R\left(1 - \frac{1}{(1 + r)^T}\right) \tag{13}$$

The income element is the difference between the current and last (discounted) net return.

17
Environmental Protection Expenditure and its Representation in National Accounts

A. STEURER, G. GIE, C. LEIPERT, C. PASURKA AND D. SCHÄFER

Introduction

The description of Environmental Protection Expenditure (EPE) is a building block in any comprehensive system of integrated environmental and economic accounting. Data on EPE have been collected since the 1970s in a number of countries. Quality[1] and availability of data has improved in recent years and the 'value-added of accounting' (Weber 1996) has become apparent.[2] However, although discussed for a long time, opinions remain divided on the interpretation of EPE accounts and, underlying that, how EPE are actually represented in national accounts.

The main focus of this paper is on the actual representation of EPE in GDP at current and at constant prices (or in value and volume, or in nominal and real terms[3]). This focus, especially on volume changes (i.e. changes in quantity or quality), is useful for various aspects of environmental-economic accounting. First, there is the idea of deducting EPE (at least those of households and government) as part of 'defensive' expenditures from GDP or other aggregates of national accounts.[4] Second, economic policy makers understand that the introduction of environmental protection (EP) regulations (as far as market producers are concerned) slows down macroeconomic capital and labour productivity development, reduces growth rates and increases inflation rates as currently measured. This understanding, which is not very comforting to the pursuit of environmental protection objectives, is based on the fact that EPE are recorded as costs in national accounts but often do not generate a separately measured output in the SNA sense.[5] Third, the SEEA (United Nations 1993a), version IV.2, proposes an extension of the SNA by using imputed maintenance cost as a means of valuation. Although these imputed maintenance costs are treated as a kind of consumption or capital, they may as well be seen as being imputed EPE-type transactions. Consequently, the considerations as presented below may be useful in this respect as well.

Views expressed are the authors' and may not represent the position of their institutions.

K. Uno and P. Bartelmus (eds.), Environmental Accounting in Theory and Practice, 309–319.

Summary of the main theses

When EPE are presented in a way consistent with Chapter XXI of the 1993 SNA, the resulting aggregate of National Expenditure for Environmental Protection, being basically the sum of all uses of environmental protection services plus environmental protection investment, is contained[6] in GDP at current prices. One could also say that EPE are a true share of GDP at current prices, i.e. the ratio of EPE to GDP is an accurate measure of the resources of an economy devoted to environmental protection activities.

As will be shown in more detail below, it is a difficult task to describe, at a conceptual level, how EPE are (or should be, following the SNA) recorded in national accounts. EPE are recorded in different ways and with different effects on aggregates of national accounts, depending on the exact details of the transactions, the actors and, finally, the information available. The question of which influence a particular item of EPE has in terms of price and volume changes (i.e. the deflation procedure) is non-trivial.

In addition to the conceptual complexity there are important practical constraints. It is time-consuming to find information on the actual treatment of the different items of EPE during the deflation procedure (i.e. when making constant price estimates in national accounts). This refers to the details of compiling price indices for individual products, the calculation of value added of non-market producers (i.e. government) in real terms and the quantitatively important payments of households for sewage and refuse disposal as a part of the gross rent paid to the 'housing rental' industry (which also includes the imputed rents of owner-occupied dwellings).

At best, EPE corresponding to final uses (investment as well as final consumption of households and government) are contained in GDP at constant prices. Based on available information for a number of countries, these final uses may represent half of total EPE, or less, with large variations among countries. Since total EPE represent 1–2% of GDP at current prices, these final uses correspond to 0.5–1% of GDP at constant prices.[7] The other half (or more) of EPE translate into price increases and do not enter GDP at constant prices.

Whether EPE actually increase or decrease GDP, and by how much, is a question completely different from calculating the share of GDP devoted to EP activities, which is statistically measurable. This new question on the effect of EP activities on the level of GDP can only be answered by making (strong) assumptions and by using modelling. Alternative assumptions concerning the utilization of production factors and, to a lesser extent, the assumptions concerning opportunity costs of government spending or the way of financing government expenditures will reverse the direction of the change in GDP at constant prices. The maximum value of change is determined by total EPE (intermediate and final uses) in GDP at current prices and by final uses of GDP at constant prices. The minimum value is determined by a zero increase of GDP at current prices and by a decrease of GDP at constant prices by intermediate uses. Potential changes may therefore range from +1% to −1% of GDP at constant prices due to the introduction of environmental policy. Note that this is the cumulated effect of the phasing-in of environmental policy over several years.

The above findings have the following implications:

(a) If deducting EPE (as part of defensive expenditures) from GDP is seen as a useful operation, then looking at the results in current prices is not enough. The results in constant prices have to be investigated as well as they would obviously differ from the results in current prices. A deduction of EPE from GDP at constant prices requires considerable knowledge on how in reality various parts of EPE (and, by extension, other types of defensive expenditures) are contained in the national accounts and how they have been treated during the deflation procedure. The point is not whether deducting EPE is a good idea with a view to welfare considerations.[8] The point is that technically deducting EPE is more easily justified when they entered GDP in the first place. It is shown in this paper that an *a priori* unknown part of EPE – this part is likely to be much more than only intermediate consumption of environmental protection services by market producers – has already been removed 'automatically' from GDP at constant prices during the deflation procedure.

(b) With a view to economic policy it is clear that EPE undertaken by market producers constitute an anomaly in the system of national accounts insofar as there is no recording of a corresponding output (as in the case of non-market producers) or, alternatively, there is no quality change of the output (apart from some consumer products).[9] Current accounting practice signals that an introduction of environmental protection, following the polluter-pays-principle, brings measured inflation up and may bring measured real growth rates down. This is due to a 'private production of a public good' (Flassbeck and Maier-Rigaud 1982) which should be included in GDP at constant prices (and would be at least partly included if the public good were produced by a non-market producer). To remedy this, adjusted productivity measures including measures of environmental productivity could be considered. See for example Nestor and Pasurka (1994). Such measures require well established links to physical data.

The above considerations send inconsistent signals for policy in general and for economic policy in particular. Fortunately, actual figures are rather small (some 2% of cumulated growth and inflation rate over the phase-in period of environmental policy). However, the findings may be confusing for environmental and economic policy and show an inconsistent presentation of environmental aspects in national accounts with a view to the expectations underlying the call for an environmentally adjusted GDP.

How are EPE represented in GDP?

Environmental protection (EP) is defined as all activities that are aimed at the prevention, reduction and elimination of pollution as well as any other degradation of the environment. The definition implies that, to be included under EP, activities or parts thereof must satisfy the end purpose criterion (*causa finalis*), i.e. that EP is their prime objective. Activities which are favourable for the environment but which serve other goals are not considered as EP (Eurostat 1994). This definition implies that EP activities are generally of a cost increasing nature.

EPE are outlays, and therefore translate into costs, on the one hand. They are revenues, and therefore count as income, on the other. They also bring about a reduction of physical pressure on the environment, or to put it in economic terms, they reduce a negative external effect. However, the system of national accounts does not explicitly record negative external effects and records the "production" of positive external effects only when they are provided by non-market producers. EP therefore has a somewhat peculiar characteristic: whereas environmental quality is a public good, the maintenance of this good is partly undertaken privately, following the polluter-pays-principle.

This section is restricted to the relation of EPE to GDP. There are at least two papers that provide detail on this issue. Vanoli (1995) and Schäfer (1995). Both papers reveal the same accounting mechanisms although their conclusions sometimes differ due to differences in their core assumptions.

Main results

The aggregate of National Expenditure for Environmental Protection has been shown as nearly fully contained in GDP at current prices.[10] It is more difficult, however, to describe the effects of EPE on GDP at constant prices. This analysis depends on the types of actors (market producer, non-market producer, consumer) and transactions (investment, marketed services, internal activities), as well as on the methods of calculating price indices, the deflation procedure used and the underlying assumptions.

Investment goods for EP are fully contained in both GDP at current and at constant prices. For EPE related to the extra cost of integrated technologies (i.e. additional costs due to changes in process) there is some dependence on actual treatment during the calculation of the price indices used for deflation. Internally produced EP services within the same unit (ancillary activities) are included in GDP at current prices, but not in GDP at constant prices if the internal measures are carried out by market producers. Internal measures by non-market producers, which are much less significant in quantity terms, are contained in GDP at constant prices to the extent that components of gross value added (wages, consumption of fixed capital etc.) are affected. External EP services are included in GDP at current prices. However, GDP at constant prices includes only services going directly to final demand or used by non-market producers. Imported investment goods and EP services have no effect on GDP. The introduction of environmental levies leads to an increase in GDP at current prices. Only if non-market producers are affected by the levies, they may also increase GDP at constant prices. Expenditure by households on goods connected with EP are included in both GDP at current and at constant prices. The case of extra expenditures incurred by households for environmentally friendly products depends on the actual treatment of such expenditures as price or quality changes when making the price indices. Conceptually, environmental characteristics of products should be recorded as quality changes. In various countries the extra cost for cars equipped with catalytic converters have been recorded both in GDP at current and at constant prices.

The observations made show that different kinds of EP measures can be expected to have different consequences for GDP at current and at constant prices. Table 1 gives an overview of the effects of EP measures on GDP at current and at constant prices. The rows show the various EP-related transactions. The columns show the various user categories.

Main assumptions and their influence

Both Schäfer (1995b) and Vanoli (1995b) assume that additional costs for EP activities are passed on in the product prices of the units bearing the costs and that producer and consumer prices rise accordingly. Additionally, the assumption must be made that there are no major changes in relative prices of ordinary products.

Whether the overall effects of EPE are a rise or fall in GDP at constant prices depends on the assumption concerning the utilization of production factors. Schäfer (1995b) assumes, that the factors of production are not utilized to the full extent whereas Vanoli (1995b) assumes that EP activities substitute (suppress) other production activities. It is difficult to judge which assumption is more realistic. Under the assumption of full utilization of production factors in a static economy the introduction of EP activities leads to a restructuring of the economy with little influence on GDP at current prices (apart from the case of financing EP via taxes) but with a fall in GDP at constant prices and in measured productivity. For non-market producers assumptions are needed concerning the sources of financing. If public budgets are kept constant, EPE must be financed by reductions elsewhere and the net effects, depending on the nature of these reductions, are likely to be small. Financing via tax increases is treated in detail in Vanoli (1995b). The other extreme – none of the production factors is fully utilized – is characterized by an increase in GDP at current and at constant prices.

Nestor and Pasurka (1995c) specified a comparative static applied general equilibrium model of the effects of environmental protection activities on the

Table 1. Representation of environmental protection measures in GDP[a] (at current and at constant prices).

Type of environmental protection product or environmental levies	Users of environmental protection products or units paying environmental levies: Market producers		Non-market producers		Final consumption		Exports	
	current	constant	current	constant	current	constant	current	constant
EP services, internally	+	○	+	+	×	×	×	×
produced for third parties	+	○	+	+	+	+	+	+
Investment goods	+	+	+	+	×	×	+	+
Environmental levies	+	○	+	+	×	×	×	×
Connected/adapted products	×[b]	×[b]	×[b]	×[b]	+	+	×[b]	×[b]

+ : included in GDP; ○: not included in GDP; × : do not occur.
[a] Assuming that additional costs are passed on in output prices.
[b] Transactions are conceivable, but are not dealt with here.

German economy. There were two cases specified. In the first case, it was assumed that the total capital stock increased by the amount of capital required for EP activities. In the second case, it was assumed that there was a fixed amount of capital stock available and that the capital used for EP activities resulted in a reduction in capital available for non-EP production.

Specific cases

This section, which is largely based on Schäfer (1995b), illustrates how EPE interact with standard national accounts procedures.

It may be helpful to describe the main principles of the deflation procedures used in national accounts. For gross value added the method of double deflation is recommended (cf. SNA 1993, Commission of the European Communities *et al.*, Ch. 16). Since gross value added is defined as output minus intermediate consumption, this means that for market producers both output and intermediate consumption are deflated independently. As the output of non-market producers is valued at cost of production, less complex methods must be used (e.g. based on compensation of employees at constant wage rates). For market producers, gross value added is thus affected when sales and intermediate consumption change, and for non-market producers, when value added components (wages, consumption of fixed capital etc.) change. Overall, it can be concluded that the quality of the data in constant prices depends on the availability of data on volume changes (changes in quantity or quality) and on the ability of price statisticians to recognize quality changes. Thus, differences among individual countries and between practice and theoretical concepts may be large.

External EP services

The services of disposal enterprises and municipalities for sewage and refuse disposal are normally provided in return for a payment (fee) related to costs. They are included on the output side of the national accounts in the enterprise or general government sector. On the expenditure side, these services are lagely included in the intermediate consumption of other economic industries. In some countries, the charges paid by households for sewage and refuse disposal are initially recorded as intermediate consumption in the 'housing rental' industry. Only the output of the housing rental industry is included in private consumption expenditure as gross rent including the refuse and sewage charges. These charges may be of the order of one third of total EPE, and the question is then how changes in the charges paid are entering the price indices for the gross rent of housing.

On the output side, external EP services are incorporated into nominal and real value added because of the value added associated with the production of such services. This also applies when no additional external disposal services are produced, but only a better, and more expensive, quality of disposal. If there is a change in the quality of the disposal service (e.g. a more frequent or less frequent refuse collection), that quality change should be eliminated in the price statistics by making an estimate of the monetary value of the quality change and include this in the volume component. The gross value added of disposal enterprises is obtained as the difference between output and intermediate consumption. The gross value added of general government units is obtained by adding costs (essential gross wages and salaries, employers' social security contributions and consumption of fixed capital). The enterprises that produce the intermediate inputs for the disposal units experience a corresponding increase in nominal and real gross value added (imports and exports are disregarded here).

Most of the output of these producers of external services is used as intermediate consumption in other industries. For these user industries, the effects of buying external EP services depend on the extent to which those costs for inputs are passed on in their product prices. If the environment-related increases in expenditure on intermediate consumption are passed on in the product prices, and if the user industries are market producers, the nominal gross value added of the downstream industries is unchanged.[11] Nominal output increases to the same extent as intermediate consumption so that real output remains unchanged. Due to the purchase of external EP services, intermediate consumption of user industries increases in real terms whereas real output is unchanged. As a consequence the real gross value added of these user industries declines. These considerations are also valid for the housing rental industry (including the imputed values for the owner-occupied dwellings of private households), as the output of both the currently rented and the owner-occupied dwellings is calculated based on the size of dwellings and the average gross rents (including disposal charges).

Of course, the situation is different when non-market producers use external EP services. In their case, real value added is unchanged, since it is determined from the cost side by adding up the value added components which are calculated independently of intermediate consumption and its external EP elements.

For disposal enterprises and enterprises buying external disposal services, the introduction of external EP services thus increases GDP at current prices in every case. On the other hand, when assuming that the factors of production are not fully utilized, GDP at constant prices increases only to the extent that external disposal services go directly into final demand or to non-market producers. Final demand contains the export of external disposal services and that portion of general government disposal services that is not financed out of fees (as government collective consumption). Both the proportion of collective consumption and the treatment of consumption of households may differ widely among countries. External services used by market producers do not increase GDP at constant prices.

Introduction of environmental levies and taxes

When market producers in levy-paying industries pass the charges on in their product prices, the introduction or increase of environmental levies produces an increase in the nominal gross value added, since environmental levies are recorded in the national accounts not as intermediate consumption but as 'other taxes on production'. Since in this case the increases in output and product prices are identical, the enterprises' real gross value added does not change. However, in the case of non-market producers an increase in the other taxes on production results in an increase in both GDP at current and at constant prices. For government units paying environmental levies (e.g. municipalities which have to pay a waste disposal tax), one has to assume an increase in revenue to cover the payment of these levies. Environment-related increases of taxes on vehicles of enterprises are also treated as other taxes on production, so that the above considerations apply. However, an environment-related increase of a value-added tax would have different effects.[12] It might also be interesting to explore how the trading of pollution permits (seen as a government-created market for environment) is or would be recorded in national accounts.

Connected and adapted products

In an environmental accounting satellite system, connected products are products, which are directly used for EP purposes; however the production of these products is of no further interest from an environmental point of view. This involves goods used by households in particular. The comments below are confined to households as consumers. Expenditure on garbage bags is one example of connected products completely attributable to EP activities. Since this expenditure by private households is included in private consumption, it is also included in GDP at current and at constant prices.

The case of a private car which is fitted with a catalytic converter may be regarded as an example of a product for which a proportion can be attributed to EP (environmentally adapted product). This expenditure is contained in private consumption and in GDP at current prices. If the price increase due to the catalytic converter is considered as a quality increase when calculating price indices and the extra costs involved are not eliminated when deflating, this expenditure is also included in GDP at constant prices. Another example may be phosphate-free detergents. Conceptually, either separate price indices would have to be calculated for phosphate-free and for 'normal' products, or there would have to be a 'chaining' when changing from one product to the other in the price statistics. In both cases, the additional costs would be contained in the volume component. GDP at current and at constant prices ought therefore to contain the full amount of any additional expenditure for purchases of environmentally friendly products by households.

Research needs

Identifying environmental aspects in the national accounts requires a clear understanding of the current treatment of enviroment-related flows in the accounts. In addition to EPE, such environment-related flows may comprise other types of 'defensive' expenditures (e.g. expenditure to remedy environmental damage) as well as flows related to the utilization of natural resources, etc. An improved understanding of the current treatment of these flows based on satellite accounts would also improve the quality of national accounts themselves, for example via improved price indices or better information on the stock of fixed capital (Steurer 1995).

This chapter clarifies some of the issues related to the recording of environmental protection expenditures in national accounts. However, several empirical and conceptual issues still need to be explored:

(a) There is a lack of knowledge about the actual importance, in terms of volumes, of the different cases described above. A comparative analysis across countries of the composition of EPE by orders of magnitude of the different transactions and actors involved would be helpful.
(b) There is also insufficient knowledge about the national practices of generating price indices and constant price estimates of environmental transactions.
(c) There is an incomplete understanding of the effects of EPE on national accounts aggregates in terms of changes induced, and greater use should be made of the experience available from economic modelling.
(d) This chapter focuses on the role of EPE in national accounts at a technical level. Broader considerations of how EPE relate to measuring welfare, how defensive expenditures are treated in national accounts or what the implications for the SEEA might be have been addressed only implicitly. These broader considerations, including the problem of constant price estimates in environmental accounting have not yet been investigated.

Notes

1. Measurement problems of course exist, e.g. with a view to quantifying in an internationally comparable way investment in integrated ('cleaner') technologies. For a critical overview see Björsell (1994).
2. Especially with regard to avoiding double counting, identifying and filling data gaps and producing meaningful aggregates. See the results from the applications of the SERIEE's Environmental Protection Expenditure Account (EPEA) in the Netherlands, France, Germany, Finland, Sweden, Switzerland or the United Kingdom (CBS 1995, IFEN 1996, Statistisces Bundesamt 1996, Statistics Finland 1996, Statistics Sweden 1996, INFRAS 1996, ECOTEC 1996, as well as US Environmental Protection Agency 1995a, or Schäfer and Stahmer 1989).
3. The 1993 SNA recommends the use of 'GDP at constant prices' or 'GDP volume'. The use of the term 'real GDP' should be avoided (cf. Commission of the European Communities *et al.* 1993, para. 16.71).
4. The idea of deducting defensive expenditure has been most clearly formulated in relation to aggregates of national accounts in nominal terms. Also, some authors have misunderstood the

concept of intermediate consumption in a sense that this means an 'automatic deduction'. EPE are more or less fully contained in nominal GDP no matter whether they are part of intermediate consumption or not. However, when EPE translate into intermediate consumption (and internal uses, which is the same) they tend to be recorded in nominal GDP only, i.e. they simply lead to inflation when occurring for the first time.

5. Of course, EP activities result in a reduction of pollution. If e.g. a system of marketable pollution permits existed, then prices could be assigned to the emitted pollution. As a consequence, pollution abatement activities would have a separately measured output.
6. Assumptions needed for this are that EPE translate into domestic final uses (either directly or via price increases in other products) and that trade in EP services and goods needed for EP (including for environmental investment) is balanced. The aggregate of national expenditure contains some double counting which occurs when EPE translate into additional costs of products which are in turn used for EP activities (e.g. a flue gas scrubber at an electric power plant translates into price increases of the electricity sold; this electricity may then be used as an input for other EP purposes). However, this double counting is very small.
7. Figures are based on the applications of the SERIEE's EPEA (see note 2 above) and supported by OECD (1996) and US Environmental Protection Agency (1995a). Note that due to differences in the financing of municipal EP activities, via fees or via taxes, the importance of collective consumption can differ substantially amoung countries.
8. This is one of the areas of work in the context of the ongoing EU-project 'Methodological problems in the construction of environmentally adjusted national income figures'; see also Radermacher (1995).
9. The treatment of current account EPE as gross capital formation in natural assets would solve the problem conceptually. See Leipert (1995) who reviews the positions of Harrison (1989) and of Carson and Landefeld (1994).
10. See Schäfer and Stahmer (1989) and personal communications from Schäfer and Gie.
11. If there are quality changes upstream, e.g. through the improved sealing of waste dumps, then a corresponding price increase (e.g. of dump charges) would be fully taken into account in the price index for downstream disposal services (e.g. refuse collection), too. When the trend in dump charges is measured, the (part of) the charge corresponding to the improved sealing of the dump would in this case be calculated separately as a quality improvement. Dump charges may not however be recorded at all in price statistics.
12. The effects depend on the deductibility of a value-added tax, as intermediate consumption is shown without the deductible value-added tax in national accounts.

18
Valuing environment in developing countries: a challenge

JYOTI K PARIKH AND KIRIT S PARIKH

Introduction

Economic valuation of environmental resources (and consequently their degradation) can help make decisions on resource utilization and allocation more meaningful. However, degradation and depletion or restoration and regeneration cannot be always valued through market transactions. This is so because, unlike material artifacts, environmental amenities (clean air, unpolluted beaches etc.) are seldom bought and sold in the market. As a result, there is no comparable estimate of the value of environmental amenities. Decisions on resource utilization, degradation of amenities and resource allocation are often made without any estimate of the value of the amenity in question. In such situations, it is observed that the resource goes unpriced; environmental amenities are often either ignored or treated as having zero value. Consider an air polluting industry, for example. If the economic value of air quality degradation can be incorporated into the cost–benefit analysis the resultant conclusions will be more holistic and comprehensive than compared to one which treats clean air as a 'free' resource.

Yet another important application of economic valuation of environmental degradation is relevant for decision making. Economic valuation helps prioritize decisions regarding sectoral resource allocation designed to improve environmental quality. While the pollution of beaches by sewage may be the most obvious environmental problem in a city, the increased incidence of respiratory diseases consequent to air pollution may be also significant. Unless a proper economic analysis is conducted a priori decisions would most likely be made to invest precious resources in improving beaches and sewage treatment rather than in improving air quality. Economic valuation helps prevent such inefficient a priori decision making.

This chapter was originally written for ESCAP, Bangkok under the title 'Accounting and Valuing Environment: A Primer for Developing Countries'. It was substantially revised at UNU/IAS. This paper draws substantially from Parikh Kirit, Parikh Jyoti, Muraleedharan T R and Hadkar Nandini, *Economic Valuation of Air Quality Degradation in Chembur, Bombay, India*, IGIDR Project Report.

K. Uno and P. Bartelmus (eds.), Environmental Accounting in Theory and Practice, 321–335.

In short, economic valuation of resources needs to be undertaken when the markets fail to generate the true price of the resource in question. Market failure leads to suboptimal tapping of the resource, and society must pay for this tendency. An economic valuation helps compute the true price of a resource, so that decision-makers can make informed choices, and society need not lose out on welfare.

Market failures are common for environmental resources. Externalities and diffused and undefined property rights lead to such failures. The same problems also pose difficulties in valuation. The impact of environmental degradation may be borne by people far away or by future generations. Such is the case with greenhouse gas emissions. Also intergenerational issues complicate the valuation problem further. How much weight should one give to tomorrow compared to today is not a difficult problem. How much weight should one give to future generations is a much more complex issue.

Once we go beyond anthropocentrism, the valuation problems become issues of deep philosophy. How much is biodiversity worth? What is the value of a habitat that is vital for the survival of some non-human species? What is the value of a pond or a lake or a pasture to the community? Finally, how do we value illness, loss of IQ and even deaths that result from pollution? What is the value of human life?

Our purpose here is a much more modest and a limited one. Some environmental resources can be valued from the limited anthropocentric perception. Such valuation can help improve the rationality of human actions. We examine here methods of such valuation.

Overview of methods for environmental valuation

Having established the importance of economic valuation and its utility, let us now turn to the tools used for valuing environmental amenities. A variety of economic techniques and models have been developed for assigning monetary values to gains or losses associated with changes in the availability (quantity) or character (quality) of environmental amenities. The aim of these techniques is to obtain an estimate of the value of an environmental amenity that would be revealed if there were a competitive market for the amenity. Figure 1 shows that either the changes can be reduced directly or indirectly.

Valuation methods can be classified as shown in Figure 2. For simplicity, they are classified into three categories:

(a) Physical linkage methods (scientific): these require assessment of loss in production, income or health and can be valued as financial transactions;
(b) Abatement cost methods (technical);
(c) Behaviour linkage methods (economic).

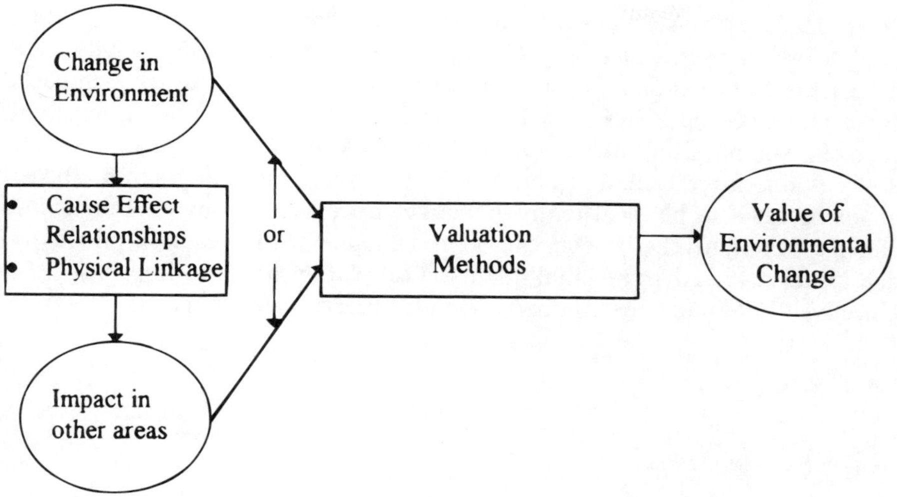

Figure 1. Steps in valuation.

Physical linkage methods

These methods are termed scientific or objective (Dixon *et al.*, 1994) because they depend on a causal connection between environmental change and its effects on other objects – processes, products or persons. In the physical linkage approach environmental values are estimated by establishing relationships between the physical effects of some environmental change on some

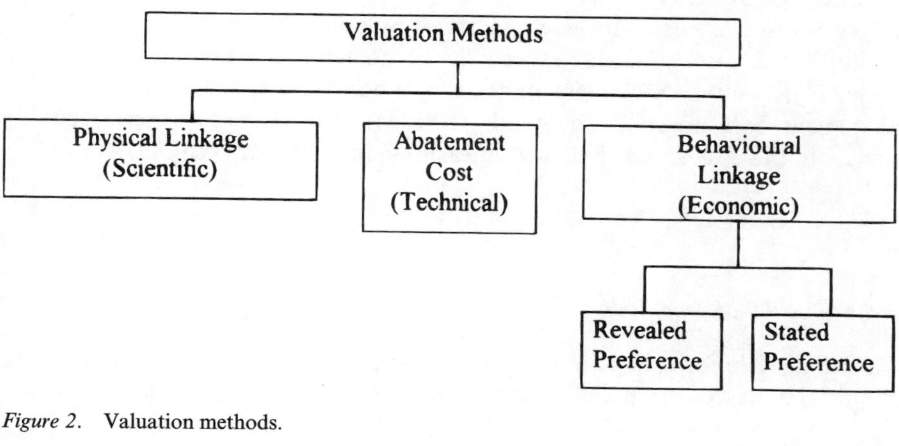

Figure 2. Valuation methods.

other things such as human health, productivity or earnings. One may also need to assess the impact of pollution on ecosystems, on productivity of a fishery, or depreciation of a physical asset. Once this is done, these impacts are quantified and then valued to arrive at the value of environmental changes. The physical linkage approach is also known as the damage cost or dose-response approach. The objective is to measure the changes in net benefits as revealed in physical terms or market prices, caused by environmental damage. Alternatively, benefits can be measured as the increased productivity due to improved environmental quality. The relationships can in principle be objectively determined based upon scientific observations (Shin *et al.*, 1993).

Abatement cost methods

Abatement cost methods approach the problem of valuation from the supply angle, as opposed to the other two methods which approach it from the point of view of demand for environmental amenity. The former are termed technological methods because they are based on the view that the costs required to abate the pollution estimate the value of damage. This method is also known as the 'maintenance cost' method, as this cost would maintain the environmental quality at a constant level. For example, one can install electrostatic precipitators to remove solid particulate matter (SPM) in a boiler or a furnace, or a desulphurization process to remove sulphur oxides (SO_x). The value of the damage due to a tonne of SPM can then be derived as the cost of removing one tonne of SPM. For this purpose, capital cost of the abatement equipment is annualized over the life time of the equipment and annual variable cost is added to that. In a developed country, where pollution control is likely to be stringent, the marginal cost of abatement is likely to be nearly equal to the marginal social cost. One can therefore use abatement cost as a proxy for social cost. However, in the developing countries, this does not apply: social costs often far exceed the abatement costs, and yet nothing is done. Thus, only if the abatement costs have been actually incurred should the abatement cost approach be used. Otherwise the damage experienced by society in the absence of abatement measures should be worked out.

Behaviour linkage methods

Valuation techniques in general assume that the value of environmental goods should be based on people's willingness to pay (WTP) to secure better environmental quality or to escape environmental deterioration. These techniques to estimate behavioural parameters can be further classified depending upon whether preferences are revealed in the market place or stated in a survey. In the revealed preference methods the value of an environmental amenity is estimated indirectly from the purchase price of a commodity whose market value at least partly depends upon the quality of the environmental amenity in question. The value of an unpriced amenity is inferred by using statistical analysis to examine how a change in the amenity affects the observed purchase price of related private goods. For example, in the hedonic price approach, preference shown for environmental amenities gets reflected in the price. Thus people are willing to pay a higher price for a house located on a beach, next to park or in a quiet area. Therefore, real estate values can be examined to detect any premium paid for location with the desired amenity (Streeting, 1990). The premium can be taken to reflect the value of the amenity.

Stated preference methods and contingent valuations assume that people respond to hypothetical market situations as if there were actual markets. The methods rely upon what people say they would buy, if the market existed. In a contingent valuation survey, respondents are presented with an opportunity to express their willingness to pay (or willingness to accept compensation) for a change in the level of environmental amenity benefits (Wilks, 1990).

The selection of a certain technique for a certain purpose is a matter of judgment as there are no standard prescriptions. The choice of a method for a given problem is a function of data availability, time, budgetary constraints, and the intended end use of the results. Physical linkage methods can be used only when scientific relationships establishing such physical linkages are available. Contingent valuation methods can be employed for valuing any sector of the environment but the accuracy will vary greatly depending upon the administration of the survey, how representative the sample population is, the description of the damage caused by the environmental degradation, etc.

The use of one or more of the above-described techniques of economic valuation calls for an extensive search for data and a scrupulous scrutiny of available data. A judicious valuation of available methods vis à vis available data needs to be done to arrive at a reasonable conclusion regarding the appropriate method for the project in question. Furthermore, one may have to use a combination of techniques in different environmental sectors for the economic evaluation. For example, physical linkage methods may be more appropriate for damage caused to human health due to air pollution whereas stated preference may be more appropriate for eliciting how much people value a national park. In addition, the synergistic effects of many environmental components acting in tandem also need to be understood before attempting to determine which method is appropriate. Having overviewed the methods briefly, we discuss each in detail.

Physical linkage methods

The valuation method that first establishes cause–effect or dose–response relationships and then values the impacts of environmental change to reflect the value of environmental change is called physical linkage. As stated earlier, the first stage is to identify the logical sequences of cause–effect relationships, with regard to the deterioration of an environmental attribute and its impact on human health and welfare (Hufschmidt *et al.*, 1983).

Estimation of the cost of the environmental degradation does not require a complete understanding of all links in the logical chain (Shin *et al.*, 1993). Information on the quality of an environmental component defined as some parameter (e.g. air quality as suspended particulate matter, SPM, concentration) and the resulting incremental incidence of an adverse impact (e.g., excess respiratory diseases) may be sufficient to obtain a first estimate. Once these relationships are determined and confirmed, one could estimate the cost of quality degradation (see Figure 3).

The practical utility of such estimates, especially for policy conclusions, often depends upon a comprehensive understanding of the events leading to a deterioration in the quality of the environment. The procedure for estimating health and productivity effects involves three steps. The first step is to determine the relationship between changes in exposure to environmental pollution

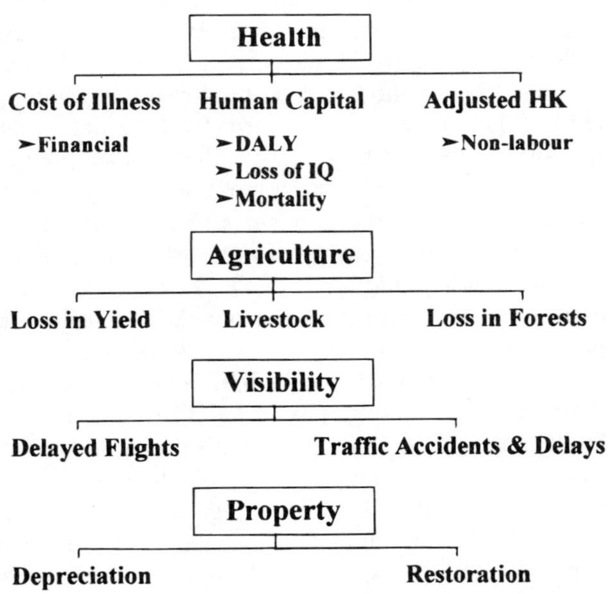

Figure 3. Physical linkage methods to assess air pollution. DALY = Disability Adjusted Life Year.

and, for example, human health as measured by mortality and morbidity rates. The second step is to use this relationship to predict the changes in mortality and morbidity associated with specific changes in environmental pollution and exposure to pollutants. The third step is to derive monetary measures of changes in health status.

The first task in the physical linkage method is to arrive at a suitable dose–response function. For human health this is arrived at through a combination of biomedical studies and statistical analysis. Damage functions could be either on mortality or on morbidity. Those specified by Lave and Seskin (1977) focused mainly on mortality, while Ostro (1983, 1987) and Krupnick *et al.* (1990) concentrated on morbidity effects. These studies seem to provide credible and consistent results. For example, Lave and Seskin (1977) estimated the elasticities between mortality and degrees of sulphate and particulate pollution. The estimates of elasticities across different data sets, model specifications and degrees of desegregation were substantially similar within the range of 0.09–0.12. That means that a 1% increase in pollutant concentration will increase the mortality rate by about 0.1%. Similarly, Ostro (1983, 1987) estimated the effect of air pollutants on morbidity using damage functions. Restricted activity days (RAD) and work loss days (WLD) were regressed to particulates, sulphates and several socioeconomic variables. The regression results showed that particulates affect both RAD and WLD significantly. The elasticity was 0.45 for WLD and 0.39 for RAD.

The estimation of mortality or morbidity effects requires separation of the effects on many confounding factors such as genetic factors and lifestyle differences. In developing countries there is an additional problem. The longevity and health status of people are generally increasing as a result of a number of factors, including better public health facilities, improvement in quality of life and improved medical facilities. Pollution levels have also been rising. To arrive at a figure of excess mortality within an overall trend of diminishing mortality is an exceedingly difficult statistical exercise.

The physical linkage methods can also be used in establishing the loss of productivity of agricultural yields, fishery or livestock or even loss in visibility and its impact on traffic accidents and delayed flights. These also pose complex statistical problems of estimating response functions. One would need to face them or find shortcuts.

Cost of illness

The cost of illness (COI) approach measures the cost of environmental damage in terms of direct outlays for the treatment of illness (hospital care, cost of medicines, cost of the services of physicians and other medical personnel) plus indirect losses in output due to illness measured by the social cost of lost earnings (Rezeler, 1993). The cost of illness approach does not account for the expenditure incurred by the individual to avert illness. The value of personal pain, suffering and inconvenience associated with illness is also not taken into consideration.

Application of COI in a developing country by Parikh *et al.* (1996) for Bombay, India is severely limited by two factors: first, many people affected by diseases do not approach hospitals for lack of awareness, accessibility and affordability; second, the entire health care system is heavily subsidized. Therefore, an estimate based on actual expenditure will be far below the true expenditure and associated opportunity cost. Nevertheless, as a lower boundary, cost of illness is quite an appropriate measure of actual illness-related expenditures.

The application of physical linkage methods faces severe difficulties in convincingly proving cause–effect relationships. For example, in the case of air pollution, it may not be really possible to establish that an increase in incidence of bronchitis is a function of increase in sulphur dioxide concentration in the ambient air. While such a medical relationship might indeed exist, it can always be questioned on counts of synergistic effects. In a locality where there is both high SO_2 and high SPM, increase in respiratory diseases cannot be attributed either to sulphur dioxide or to SPM exclusively. Computing the economic cost is even more complicated when one has to establish a numerical relationship between levels of one pollutant and incidence of a certain disease. Different pollutants and their various combinations cause a variety of diseases, and to establish a convincing numerical link between given levels of pollutants and loss of human capital as its consequence is, to say the least, statistically arduous.

It is, however, possible to arrive at reasonable policy conclusions using physical linkage methods by making some simple assumptions. Consider the example of air pollution. The first task is to identify the dominant pollutant in a given city. A dominant pollutant can be defined as the one which has demonstrated health effects (supported by medical literature) and which is widely prevalent and increasing at a rate faster than other pollutants in the city. Once we have identified dominant pollutants, we could take these pollutants in isolation and estimate the health damage due to them. By a careful selection of the dominant pollutant, we can obtain a reasonable lower bound on the estimate of environmental degradation.

Human capital approach

It is not enough to account for expenditure on illness. Illness and disability, whether temporary or permanent, have private and social costs that need to be accounted. The human capital (HK) approach for valuing morbidity assumes that the value of an individual is the potential of his 'production'. Once the damage function is obtained, the essential factor is the unit economic value of the physical damage (mortality/morbidity). A product of the numerical values of both the number of people affected/dead and the unit cost of treatment/death will provide the monetary value of health damages (OECD, 1989). It is therefore important to arrive at numerical values for the value of statistical life as well as the cost of WLD/RAD consequent to morbidity and mortality. The following methodologies to arrive at such values have been used.

According to Mishan (1982), the present value of a person's expected future earnings may be calculated as

$$L_1 = \sum_{t=j}^{\infty} \frac{Y_t P_j^t}{(1+r)^{t-j}}$$

where Y_t = the expected gross earnings of the person in t^{th} year; P_j = the probability in year j of the person being alive during the year t; and r = the social rate of discount expected to prevail during the year t.

Alternatively the net income can be specified by subtracting C_t, the individuals' expected expenditure during the year t from Y_t.

Among the drawbacks of the HK approach are the following:

(a) Non-market productivity is usually excluded in the valuation. Thus persons not being paid for their services have zero economic value. For example, work within the household, by women and children, is not included.
(b) Other dimensions of illness and death such as pain, suffering, aversion to risk, loss of leisure and adverse effects to others are not included (Dalvi, 1988).
(c) The final values thus generated are very sensitive to the selection of an appropriate social discount rate.

The adjusted HK approach addresses some of these problems. It calculates the value of human capital as follows.

$$L_2 = \left[\sum_{t=1}^{T} \frac{Y_t}{(1+r)^t}\right]\alpha$$

where T = the remaining life time; Y_t = after tax labour and non-labour income; r = individual's opportunity cost of investing in risk reducing activities; and α = a risk aversion factor. Both r and α are assumed to remain constant over time.

This method also fails to account for intangibles like pain and suffering. It is suggested that the adjusted HK method is the most appropriate method for evaluating environmental policies that involve risk to human life.

If deaths occur due to pollution, human life has to be valued beyond the 'cost of illness' and disability adjustments. It should be emphasized that a specific person's life is not valued, which is always infinite to him/herself and her relatives. Statistical value of life is a concept that captures the payments made either by insurance companies or workmen's compensation acts, compensation awarded by courts, and so on.

Statistical value of life

An alternative method of approaching the delicate question of valuing human life is to analyse the behaviour of people in paying for reduction in risk to their life or accepting compensation for undertaking risky jobs. The willingness to accept compensation for one's life will be meaningless, because the money is of no use without life. However, in valuing the impact of environmental degradation on mortality, we are not analysing the value of a specific person's life but rather the value of a statistical life. A wage differential for risky jobs can be used to estimate individuals' willingness to accept money for a change in risk of death. The underlying assumption is that workers will accept risk up to the point where the marginal benefit of compensation equals the marginal cost of taking the jobs. A risk premium is obtained by the partial derivative of the wage function with respect to risk, where the wage function is specified in terms of job characteristics and factors affecting workers' productivity.

The aversive behaviour (or defensive expenditure) approach infers the value of risk reduction from the observation of people's voluntary purchase of certain risk reducing goods or avertive consumption behaviour. According to this approach, people use life saving consumer goods such as seat belts or smoke detectors until marginal cost is equal to the benefit of reducing the probability of death. This avertive expenditure is an approximation of individual WTP to avoid risk (Gerking and Stanley, 1986). By dividing the annual cost of avertive behaviour by the reduced risk of death (illness), we can estimate the value of life/loss of time (Blomquist, 1979).

The approach assumes that the labour market is free and is in equilibrium, that the workers perceive the risk correctly and, most importantly, the workers have a range of choices from which to choose. In addition to the obvious violations of these assumptions in most real life situations, there are further objections to the application of this method for calculations in environmental economics: first, the wage differential is for a voluntary risk whereas risks associated with environmental degradation are involuntary and, second, the wage differential relates to compensation received for an increase in risk over the 'normal risk'.

Behaviour linkage methods (Figure 4)

Here, we assume that the value of environmental goods and services is based on people's willingness to pay to obtain them or to avoid the impacts of

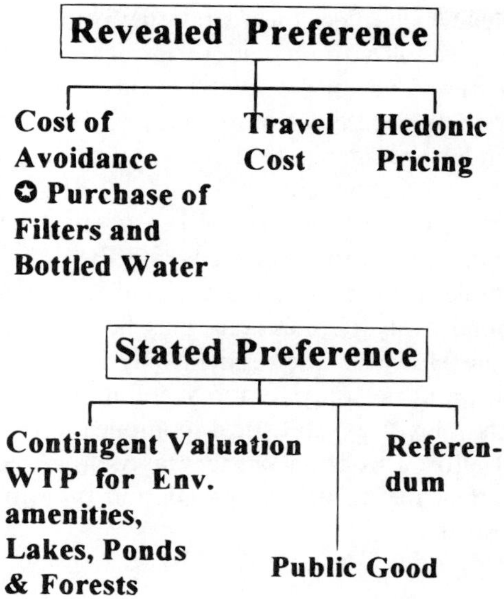

Figure 4. Behaviour linkage methods.

degradation. These preferences may be merely stated or revealed directly or indirectly. Examples of revealed preference for a resort or clean water are how much do people pay to go to a resort or to buy water filters?

Revealed preference methods

These methods are based on actual information revealed in the market places.

Travel cost method

The travel cost method (TCM) is widely used to estimate amenity values for outdoor recreation sites (US Water Resource Council, 1979). The basic philosophy of this method is to use the cost of travel as a surrogate for the willingness to pay for using the recreation site. Travel costs could include actual transportation cost, fees paid at hotels and at times the opportunity cost of travel time spent on the journey. People staying far from the site will be paying higher travel cost than those who stay nearby and accordingly the visitation rate of the former will be smaller than the latter.

The first step in applying the TCM is to collect data through a visitor questionnaire at the site. The required data include transportation expenditure, hotel expenses and park entrance fees, amount of time spent on travelling, and various socioeconomic characteristics of visitors. For economic approaches, the second step is to derive an equation that relates the visitation rate and the independent variables which affect the visitation. Examples of these variables include total travel cost of visitors and their income, age or other socio-economic characteristics. The third step is to construct a system of demand equations to obtain an aggregate demand curve for the site. The last step is to measure the area under the aggregate demand curve to quantify the benefit visitors enjoy from the recreation site.

Travel cost of visiting the resorts may not be easy to determine. The visit to the resort may be combined with other visits and there may be other reasons for travel. For example, the travel cost associated with commuting for those who prefer to live in far away, cleaner areas may be considered an expenditure to avoid air pollution. However, the resident may be also paying for a larger house in the suburbs, and not only to avoid pollution. Thus, the TCM valuation procedure also involves statistical complications. Moreover, TCM requires substantial effort to obtain primary as well as secondary data, and the method is less applicable to urban degradation than to amenities such as national parks or resorts.

The hedonic pricing method

The hedonic pricing technique uses a related market approach to obtain the value of an environmental amenity from indirect observations (Rosen, 1974). The basic premise is that for many environmental goods, it is often possible for individuals to choose their level of consumption through their choice of related market goods. For example, in the decision to purchase a house in a residential area, one could choose the levels of noise, air quality, water supply and sanitation facilities etc. (Anderson and Crocker, 1971; Wieand, 1973; Schnare, 1976). The value of an apartment facing the sea may be greater than that of the same size apartment on the other side of the block. Such differences in prices may be attributed to the value of enjoying the sea view.

In a decision to buy a family home, there is an implicit market in environmental quality, and the demand for non-market environmental goods such as air and noise pollution contributes to the observed prices and consumption of market goods (i.e. houses in this case). For example, the typical preference for living in quieter residential areas might be reflected in willingness to pay more for a house that is not adjoining a major freeway or an airport (Burns, 1989; Nelson, 1980). This rationale led to the development of the hedonic pricing technique as a means of describing valuations of non-market environmental goods.

In essence, the hedonic pricing method employs statistical techniques to isolate environmental values which contribute to an observed difference in product prices. Typically the composition of real estate values has been analysed to draw out these environmental values (Streeting, 1990). The key initial task of the hedonic price technique is estimating the implicit price function (Streeting, 1990). The construction of the implicit price function is based on the idea that goods and services are composed of a number of attributes, and that the relative amounts of these attributes contribute to the total value of any particular good. For example, in the housing market, the price of a house is related to structural aspects (reinforced cement concrete/wooden/pre-cast, etc.) and physical attributes (living area, number of bed rooms, etc.), locational and neighbourhood characteristics (shopping centres, school quality, population density, etc.), together with environmental features (air and noise pollution, etc.). In this sense, the value of each attribute is implicit, or reflected in the total house price.

In mathematical terms, the generalized form of the implicit price function can be expressed as follows:

$$P = F(X_1, \ldots, X_n)$$

where P is the price of the product under consideration and $X_1 \ldots X_n$ are the attributes of the product.

Regression analysis is used to estimate the equation, and the coefficients of each of the independent variables ($X_1 \ldots X_n$) represent the implicit prices for each of the attributes. This then enables the price differential within the product class (P) to be assigned quantitatively to each attribute.

Stated preference methods

These methods are not as direct as revealed preference methods. Among these is the contingent valuation (CV) method.

Contingent valution methods

In many respects, CV is similar to market research surveys that estimate consumer demand for a new product. The CV method uses surveys to elicit people's valuation of increases or decreases in the provisions of environmental amenities by constructing a hypothetical market. This market is outlined to the respondent in a scenario describing the amenity, the actual or likely change in provision of the amenity, the organization providing the amenity and the method of payment. Respondents are then asked for their valuation, contingent on the scenario described to them.

In a CV survey, individuals may also be asked about their attitudes, expectations, needs and opinion related to the amenity in question. These supplementary questions provide useful insights into the attitudes and behaviour of different people with respect to the environment. In most applications, the CV procedure involves asking people what they are willing to pay for an environmental benefit, or what they are willing to accept for an environmental loss. The aim is to reveal the price at which the respondent is no longer willing to purchase the environmental amenity, thereby revealing the individual's maximum willingness to pay. CV can also be applied to reveal a potential loss by asking people what they are willing to accept by way of compensation for environmental degradation or loss of environmental amenity (Imber *et al.*, 1991). However, questionnaires asking how much people will pay to acquire a benefit do not necessarily yield the same as similar questionnaires asking the amount of compensation demanded to give up the benefit (Knetsch, 1993).

CV surveys can be administered by personal interview, over the telephone, or by post: no specific method of survey administration is preferred. The advantage of personal interviewing is that the interviewer can carefully probe the respondent, repeat questions, or use visual aids to clarify any ambiguous response. The questions can be put in a sequence, which cannot be ensured in a postal survey. However, it is expensive to use personal interviewers for large samples. An ill-trained interviewer can do more harm than an improperly designed questionnaire. Because of these, it is often necessary to resort to telephone or postal surveys or else train the interviewer very carefully. A contingent valuation study by Muraleedharan *et al.* (1995) of Borivali National Park in Bombay highlights the needs for bias removal in the context of a developing country.

CV surveys can replicate either a private goods market or a referendum to obtain benefit estimates. The advantages and limitations of the two models are briefly reviewed below.

Private goods market model

In a private goods market model, a scenario is usually described to the respondent where the opportunity to obtain the benefits of amenities is offered at a range of prices. This model is best suited to quasi-private goods such as access to beaches, fishing rights or any other amenity where a permit for access (to the exclusion of others) is feasible. Application of this model to public environmental amenities such as clean air tends to be less successful as the procedure usually requires the respondents to imagine that they can own and use the amenity to the exclusion of other persons.

The elicitation method often used in private goods models is a bidding game technique designed to resemble an auction. The interviewer raises or lowers the bid until the respondent decides to make a purchase, thereby revealing his maximum willingness to pay. In an open-ended question, the opening bid or the final valuation can be stated by the respondent without any prompting. In a close-ended format, respondents are asked to answer 'yes' or 'no' to a proposed payment.

The referendum method

In the referendum method, respondents are asked whether they would be willing to pay or sacrifice a specific amount of money in order to preserve the environmental amenity in question. A referendum can also be held about a specific action where there is a discrete choice, and the answer can only be 'yes' or 'no'. A range of amounts is put to a number of subsamples. The referendum model can be used to elicit respondents' votes on a tax level, or the tax rate for the provision of a public amenity. It has been extensively used in the developed countries in practice. For example, in Austria an already built nuclear power plant at Zwentendorf was mothballed following a referendum. The value of forgone benefits and additional expenses for alternative electricity could be considered as the cost of avoiding the risks of nuclear power.

Willingness to pay/accept from CV studies

Once the survey responses are available, obtaining the numerical value of willingness to pay/willingness to accept (WTP/WTA) is rather easy. One could either take the mean or the median value of the reported WTA/WTP: there is no norm about which should be used. While the median has been the preferred mode in the referendum model, experts have questioned the validity of this practice. Haneman (1991) has suggested that the choice between mean and median be derived from a prior choice of a social welfare criterion.

How acceptable are contingent valuation techniques? An influential report of a blue-ribbon committee having among its members two Nobel prize winning economists (Arrow *et al.*, 1993) guardedly approved the CV approach, if it is correctly used and if certain conditions are met. In some cases some comparisons with conventional valuation methods are also available.

Concluding comments

A variety of valuation techniques are needed to value environmental change. In applying many of these, some shortcuts on simplifying assumptions would be unavoidable. In making these assumptions, however, one should keep in mind the purpose for which valuation is done and be careful about obtaining either a lower bound or an upper bound estimate.

19
Greening the national accounts: valuation issues and policy uses

KIRK HAMILTON AND MICHAEL WARD

The 1993 revision to the SNA and the simultaneous publication of UN guidelines for integrated environmental and economic accounting (the SEEA) both raise important issues concerning the valuation of environmental stocks and flows. The issues in the SNA include the treatment of resource assets where property rights are not established, valuation of non-market assets, and the scale of informal sector activities with environmental spillovers. While the SEEA provides an accounting structure, it does not offer clear options for valuing environmental assets. Concrete suggestions are presented for valuing living resources, exhaustible resources and pollution emissions. For pollution, marginal social costs, rather than maintenance costs as suggested in the SEEA, provide the correct basis of valuation. The policy implications of green national accounts are discussed, with an emphasis on policies for sustainable development.

Introduction

The interaction of economic strategy with the environment has become a major policy concern, not just because of the implications of macroeconomic decisions, but also because of the more immediate impact which official actions and their commercial consequences have on the quality of life. For the longer term the concerns are about whether current decisions regarding economic development and the exploitation of the environment will reduce the well-being of future generations – in other words, is development sustainable? Related questions include: what policies will be required to achieve sustainable development? and what indicators exist of the sustainability of economic development?

The growth rate of gross domestic product (GDP) has long been a key indicator in macroeconomic policy-making, but GDP ignores explicit environmental and resource impacts. As a result, 'greening' the national accounts has

---

The opinions expressed are those of the authors and not necessarily those of the World Bank.

K. Uno and P. Bartelmus (eds.), Environmental Accounting in Theory and Practice, 337–354.

been urged as a step that is necessary to guide both economic and environmental decision-making into sustainable pathways. This viewpoint is perhaps most eloquently argued in Repetto *et al.* (1989). An important response to this demand to green the accounts was the publication of the guidelines for an integrated System of Environmental and Economic Accounts (SEEA; United Nations, 1993a). This is designed as a satellite system that augments, but does not directly alter, the System of National Accounts (SNA). The SEEA complements the newest revision to the SNA (Commission of the European Communities *et al.*, 1993), which introduces substantial conceptual changes that facilitate the linkage of the SNA to other resource and environmental information. Developing elaborate accounting systems such as the SEEA can be an expensive business. An important concern, therefore, is what additional policy insights are provided by greening the national accounts.

This chapter has three purposes: (1) to explore the valuation issues raised by the revised SNA, with an emphasis on links to resources and the environment; (2) to offer suggestions for many of the valuation issues that are ignored or imprecise in the SEEA; and (3) to examine the policy implications of greening the national accounts, with an emphasis on policies for sustainable development.

Valuation issues from an SNA perspective

It is not possible at present, for practical and conceptual reasons, to place a meaningful economic value on the use of all environmental assets (and to assess the cost of the damage sustained by such use) and to do so without affecting the level and structure of relative prices in general. However, because many important environmental assets remain unpriced, or underpriced, these resources tend to be used indiscriminately and, consequently, inefficiently. The resulting misallocation of resources and market failures associated with not attaching appropriate values to scarce goods is wasteful, and the lack of suitable pricing sends the wrong signals to policy makers, particularly in formulating strategies for long-term sustainable development or in choosing between competing social needs. The absence of such valuations also leads to a distortion of asset use in favour of the present rather than future generations, and to benefit the rich rather than the poor. The failure to cost the use of the environment also tends to underestimate the growing divergence between social and private costs. Some form of valuation is necessary, therefore, to develop environmental policies sensitized to the overall interests of the public good. If the market is to occupy a position of pre-eminence in determining the most efficient allocation of all scarce resource, then valuation procedures must be extended to non-traded assets to ensure a balanced and comprehensive treatment of scarcity and social concerns.

The extension of the SNA's conceptual framework to incorporate asset classification and asset balances supports environmental accounting initiatives but, simultaneously, it requires the implementation of clear rules of accounting relating to principles of valuation, aggregation, net worth assessment and the more precise definition of relevant time horizons.

Conceptually, the revised United Nations SNA (Commission of the European Communities *et al.*, 1993) marked a significant step because it permitted the potential integration of environmental concerns into macroeconomic accounts and policy decisions through a satellite system. By broadening the definition of the production boundary and putting the emphasis on the comprehensive principle of 'exchange' as opposed to the simpler but narrower concept of a 'market' (in which there are identified transactions where a price system prevails), the 1993 SNA opened up the possibility of a more constructive valuation of the voluntary and involuntary use of natural assets by humans. Such assets comprise land, including soil fertility and subsoil resources; air and the atmosphere surrounding the planet; and water, including the world's oceans and their resources. The drawback of this theoretical advance, however, was the need to introduce a much more extensive range of imputations into the core national accounts, an imperative that automatically extended to the assessment of environmental effects within (in value terms) the same macro-economic context. It also introduced a new scale of measurement problems. Whereas questions of current market valuation are relatively straightforward (although no less difficult to resolve) in matters of economic exchange where the market transaction is implicit, the utilization of natural assets through pollution, depletion, congestion, etc. raises an entirely different range of valuation issues. These have intergenerational consequences and differential implications for various sections of the community, depending upon the existing level of well-being of the community.

For these reasons, and to ensure inter-temporal and cross-country comparability, it is essential to establish some well recognized and universally accepted principles of quantitative imputation as well as appropriate 'pricing' conventions. There is no doubt that unless some standard valuation procedures are followed, the new SNA, despite its theoretical superiority over previous frameworks, and the related system of integrated environmental and economic accounts that can be based upon it, will pose even more difficult questions when assessing the economic impact of economically induced environmental deterioration.

An important further point must be made; the national accounts are country specific and concerned with macroeconomic and government policies. Though it is tempting to see the environment as a 'sector' cutting across different activities within this context (and the satellite approach tends to reinforce this view), environmental concerns are not only macro and related to the micro-incentive structure but also supra-national. Furthermore, for many environmental concerns, it is very costly and time consuming, if not impossible in practical terms, to quantify some critical issues such as the impact of rapid population growth and loss of biodiversity. It is difficult to assess the impact of such phenomena on an economy from the environmental perspective because the issues may not relate directly to traditional macro-analysis.

This section attempts to identify, in the absence of appropriate and recent case study experience, those areas in the new SNA where imputations at the macro-level are most complex, tenuous and difficult for developing countries. The problem in practice is one of compounding estimates of non-directly measurable environmental phenomena on national accounts estimates which predominantly reflect a current supply and demand perspective and its associated price system. Because time and relative position are important, it underlines the need for treating each concern separately. Current environmental valuations have to take account of the historical legacy of past environmental damage, current concerns, and possible future influences and strategies.

The SNA has always encompassed the twin concerns of comparability of data and the standardization of international guidelines (frameworks) and statistics. As the definitions have become broader and the boundaries of production extended and grown less precise, it has become increasingly difficult to satisfy these two measurement concerns. The continuing thrust for a comprehensive approach reflects a desire to identify and understand the complex processes at work rather than a wish to come up with a single national number such as an adjusted 'national income' measure (however defined) or 'green GDP'. Why (then more specifically) is it necessary to link such environmental issues to national accounts? Briefly, the main reasons are:

(a) to take long-term environmental costs into consideration in formulating overall development strategies and investment programmes;
(b) to evaluate the impact of economic measures, such as subsidies, taxes, transport industrialization, energy production and pricing policy, on the environment;
(c) to help manage existing natural resources such as forests, mineral deposits, soil, and water more effectively, and to ensure coherence with economic and social objectives;
(d) to assess in economic terms the sectoral and distributional impact of environmental change.

From the earliest time, the main focus of the various versions of the SNA was on the articulation of monetary flows and the interaction of production, income, expenditure and capital formation. Such a systematic assessment of economic activity was intended to help chart the course of economic development and, consequently, provide recommendations concerning the adjustments needed to guide current government policy and keep economies on an equilibrium growth path. In the manner in which it is constructed, the SNA determines the kind of analysis that can be carried out and, therefore, it also strongly influences the way economic and social issues are perceived. Extending the SNA to incorporate environmental values enhances its capacity to inform on matters of longer-term economic strategy but, integrating all environmental issues into the existing national accounts system is complicated by the absence of markets and also by the intrinsic limitation of current GNP measures which distinguish only roughly between productive and unproductive activities and have no facility to incorporate previous losses and irreversibility (exhausted resources) into current valuations.

Equally critical to these assessments is the interpretation of the boundaries for production, consumption and assets. Production is a physical process carried out under the responsibility, control and management of an institutional unit in which labour and assets are used to transform inputs of goods and services into outputs of other goods and services. Under SNA definitions, for example, the natural growth of the fish stock in open seas and of wild uncultivated forests are not production, but fishing (i.e., the depletion of fish stocks) and felling (deforestation and the depletion of wood stocks) are. This clearly has implications for the meaning of consumption and asset utilization. Assets, similarly, are entities owned by units that derive economic benefit from their use over time. Assets that occur naturally, as opposed to those that are physically produced, include land, subsoil deposits of minerals and fuel and uncultivated forests. These are incorporated within the SNA only where an institution exercises effective ownership rights that enable it to benefit economically from the sale or use of such assets. In the national balance sheet, the use of an asset in production is assessed as depletion and the discovery of new assets is treated as an accretion. Unfortunately, this appears to introduce an element of asymmetry with the production accounts in the SNA and ignores the fact that in the ultimate analysis the stock of global assets is given and finite.

How natural assets are valued in the SNA clearly has a critical impact on how people observe the way in which they are used. Assets (and liabilities) are valued using a set of prices specific to the assets that are current on the date to which the balance sheet relates. For most man-made assets, this creates few problems; either existing prices observed in the market can be used, or written down replacement cost values can be determined on an assumed amortization or proper depreciation basis reflecting their reduced service value. However, where assets are directly used as part of final production, a rate of discount must be applied to compute the present value of the expected future returns from those assets remaining, because each asset varies in value and lifetime. This rate of discount relates to each particular asset, including its location and use, and may not necessarily correspond to the general financial rate of interest. The basic value of such assets, furthermore, is linked to how much of them the economy possesses and the level of well-being in the community at large.

In trying to meet a wide range of analytical and policy needs, therefore, the national accounts system has become stretched. It appears to have become stuck between the competing desires to have a comprehensive framework for all activity and one that prevents the sort of misinterpretation of economic behaviour that arises from swamping records of well defined market transactions by a substantial array of non-monetary imputations. So much of the impact of the economy on the environment may be potentially confused and obfuscated therefore, by the initial imprecise valuation of the basic economic activity being undertaken.

The most important areas and transactions where these 'rough' monetary measurements are applied, and where the associated environmental implications and resource use are often far from minimal, especially in low income countries, are as follows (though not necessarily in order of importance):

(a) illegal activities,
(b) household production for own consumption,
(c) informal production and construction,
(d) housing services from home ownership,
(e) domestic and personal services performed within the household,
(f) consumption of fixed capital,
(g) activities remunerated in kind.

Some of these transactions are highly valuable, while others are not denominated in monetary terms. In a not insignificant number of countries, the sum total of such vaguely defined and measured activities could amount to close to 50% of GDP. Each of these activities has some specific, as well as general, association with the environment. Thus, illegal activities may include unofficial logging and mining or drug trafficking, while rural household production could involve soil degradation and erosion as well as the indiscriminate gathering of firewood and contamination of safe water. Informal production (and food processing), because it is uncontrolled and unlicensed, is frequently associated with hazardous materials, pollution and untreated waste disposal. Other activities similarly pose serious health risks to employees, individuals and the community. When the base data are so poorly and inconsistently assessed, both over time and in cross-country comparisons, it is consequently difficult to produce a robust environmental impact assessment in monetary terms, however well defined the theoretical valuation process may be.

Valuation issues in the SEEA

The System of integrated Environmental and Economic Accounts (SEEA; United Nations, 1993a) provides an essential framework for resource and environmental accounts as a satellite system of the SNA, but it does not resolve many thorny issues in valuation. Its contribution is therefore more at the level of structure rather than of content. Some of the key issues in valuing environmental stocks and flows will be explored below.

Valuation is clearly one of the cornerstones of green national accounting. While many policy uses exist for physical accounts of environmental resources linked to the SNA, when it comes to the key issues of measuring sustainable income and broadening the measure of wealth, valuation of environmental stocks and flows is an essential ingredient. This section examines the issues and some suggested solutions for valuing living resources, exhaustible resources and pollution in an extended national accounting framework.

Living resources

The SEEA is notably silent on the issue of valuing stocks and flows of living resources such as forests or fish. The standard models of natural resource economics (see, for instance, Clark, 1976) may be applied to suggest some possible approaches. In the canonical model of a living resource, growth $g(S)$ is assumed to be a function of the stock size S. Growth as a function of S is assumed to have an inverted-U shape, with the maximum occurring at the maximum sustainable yield. For harvest q (all variables are assumed to be functions of time, unless explicitly stated otherwise), the equation for the evolution of the living stock is therefore:

$$\dot{S} = -q + g(S).$$

For a fixed discount rate r the efficient path for unit resource rent (i.e. price minus marginal harvest costs) n is given by,

$$\dot{n}/n = r - g'.$$

The steady state equilibrium for the resource is therefore characterized by a constant unit resource rent n^*, a constant harvest rate $q^* = g^*$, and the equality of the slope of the growth curve and the discount rate, $r = g'(S^*)$. Profits in the steady state, revenues minus costs of harvest, are given by $f(q^*)$. Since the steady state harvest rate can be sustained indefinitely, the value of the stock of living resources is therefore:

$$(t) = \int_t^{\infty} f(q^*)e^{-rs}\,\mathrm{d}s = \frac{f(q^*)}{r}. \tag{1}$$

As long as this equilibrium is attained, the value of depletion is zero, since growth equals harvest. Away from equilibrium the value of depletion was established in Hartwick (1990) to be:

$$\dot{V} = n\dot{S} = n(-q + g). \tag{2}$$

This can be positive or negative depending on the initial conditions: if the initial stock is less than the steady-state equilibrium stock, then it will be positive (i.e. there will be appreciation rather than depreciation, which would increase eco-domestic product (EDP) in the SEEA); there will be depreciation if the initial stock is greater than the equilibrium value.

An important issue for forest resources that is not considered in the SEEA is that of deforestation. Deforestation is essentially the conversion of land from one use (as a forest yielding a potentially sustainable stream of products) to another (e.g. agriculture, which may or may not be sustainable). This suggests that deforestation could be treated as a transfer of land from one form of economic asset to another, with the appropriate additions and subtractions showing up in the asset accounts. In the standard SNA this would be treated as land improvement, a type of investment. Hartwick (1992) shows that a critical issue is whether the new land use yields a stream of income whose present value is less or greater than the present value of the stream of forest products when the land is under forest; if the present value is less then there should be a depreciation charge to reflect this.

One complication in considering deforestation is that standing forests provide positive externalities in the form of water-shed protection that benefit agricultural activities downstream. As an extension of the Hartwick (1992) model, therefore, it would seem logical to show the reduced downstream benefits as a deduction from the income stream of deforested land that is converted to other uses.

Exhaustible resources

The SEEA considers the valuation of exhaustible resource stocks in terms of an expanded asset account, but is less than precise in its consideration of the valuation of depletion, which is a critical input to the calculation of eco-domestic product. As one approach to valuing depletion it mentions the method of El Serafy (1989). Hartwick and Hageman (1993, Appendices 3 and 4) provide a useful analysis of the issues, but this work has received little recognition.

The theoretical approach of Hartwick and Hageman can easily be extended and provides critical insights. Consider a stock of resource S that is extracted in amount q up until the time of exhaustion of the deposit T, so that,

$$\dot{S} = -q \quad \text{and} \quad S(t) = \int_t^T q(s)\,\mathrm{d}s.$$

For resource price $p(q(t))$ and extraction cost function $c(q(t))$, profits from extraction are given by,

$$f[q(t)] = p[q(t)]q(t) - c[q(t)].$$

If the discount rate is fixed at r, then the asset value of the resource deposit is just

$$V(t) = \int_t^T e^{-r(s-t)} f[q(s)]\,\mathrm{d}s \tag{3}$$

This is what the SEEA terms the 'present value' approach to valuing resource assets. It requires forecasts of prices, quantities and costs in order to arrive at a unique value, a formidable obstacle in the minds of most national statisticians (see, for instance, Aaheim and Nyborg, 1994).

Among many choices of the paths of p and q, there are two practical options that should be considered. One is to assume an efficient path for resource rents, the Hotelling rule; the other is the method of El Serafy (1989). The Hotelling rule stipulates that unit rents should increase at a percentage rate equal to the discount rate. Under these conditions investors will be indifferent between holding natural resource assets and other financial or productive assets. If this is assumed, then for $s > t$,

$$\frac{df}{dq(s)} = e^{r(s-t)} \frac{df}{dq(t)}.$$

Noting that e^{-rt} can be factored out of expression (3), depreciation of the stock is given by

$$-\dot{V}(t) = f[q(t)] - rV(t). \tag{4}$$

Assuming that $q(T) = 0$ and $f(0) = 0$, integrating by parts and applying the Hotelling rule yields:

$$\begin{aligned} rV(t) &= \int_t^T r\, e^{-r(s-t)} f[q(s)]ds \\ &= -e^{-r(s-t)} f[q(s)]\bigg|_t^T - \int_t^T e^{-r(s-t)} \frac{df}{dq(s)} \frac{dq(s)}{ds} ds \\ &= f[q(t)] - \frac{df}{dq(t)} q(t) \end{aligned}$$

For efficient resource extraction, therefore, depreciation is calculated as the unit rent times the quantity of resource extracted,

$$-\dot{V}(t) = \frac{df}{dq(t)} q(t). \tag{5}$$

This result is the same as that derived in the formal model of national accounts in Hartwick (1990). Hamilton (1994) extends this model to the case of multiple resource deposits with heterogeneous extraction costs (e.g. onshore and offshore oil). For profit functions:

$$f_i[q_i(t)] = p[\Sigma q_i(t)]q_i(t) - c_i[q_i(t)], \quad i = 1 \text{ to } n,$$

the total value of resource depletion is given by

$$-\dot{V}(t) = \sum_{i=1}^{n} \frac{df_i}{dq_i(t)} q_i(t).$$

If efficient resource extraction is assumed and marginal costs of extraction can be measured, then the foregoing derivation for $rV(t)$ yields a simple formula for the value of the stock,

$$V(t) = \frac{f[q(t)]}{r}\left\{1 - \frac{df}{dq(t)}\, q(t) \middle/ f[q(t)]\right\}.$$

The value of the resource stock is therefore the present value of an infinite stream of the current profit times a factor equal to one minus the ratio of current rent to current profit. This clearly requires there to be increasing marginal extraction costs, which is a reasonable assumption. Note that although this expression seems to be independent of the terminal time T, this is captured implicitly in the ratio of scarcity rents to average profits.

The approach of El Serafy (1989) simply assumes that profits are constant at each point in time, $f[q(t)] = R$, for constant R. Under this assumption it is straightforward to derive

$$V(t) = R\int_t^T e^{-r(s-t)}ds = \frac{R}{r}(1 - e^{-r(T-t)}) \tag{6}$$

Now the value of the resource is equal to the present value of an infinite stream of the current profit times a factor reflecting the finitude of the resource. Other than the fact that the right-most factors in this expression and the preceding one will converge to zero as the resource is exhausted, there is little linkage between these expressions in general. This result suggests that the rough equivalence of the optimal depletion and El Serafy approaches simulated by Hartwick and Hageman (1993) is strictly an artifact of the assumed functional forms.

From expression (6) it follows directly that

$$-\dot{V}(t) = R\, e^{-r(T-t)} \tag{7}$$

El Serafy (1989) does not derive the formula for the asset value implicit in his approach, expression (6), and derives expression (7) in its more familiar discrete-time form: i.e. if

$$T - t = \frac{S(t)}{q(t)} = N$$

then depreciation is given by

$$V(t) - V(t+1) = \frac{R}{(1+r)^{N+1}} \tag{8}$$

If profits are assumed to be constant over the life of the resource, therefore, the El Serafy method offers a tractable means of calculating both asset values and the value of extraction. From the perspective of the national statistician, constant profits may be viewed as the most neutral forecast of prices, quantities and costs.

There remains the question of the choice of discount rate in the El Serafy approach. The standard derivation of the rate of interest from growth theory yields, for the pure rate of time preference ρ, elasticity of the marginal utility of consumption $\eta(C)$, percentage rate of growth in per capita consumption $\dot{C}/C$, and marginal product of capital F_K,

$$\rho + \eta(C)\frac{\dot{C}}{C} = F_K.$$

This equates the rate of interest in the economy both to the marginal product of capital and to the social rate of return on investment (SRRI, the left hand side of this expression). In the real world (and in more complex growth models) risk and taxes drive a wedge between the rate of return private producers require and the social rate of return on investment.

For resources that are owned by governments the tax rate is not an issue. If the quantity of resource at issue is proven plus probable reserves, then risks concerning the likely quantity of resource to be exploited are minimal. There are risks concerning future prices of resources, with both upward and downward potential identifiable. On balance, there is a strong argument for using the social rate of return on investment as the relevant interest rate. Pearce and Ulph (1995) offer evidence that the SRRI lies in the range of 2–4% for the UK, values that should be applicable to most developed economies.

A number of conclusions follow from this analysis. First, valuing exhaustible resource assets requires assumptions or forecasts concerning the future level of resource prices, quantities extracted and costs. Secondly, the value of depletion is, in general, itself dependent on the value of the asset (equation 4). If efficient production is assumed, then the value of depletion is equal to the current unit rent times current extraction; resource rents are generally difficult to calculate, however, because they involve the measurement of marginal costs of extraction. Assuming constant profits from extraction permits straightforward calculations of both asset values and the value of depletion (equations 6, 7 and 8), but is inconsistent with any intertemporal optimization by resource producers. Assuming constant profits requires a choice of discount rate, with the social rate of return on investment being a defensible choice.

As long as one is willing to abandon notions of optimality, therefore, the El Serafy approach offers a practical methodology for statisticians to use in valuing resource stocks and depletion. The resource economics literature is filled with examples of studies that fail to find empirical support for the Hotelling rule, with that of Miller and Upton (1985) being one of the few exceptions.

Valuing environmental degradation

In dealing with environmental damage the SEEA introduces the concepts of 'cost caused' and 'cost borne', suggesting that both may be related to measures of product. Transboundary pollution makes clear the issues regarding costs caused and borne.

Consider two countries, A and B, that each emit pollutants as a result of their own economic activities, and for each of which there is some export of pollution to the other country (there could be, for instance, different prevailing winds at different times of the year). Then the cost borne by country A is the damage that is done to its economy and citizens both as a result of its own economic activity, i.e. that portion of its pollution emissions that falls on itself, and as a result of economic activity in country B, i.e. that portion of country B's total emissions that are exported to country A. The cost caused by country A is, again, the damage that is done to its economy and citizens as a result of its own economic activity, plus the damage that is done to country B as a result of the export of pollution emissions from country A. The situation in country B is the exact mirror of this.

It is clear that 'cost borne' answers the question 'What is the total pollution damage incidence within the country, both as a result of domestic and imported pollution?' This is an important question, particularly when it comes to international negotiations on transboundary pollutants, but it bears no relationship to national accounting in the sense that charges should be made against income or product to reflect these costs. This is because the costs borne by country A are not linked or related to the level of national product in country A. In contrast, 'cost caused' answers the question 'What is the total pollution damage caused by economic activity in the country, whether the damages occur within the country or in other countries where the pollution is incident?' In this case there is a clear logic associated with charging these costs against income or product, since the costs are associated with the level of national product and are either incident domestically or are notionally owed as compensation to other countries. This argument applies to all pollutants crossing national boundaries, either ones that are regionally important such as acid emissions, or pollutants with global social costs, such as greenhouse gases. Setting aside transboundary pollutants, the same logic applies when attempting to measure the net product of individual sectors in a national economy. Again, there is no good argument for adjusting the income or product of sectors affected by pollution (costs borne), but ample justification to make a deduction from the product of emitting sectors (costs caused) because compensation is, at least notionally, owed to the affected sectors.

The next issue is the actual valuation of environmental damage. The SEEA suggests that 'maintenance costs' are the preferred approach: the total cost of maintaining environmental quality at the same level at the beginning and end of the accounting period. While this approach has the flavour of maintaining capital intact, the conception of sustainable development favoured by Pearce *et al.* (1989), it presents practical as well as conceptual difficulties. At the practical level, suppose it is assumed that some level of environmental protection costs were incurred in a country and that environmental quality deteriorated over the accounting period. Then the maintenance cost is the sum of actual costs incurred and the costs that would have been required in order to achieve the target of a constant level of environmental quality. Both costs are difficult to measure, with the obvious greater difficulty being associated with the hypothetical costs. The only solution to the latter problem is some sort of modelling and, to be accurate, general equilibrium effects would need to be considered: simply applying abatement cost functions to the 'excess' emissions would not measure the economic cost of achieving the target.

The problems with maintenance costs at the conceptual level are, if anything, even more formidable. This is best explored with a formal model, as presented in Hamilton (1996). Here welfare, or utility, U is assumed to be a function of both consumption C and the flow of services provided by the environment B. We wish to model a pollutant whose effects are cumulative. The level of the flow of environmental services is therefore related negatively to the cumulative stock of pollution emitted, X, so that $B = B_0 - \beta X$. We assume no abatement expenditures, so pollution emissions resulting from production F are given by $e = e(F)$, while pollutants accumulate according to $\dot{X} = e$. For a fixed pure rate of time preference r, the objective for a social planner in this simple economy is to maximize the present value of welfare. For production function GNP $= F(K,L)$ (K is produced capital, L is labour, assumed to be fixed), the model is therefore,

$$\max \int_0^\infty U(C, B)\, e^{-rt}\, dt \text{ subject to:}$$

$$\dot{K} = F - C$$

$$\dot{X} = e.$$

Here C is the only control variable. The current value Hamiltonian for this problem is,

$$H = U + \gamma_1(F - C) + \gamma_2 e$$

for shadow prices (in utils) γ_1 and γ_2, and the first order condition for a maximum (ignoring the dynamic conditions) is,

$$\frac{\partial H}{\partial C} = 0 = U_C - \gamma_1 \Rightarrow \gamma_1 = U_C.$$

Note that $\gamma_2 < 0$, since increases in the accumulation of the pollutant decrease welfare. The current value Hamiltonian can therefore be written as,

$$H = U + U_C(\dot{K} - \sigma \cdot e),$$

where $\sigma \equiv -\gamma_2/U_C$ is the marginal damage, or marginal social cost, associated with a unit of pollution emissions. The maximum amount of produced output that can be consumed while leaving wealth instantaneously constant is therefore,

$$\text{gNNP} = C + \dot{K} - \sigma \cdot e \qquad (9)$$

This is the measure of 'green' national income. Note that σ is also the level of the Pigovian tax required to maximize utility.

Abatement expenditures, a, are introduced into this model as the use of current production to reduce the level of emissions, so that the emission function is re-defined as follows:

$$e = e(F, a), \quad e_F > 0, \quad e_a < 0.$$

The maximization problem is now specified as:

$$\max \int_0^\infty U(C, B)\, e^{-rt}\, dt \text{ subject to:}$$

$$\dot{K} = F - C - a$$

$$\dot{X} = e.$$

The control variables are C and a and the current value Hamiltonian is as specified above. The first order condition for γ_1 is again that it should equal the marginal utility of consumption. For γ_2 we now have,

$$\frac{\partial H}{\partial a} = 0 = -\gamma_1 + \gamma_2\, e_a \Rightarrow \gamma_2 = \frac{U_C}{e_a}. \tag{10}$$

It will be useful in what follows to define $b \equiv -1/e_a$; this is just the marginal cost of pollution abatement. The current value Hamiltonian therefore becomes,

$$H = U + U_C(\dot{K} - b \cdot e),$$

and the expression for green national income is,

$$\text{gNNP} = C + \dot{K} - b \cdot e. \tag{11}$$

Equation (10) implies that $b = -\gamma_2/U_C$. The marginal cost of abatement is identically equal to the marginal social costs of emissions and to the value of the optimal unit emissions tax.

Equation (11) yields another interpretation. First, GNP $= F = C + \dot{K} + a$. This implies that

$$\text{gNNP} = \text{GNP} - a - b \cdot e.$$

We conclude that, in order to arrive at a greener income measure, abatement expenditures should be subtracted from GNP – they become, in effect, intermediate consumption. Of course, most abatement expenditures are already treated as intermediate consumption in the national accounts. Therefore it is only expenditures on pollution abatement in final demand that need to be deducted in measuring green national income.

Measurements of the marginal social costs of pollution emissions are increasingly available as a result of work on the social costs of fuel cycles, particularly in the USA and Europe (see Commission of the European Communities and US Environmental Protection Agency, 1993 for recent estimates, and Hamilton and Atkinson, 1996 for an application). While much effort has gone into valuing the social costs of air pollution, less is known about water pollutants.

From the standpoint of the SEEA treatment of environmental degradation, the key point is that formal models of green national accounts lend no support to the use of maintenance costs as the basis of valuation. This conclusion is quite robust: Hamilton (1996) shows that similar results hold for flow pollutants, stock pollutants, acid rain and emissions of greenhouse gases.

It is important to ask whether optimizing models can yield useful results for practical national accounting. As shown in Hamilton (1996), valuations of pollutants based on marginal social costs will in general be higher than the optimal value, but the amount of bias will approach zero as the optimal level of pollution emissions is approached. Such a measure, with known direction of bias and decreasing amount of bias as the optimum is approached, can therefore be effective for guiding policies that aim for the optimum.

Policy uses of green accounts

Satellite accounts based on the SEEA can be used effectively in policy analysis. The range of applications includes the measurement of physical resource scarcity, valuation of depletion, measuring the incidence and burden of existing and proposed regulations and taxes, estimating emission tax rates, modelling the effects of structural change, and providing environmental components for existing macro-policy models.

The promise of measuring EDP (or a 'green' NNP) is that it can provide a highly integrative measure of economic activity and its effects on the resource base and the environment. The gap between GDP and EDP quantifies the extent of depletion and degradation and therefore serves as a signal of the importance of environmental effects. The composition of the gap can be used to set priorities for dealing with environmental problems, since valuation provides the numéraire for comparison of disparate environmental phenomena. Measuring EDP can provide an enabling environment for policy makers to consider the linkages between the economy and the environment.

Defining a true measure of the level of income, one that does not treat the liquidation of assets (natural resources) or the creation of liabilities (pollution) as income, is clearly important. Development strategies that lead to excessive resource depletion or pollution emissions will not lead to the same growth in EDP as in GDP. But the question of the sustainability of development cannot be answered directly by an EDP measure. For this we turn to greener measures of saving and wealth.

Given the centrality of savings and investment in the economics of development, it is perhaps surprising that the effects of depleting the environment have not, until recently, been considered in the measurement of national savings. This omission may be explained both by the models used by economists, which typically rely on gross measures of activity, and the fact that the System of National Accounts (SNA) ignores the degradation of the natural environment in the standard measures of income and product. To correct this, genuine saving is defined as net saving less the value of resource depletion and the value of environmental degradation (cf. Pearce and Atkinson, 1993; Hamilton, 1994). The policy implications of measuring genuine saving are quite direct: sustained negative genuine savings must lead, eventually, to declining welfare. There appears to be ample evidence for negative rates of genuine savings in a range of developing countries (World Bank, 1995). The key question, therefore, is what policy implications follow from the measurement of negative genuine savings.

A basic determinant of genuine savings rates for developing countries is the value of resource depletion. However, it would be wrong to conclude that the policy response regarding savings and natural resources is to boost genuine savings by restricting resource exploitation. One of the key lessons from growth theory, alluded to in Weitzman (1976), is that the discovery of a natural resource, properly managed, leads to a permanent increase in the sustainable stream of income for a country. The question with regard to natural resources is therefore one of what constitutes 'proper management'. Clearly, an important policy concern is the achievement of efficient levels of resource exploitation. The policy considerations are therefore:

(a) Do tenurial regimes encourage sustainable exploitation?
(b) Are royalties set correctly, to capture resource rents while leaving the exploiting firms with adequate rates of return?

Another element of proper management of the resource endowment is to ensure that royalties on natural resource exploitation are invested in other productive assets. It is this rather simple concept of 'preserving capital' that is captured by genuine savings measures. Basic questions for countries with natural resources are, first, are the royalties from natural resources invested or consumed? What kinds of investments are made? The export of natural resources necessarily involves the liquidation of some amount of the natural resource base. From the perspective of genuine savings an important question is, therefore, do policies to promote natural resource exports also embody plans for the investment of the resource royalties? It is clear that in many developing countries rapid urbanization and industrialization is leading to major problems of environmental quality; this is particularly so in the metropolitan areas of Asia and Latin America. The policy questions this poses are do policies with respect to pollution emissions aim for the economic optimum, where total social benefits and total abatement costs are equated at the margin?, and even if the optimum is achieved, are sufficient savings being made to offset any increments to pollution stocks that this may entail?

Finally, it should be obvious that the gross savings rate is a basic determinant of the genuine savings rate. This leads to consideration of the whole range of micro- and macro-economic policies that affect savings behaviour by individuals and institutions, including:

(a) Is the level of government current expenditure appropriate and sustainable?
(b) Does the tax system penalize or encourage saving?
(c) Does monetary policy set positive real interest rates?
(d) Do government policies support a viable financial sector?

Some important caveats need to be added to this discussion of genuine saving. First, not all saving is the same, in the sense that savings sitting in foreign bank accounts belonging to a small segment within a society may not lead to development. In other words, there are distributional issues to be considered. Second, not all investment is the same, in the sense that there are both productive and wasteful investments. A key concern that follows from the consideration of genuine savings, therefore, is the quality of investment. Investments in human capital, especially in primary education in developing countries, are likely to be important in this regard.

Conclusions

The 1993 revision to the SNA and the simultaneous publication of the SEEA have given an important impetus to efforts in many countries and organizations to produce greener national accounts. The implementation of these systems will raise many of the issues discussed above, particularly on valuation.

Among the accounting issues specific to the SNA is the definition of economic assets and the distinction between wild resources and those over which property rights are established. For natural assets, the lack of an existing market valuation implies that the present values of their services must act as a proxy, with difficult decisions therefore required on discount rates and service lives. Implementing the SNA for developing countries entails valuing many difficult to measure activities, including household production for own consumption, informal production, illegal activities and activities remunerated in kind. For these countries the starting point in green accounting, conventional GDP, is known only imprecisely because of the scale of these activities in the economy.

The SEEA provides a useful framework for satellite accounts on the environment and natural resources, but leaves many of the details on valuation imprecise. For living resources, not discussed explicitly in the SEEA, the approach suggested by theory is to measure depletion as harvest minus growth, valued at the unit rental rate. For exhaustible resources, valuing depletion requires making assumptions about the future path of extraction profits and quantities, as well as the discount rate. Two practical alternatives exist for this: assuming efficient extraction, in which case depletion equals the current quantity extracted times the unit rental rate; or assuming constant profits over the life of the resource (El Serafy, 1989), in which case depletion equals the present value of the profit taken in the final period of extraction. Given the difficulty of measuring resource rents (marginal extraction costs must be gauged), the second method may be more practical.

The treatment of pollution in the SEEA is particularly problematic. Only 'costs caused' have a place in green national accounts, because only these costs can be unambiguously associated with production activities in the country in question. These costs include damages imposed on other countries. Pollution emissions should be valued at their marginal social costs, rather than the SEEA notion of maintenance costs. And environmental protection expenditures in final demand should be deducted from any measure of EDP.

Turning to policy implications, it is clear that measuring EDP will produce a 'truer' measure of income than GDP if all the valuation issues are solved satisfactorily. However, it is likely that it is the components of the deductions making up EDP that will have the greatest policy resonance. When it comes to an indicator of sustainability, genuine savings would appear to be the superior choice. Policies to raise the level of genuine savings span the full range of macroeconomic, resource and pollution control measures. The analysis of genuine savings therefore provides a natural point of linkage between the interests of ministries of finance, planning, natural resources and environment.

More broadly, accurate national accounting for income and wealth is important. It may indirectly encourage policies and a mindset for politicians, statisticians, planners, and others that encourages sounder economic management. However, environmental accounting will not, by itself, result in improved environmental policies; the latter can be expected to be encouraged only indirectly. Better accounting should be seen as one element along with other tools in a multi-pronged strategy which includes environmental impact assessments at the project level, and integrated environmental and economic analyses for policy work at the sectoral and macroeconomic levels.

20
Modelling and accounting work in national and environmental accounts

ANDRÉ VANOLI

Modelling plays a limited role in the central framework accounts of the System of National Accounts (SNA), which are based essentially on observations. By contrast, satellite environmental accounts in money terms, with the more ambitious objectives advanced in the United Nations System of integrated Environmental and Economic Accounting (SEEA), seem largely a modelling construct. Modelling is used in a broad sense, covering valuation relying on economic theory. Such an approach requires careful explanation of what types of flows and stocks are intended to be measured, what seems to be measurable, what is actually measured (or measurable) and what is the meaning of the aggregates that are proposed. This chapter discusses some issues in relation with the interim version of the SEEA. They include mainly the difficulty raised by the advocated combination of exchange values (market prices) and use values (based on analysis of the consumer surplus) in the welfare approach. Ex post modelling used for the estimation of the hypothetical maintenance costs also raises some questions as to the meaning of the results, based on partial and constrained modelling, and their inclusion in an accounting framework.

Introduction

One of the main uses of national accounts is to serve as a basis for modelling. Economic modelling is normally future orientated, and aims to forecast what will probably happen, taking into account the past and present, what can already be known about the future (the fiscal policy for instance), economic inter-relations and various assumptions about the price of crude oil, the exchange rate, etc., or simulating the consequences of various phenomena (a sharp rise in the price of oil) or of economic policy measures (e.g. the introduction of an eco-tax). In general, modelling tries to analyse 'what would or could happen if . . . ?'. It is conditional and hypothetical. Modelling by nature cannot deliver a single answer. Differences in basic structure, parameters or exogenous assumptions lead to different answers. Different models give different answers and a given model may produce different variants.

K. Uno and P. Bartelmus (eds.), Environmental Accounting in Theory and Practice, 355–373.

Accounting for a past period (ex post accounting), on the other hand, aims to deliver a single, straightforward answer to the question 'what has actually happened?', for instance, 'what has been the current value of household final consumption expenditure in year *t*?', supposing the object to be measured was defined unambiguously. Ex post accounting is based on observations, facts or actual data. *A priori*, the object to be measured exists, or has existed. It is, or it was, observable. This is certainly not the case with accounting for the future or simulating various states of the economy. The object to be measured then is conditional and hypothetical.

Thus there is, at first glance, a sharp distinction between ex post accounting as an observation system and modelling. However, if it is obvious that modelling the future or alternative states of the economy cannot be based on future or alternative observations,[1] it is worthwhile investigating whether past accounting can be based on observations only.

Modelling in ex post central accounts

We can leave aside the fact that what is observable in principle is not necessarily observed, or easy to observe in practice. Statisticians have to work in most cases with partial and imperfect data. They apply techniques in these situations which use statistical models as estimation tools. National accountants often work with incomplete data, different data related to the same item, or conflicting data for inter-related items. They make estimates, choices and adjustments, to obtain consistent figures. In practice, either a single result is not obtained or, often, we cannot be sure whether a single result is actually the 'true' figure. However, we assume generally in ex post accounting that there exists such a true figure, which we try to approximate as best as we can, knowing that there are errors of observations (including sampling errors, underestimation by respondents, imperfect data processing techniques, etc.). Nevertheless there are at least three types of issues for which the pre-existence of a true value to be approximated by observation is not evident.

Non-monetary transactions and certain aspects of monetary transactions

The first type of issues refers to the broad range of non-monetary transactions or certain aspects of monetary transactions. Monetary transactions are the basis of economic accounting, business and national accounting. Normally, for monetary transactions (exchanges or transfers) there exists a true figure that is observable regarding all aspects of these transactions. At the root of the monetary value are exchanges, both of products and factors of production, and the circulation of economic value derived from them by means of payments.

Beyond monetary transactions, the central national accounts framework extends to non-monetary transactions which are closely connected with monetary transactions and from which their value is derived. Costs of non-market production activities by government are monetary and observable. National accounts equate the value of non-market government output with its costs, assuming a zero net profit. It can be debated if all costs of government non-market output are actually monetary. The question is whether to include the opportunity cost of the capital used in government productive activities thus increasing the value of output. The negative conclusion of the last revision of the SNA illustrates the reluctance of national accountants to depart from observations. Thus the assumption is not only zero net profit (beyond costs) but zero net operating surplus resulting from governmental non-market activity.

The 1993 System of National Accounts (SNA) recommends the valuation of output by costs because there are in general no representative markets of services similar to government services (Commission of the European Communities *et al.*, 1993, paras 6.90 and 6.91). Preference is thus given in principle to the valuation of non-market output at market prices of similar goods and services, provided such markets exist. This is the case, for instance, when the non-market services of owner-occupied dwellings are valued by the market prices of renting similar dwellings, or when 'reliable market prices' are applied to output produced for own final use (Commission of the European Communities *et al.*, 1993, para. 6.85). However, the central framework does not recommend simulating markets to derive a shadow price when no reference markets exist. In this case, costs are preferable, because they are observable. As to measuring costs, there is no problem, but their use for measuring output inevitably implies certain assumptions.

Using costs or market prices for similar goods and services in order to value non-market output implies that the values and the quantities involved would not be significantly different if non-market activities were converted to market activities. There is, however, no observable true value in these cases, even though the underlying physical quantities are observable. The valuation of this output is therefore subject to assumptions which are in the nature of modelling.

Sometimes there is no direct correspondence between actual economic and national accounts transactions. Obvious cases are insurance transactions and financial intermediation services indirectly measured (FISIM). In both cases, the value of output is usually not strictly observable, unless a sharp distinction would be made by the enterprises between the management of their own funds (and the corresponding investment income) and the funds that are the property of policy holders or victims, in the case of insurance, or intermediated funds in the case of financial intermediaries. As far as total output is concerned, the problem is the difficulty of measurement and not so much non-observability. The problem increases when looking at the allocation of FISIM between users. For instance, the allocation of FISIM to depositors is based on the opportunity cost of keeping deposits instead of investing the funds in interest earning assets. There is an element of 'what if?' involved. Thus, while total

FISIM can be deemed observable, its sharing between borrowers and depositors is not directly observable and requires modelling. It is not clear at the moment to what extent changes of bank practices in invoicing their services could drastically modify the issue.

The above review of non-monetary transactions and certain aspects of monetary transactions touches upon issues that have been extensively debated in the history of national accounts. These debates reflect tensions between observation and modelling on the one hand, and reality and appearance on the other hand. National accounts constitute an information system which ideally should be based on observations rather than on assumptions. At the same time, national accounts try to describe and measure reality in a meaningful way. As certain aspects of reality are not directly observable and their exclusion from the central national accounts would give a distorted picture of the economy, there is no alternative but to accept some dose of 'soft modelling'.

Accounting at constant prices

An issue which inevitably implies modelling is accounting at constant prices, i.e. in volume terms. As is well known, there is no single way of assessing the change in volume of any aggregate between two periods of time, unless there has been a homothetic change of all prices. Accounting at constant prices is inevitably a 'what if?' exercise: what would have been the value of a given set of goods and services, using a price system different from the actual one? Since a different price system would have brought about a different set of quantities, combining present quantities with a different price system is by definition hypothetical. Thus there is no single actual true volume change that could be approximated.

Price index theory analyses the differences between alternative methods. Empirically we try to avoid applying a price system that is too different from the original one associated with the set(s) of quantities in question. Chaining, under certain conditions, provides a relevant solution. At the detailed level of specific products whose qualitative characteristics have changed, modelling can be the most efficient way. Hedonic techniques can be applied as a means of distinguishing an actual price change from a volume change in the apparent price change of a unit of the product.

Consumption of fixed capital and the value of assets

A third issue is the consumption of fixed capital, an item which plays such an important role in deriving net figures for value added, product, income, saving and capital formation. In practice, consumption of fixed capital is not observed as business depreciation allowances are not relevant in this context.

Capital consumption is generally calculated using the perpetual inventory method. Even supposing that all other elements entering the calculation, such as the service lives of fixed assets, are known, assumptions are necessary about the efficiency profiles and the rates of depreciation of the assets.[2]

There are further conceptual and substantive problems. According to the Commission of the European Communities *et al.* (1993), 'the value of a fixed asset to its owner at any point of time is determined by the present value of the future rentals (i.e. the sum of the discounted values of the stream of future rentals) that can be expected over its remaining service life. Consumption of fixed capital is therefore measured by the decrease, between the beginning and the end of the current accounting period, in the present value of the remaining sequence of rentals' (Commission of the European Communities *et al.*, 1993, para. 6.182) and 'the calculation of consumption of fixed capital is a forward looking measure that is determined by future, and not past events' (Commission of the European Communities *et al.*, 1993, para. 6.183). While the basic principle is not debatable (the price a purchaser accepts to pay for an asset depends on what he expects to derive from the ownership of this asset), there are questions about the assumption of rationality, limited information, uncertainty and imperfect expectations. Even for new assets, the assumption that 'as a result of market forces, the purchaser's price of a new fixed asset should provide a good initial estimate of the present value of the future rentals which can be derived from it' (Commission of the European Communities *et al.*, 1993, para. 6.184) is not always realistic as we can see from the appearance and quick disappearance of many small enterprises.

Generalized markets for existing goods would provide useful information about the adjustments of expectations. Unfortunately, such markets are limited in scope, with few exceptions. Also, the value of individual assets belonging to an enterprise thought of as a going concern is usually not separable from the total value of the enterprise. Of course, we use prices of assets as they are observed. In the case of existing assets we thus apply to all assets of a particular type prices revealed in actual transactions, though they might represent a limited share of the total stock only. Those figures are not adjusted even if the market values seem to be unrealistic, e.g. in the case of speculative bubbles in financial or real estate markets. Furthermore, the SNA does not record all the change in the real value of (fixed) capital under consumption of fixed capital. Those changes include real holding gains or losses recorded in the revaluation account and unforeseen obsolescence entered in the 'other changes in volume of assets' account.

The conclusion of this brief discussion is that a certain dose of explicit or implicit modelling is unavoidable when the national accounts come into contact with the future through the valuation of assets and the measurement of consumption of fixed capital. The principle of basing the accounts on observations cannot be held up when measuring capital and addressing the concept of 'keeping capital intact'. As a result, net value added, product, income, etc. do not seem to be completely observable items and aggregates (business accounting faces similar difficulties). A certain amount of modelling, in the sense of measurement partly relying on assumptions, is thus required, especially for the measurement of non-monetary transactions, accounting at constant prices, and the measurement of assets and consumption of fixed capital.

Environmental accounting and welfare analysis

The extension of the accounts to measure non-monetary transactions, on the one hand, and the extension of the scope of assets, on the other hand, are two main dimensions along which satellite accounting extends beyond the limits of the central framework. However such extensions are not free to use any kind of modelling, but should be based on clear answers to the following questions: what do we intend to measure? What do we actually measure? Is the result of the measure acceptable? Similar questions need to be asked again when combining the measure of a particular phenomenon with the measure(s) of other(s). In macroeconomic satellite accounting we often face situations where, on the one hand, sophisticated theories and rigorous techniques are used for micro-measurement, whereas, on the other hand, rough reasoning has to support conclusions at the macro-level.

Ex ante environmental modelling

Ex ante modelling of the future or of alternative states of the economy tries to answer the question 'what could happen if... ?' In the field of the environment, various objectives or constraints have been introduced into models of environment–economy interaction.

Environmental objectives (constraints) include changing the rate of depletion of exhaustible or renewable market natural resources, substituting exhaustible resources with renewable ones, lowering the rate of degradation of non-market natural assets used as sinks for waste disposal or for supplying final services (amenities), halting additional degradation, or restoring natural assets to a desirable level. Such objectives imply choices that can be theoretical, ethical or ideological. Sustainability in achieving these objectives can be more or less weak or strong.

The ways and means of achieving the objectives in question are diverse (administrative or economic). They would cause changes in the structure and size of economic activities. Uncertainties about the nature and rate of technical progress and about the expected prices and reserves of market natural resources play a significant role in environmental models. So does the time horizon specified for reaching the (model) objectives. For instance, in the Netherlands, an attempt was made to estimate what net domestic product would have been in an environmentally sustainable economy (de Boer *et al.*, 1994). In this model, the whole rebalance of the economy is to be achieved in one period. This represents a catastrophic scenario with very strong constraints, aiming at illustrating the magnitude of the problem, rather than at measuring what would actually be the sustainable income of the Netherlands. The difference between the hypothetical and actual net domestic product can be interpreted as an indicator of the costs of achieving sustainability (Faucheux and Froger, 1994; Faucheux *et al.*, 1994), in terms of the global costs which indicate the magnitude of the re-orientation of economic activity that would be required (Faucheux and O'Connor, 1995).

Modelling the future and alternative states of the economy does not aim at providing a new measure of ex post net domestic product, even if this point has been confused sometimes. The results of such a modelling are significant. They can be presented in the national accounts framework as is with the relevant items shown separately when required.

Ex post environmental accounting

The objectives of ex post environmental accounting in money terms are different from ex ante modelling. Environmental accounting aims at answering the question 'what happened really during the accounting period?' It proposes alternative measures of net domestic product, final consumption and other aggregates. The purpose is to give a different picture of the actual economy.

What do versions IV.3 and V.5 of the SEEA intend to measure?

Version IV.3 of the SEEA (environmental costs at market and contingent values) deals with the valuation of the repercussions of a deteriorated natural environment on households (United Nations, 1993a, para. 320).[3] The version is based on the contingent valuation method, that is the willingness to pay (WTP) for improving the natural environment (para. 321). These imputed repercussion costs (Table 4.87) are recorded as a reduction in individual consumption (column 3, rows 6 and 7) and as additional costs of different economic activities of households (columns 1 and 2). If we interpret the WTP as the willingness to pay of households for avoiding the actual deterioration of the environment during the accounting period, it is difficult to understand why this value is deducted from actual household final consumption in the SNA sense. The SEEA does not explain the meaning of the implicit new aggregate of final consumption.[4] The explanation 'adjusted for the environment' is not specific enough. Nor is the recording of the imputed repercussion costs as additional costs of industries (to get an eco-domestic product) more easily understandable. The allocation of these costs between industries in proportion to the work time spent in each industry is also not illuminating. The use of the word 'costs' in the expression 'repercussion costs' can be misleading in this respect. What the contingent valuation method tries to measure is the value, based on the WTP, of the losses of environmental (consumption) services due to the degradation of natural assets. As version IV.3 does not take explicitly these services into consideration, it is not clear what is proposed.

Version V.5 is fortunately more explicit because it introduces environmental consumption services (called consumer or consumptive services in the SEEA). 'The consumer services of the natural environment are considered to be the result of a "productive" activity of that environment' (para. 366). When introducing such services, provided they can be valued, one expects to find a positive value that could be added to the central national accounts household final consumption. Household final consumption with environmental consumption services would equal conventional household final consumption plus the value attributed to the (free) environmental (consumption) services. However, this is not what version V.5 presents. Consumption services (in Table 5.10) have a negative value of 88 (75.3 plus the actual repercussion costs 12.7), explained as 'the description of consumer services is limited to recording the decrease of these services' (para. 366). What is supposed to be measured through the use of contingent valuation is actually the value of the losses of environmental (consumption) services due to the degradation of natural assets. However, version V.5 combines these losses with the total value of conventional household final consumption (neglecting that version V.5, compared with version IV.3, extends the boundary of production to household activities beyond what is already covered in the central framework). Thus, implicitly, version V.5 defines household final consumption with environmental consumption services as conventional household final consumption less the decrease in environmental consumption services.

There seems to be confusion about the absolute level in the period under review and the change from the level in the previous period. The absolute level in a period of time of the household final consumption with environmental consumption services cannot be lower than the absolute level of conventional household final consumption unless the total value attributed to the (free) environmental (consumption) services is negative, which is certainly not the case.

Apparently, the SEEA is not interested in measuring the total amount of environmental (consumption) services as the issue is not raised in the Handbook. The focus is only on the losses incurred in these services because of the degradation of natural assets. This is an easy to understand priority. The SEEA wants to show that the change in the level of living (or welfare) is over-estimated when referring to conventional household final consumption while there is a decrease in environmental (consumption) services. The conclusion is obvious: if there are two subsets in a basket of goods and services, one increasing and the other decreasing, the change in the total basket itself is certainly lower than the change in the first subset. However, in order to show this, it is not justified to say that the value of the total basket is lower than the value of the first set. What is true in this case is that the change in the value of the total basket is lower than the change in the value of the first set. If we want to calculate an indicator of the rate of change in household final consumption with environmental consumption services, it seems necessary to give approximately correct weights to the various components. Perhaps, this is not what the SEEA aims at. In this case, what is it that version V.5 tries to measure?

Above, it was assumed that a value is attributed to the whole of the (free) environmental consumption services provided during a given period by natural assets, and not only to the decrease of these services. However, the possibility and meaning of such a valuation is a real issue.

Measuring consumer surplus or changes in consumer surplus

The measurement in money units of the benefits provided by non-market natural assets relies on the concept of consumer surplus. As those benefits have no market price by definition, the idea is to estimate demand functions either indirectly (travel cost method, hedonic prices method) or directly (contingent valuation through direct surveys). Then, changes in consumer surplus are analysed, following actual or hypothetical changes in the quantitative or qualitative flows of services provided by natural assets. This type of approach in valuing the non-market services of natural assets refers to basic assumptions of welfare economics, namely that individual preferences are the basis for valuing environmental benefits and individuals are the best judges of their preferences.

The existence of a consumer surplus means that the value attributed by a consumer to the units of a good or service bought, other than the last one corresponding to the market equilibrium, is higher than the market price. The expression 'exchange value' should be used when referring to the market price and 'use value' for the value the consumer attributes to each unit it consumes. Exchange value and use value coincide, under certain conditions, for the last unit consumed, but not for others. In the case of market goods and services, the total benefits derived by the consumer from buying a number of units of a given product are thus equal to the expenditure on this product (quantity times the market price) plus the consumer surplus. In the case of non-market environmental services, as there is no market price, the total benefits derived by the consumer are supposed to be equal to the consumer surplus. Of course, the application of welfare economics to the environment is complex.[5] In this context, it is sufficient to remember that the purpose is to measure consumer surplus (or changes in consumer surplus) and not to estimate the equivalent of a market price.

When advocating the use of this approach at the macro-level, as in versions IV.3 and V.5 of the SEEA, we are obviously in the field of modelling, even if the methods in question use indirect observations (travel cost, hedonic prices) or direct observation of the WTP that are, however, generated in the context of the experiments themselves. Modelling is unavoidable, provided we do not forget what it means when we insert its results in an accounting framework.

There are at least two different objectives when considering these methods at the macro-level. The first is measuring the value attributed to the whole of environmental consumption services or the natural assets themselves. The second is measuring only the change in the value attributed to the environmental consumption services or to the natural assets.

VALUING TOTAL ENVIRONMENTAL CONSUMPTION SERVICES OR NATURAL ASSETS

Intuitively, the valuation of total environmental services seems more difficult and problematic than measuring their change in value. The problems are both practical (can we expect enough studies?) and theoretical (aggregation of individual surpluses, value of pure public goods etc.); there would also be a problem when combining aggregated consumer surplus for environmental consumption services provided by natural assets and the benefits derived from market goods and services. The latter cover both known expenditures, and consumer surpluses which we do not measure and do not intend to measure. Is it because of these difficulties that the SEEA does not propose measuring the amount of environmental services itself, but only the changes therein? It would be interesting to discuss the feasibility, even in the long run, of measuring at the macro-level the value of natural assets and the services provided by them. Of course, the answer can be different for various types of natural assets or environmental services. In case the answer is largely negative, is it desirable to develop physical measures of the services in question?

VALUING THE LOSSES IN ENVIRONMENTAL SERVICES

Measuring the losses in environmental services, that is the negative changes in the corresponding consumer surplus, seems to be easier. Recent studies have focused on measuring the negative or positive changes of consumer surplus, associated with changes in the flow of non-market environmental services. At the level of specific projects or situations, progress has been made, especially in cost-benefit analysis of projects involving natural assets. It remains unclear if this approach facilitates the measurement, during a given year and repeatedly over time, of the total actual (net) losses of environmental services associated with the deterioration or improvement of natural assets. This is a prerequisite for dealing with annual aggregates. Obstacles to this measurement may include:

(a) the difficulty of observing during the accounting period the losses of environmental services even in physical terms. This is an issue which is connected to the difficulty of observing the physical changes in the natural assets themselves;

(b) an insufficient number of studies and projects measuring in money units the losses or gains of services, including the issue of applying a WTP or WTA approach;

(c) the feasibility or non-feasibility of transferring valuations, that is, of using values estimated in a given context, but applying them elsewhere in a context relatively similar to the first one. Such transfer may be necessary in order to save on estimation costs, but it is a complex and delicate procedure;

(d) the acceptability of using such values at the macro-level where the requirements can be, on the one hand, less demanding than for specific project cost analysis (but are the same in nature) and, on the other hand, more demanding because of the issue of aggregation.

If the measurement in money units of the use value, attributed by households to the actual (net) losses of environmental (consumption) services were feasible, it would represent a very meaningful result just as it is in economic and environmental analysis.

COMBINING EXCHANGE VALUE AND USE VALUE

The possible combination of use values of environmental services with the monetary value of market goods and services is an unsettled issue. Applying average WTP or WTA for the estimation of the change in the consumer surplus does not seem to be equivalent to estimating market prices, i.e. exchange values, even if quantity and/or quality changes of environmental services are deemed equivalent in terms of the variation of surplus to a price change for market goods and services. As to nature, there is no supply curve being a function of prices. On the demand side, the estimated demand functions are contingent upon specific scenarios. They do not result from a general trade off in which all resources, both monetary and in kind (the free gifts of nature), would be taken into account. There is no enlarged budget constraint for consumer choices, and possibilities for substitution are limited.[6]

The issue of accepting or not the combination under review persists even in the case of (net) losses of environmental (consumption) services and the changes in conventional final consumption of market goods and services. The latter, i.e. the change in expenditure for these goods and services, does not measure the change in consumer surplus associated with these products. We do not know whether, from one period to another, the rate of change in expenditure and the rate of change in (supposedly aggregated) consumer surplus are the same or are different, depending for instance on demand elasticities. This justifies the claims of national accountants that the national accounts, recording exchange values, do not aim at measuring welfare. As long as we accept the idea that the value attributed by consumers to non-marginal units of goods and services is greater than the (equilibrium) market prices, this conclusion seems unavoidable.

Other goods and services also contribute to the welfare of individuals. The central framework does cover certain non-market goods and services, such as those of owner-occupied dwellings produced by households and consumed by members of the same households, and goods and services produced and supplied on a non-market basis by government bodies and non-profit institutions. As mentioned in the first part of this chapter, national accounts value these products by the market prices of similar goods or services or, when not available, by their costs of production, most of which are monetary costs. This is an approximation of the exchange value and not of the total value (including expenditure plus consumer surplus).

In the context of satellite accounting, other extensions of the boundary of production and consumption have been made, for instance, to other household services or leisure activities. Even when the primary concern was welfare, the valuation of these services always tried to approximate exchange values. In the case of household services, the alternative values applied are either the prices of similar services in the market (households are deemed to value their output at least at the prices they would have to pay for it), the wage rates of domestic servants or the wage rates members of the households would earn in the market. The last method is also used for valuing leisure. Leisure is an interesting case, because its valuation through the opportunity cost of the time spent in leisure activities has been extensively discussed in economic theory.[7]

The issue of the valuation of government non-market goods and services, household services, leisure, in terms of both quantities and values, is extensively debated. If one accepts this valuation as a reasonable approximation of an exchange value, those goods and services can be combined with market goods and services. Even if no other goods and services contribute to welfare, the resulting aggregate is not a measure of economic welfare as it does not cover the consumer surplus. Nor is it an indicator of the change in welfare, since exchange-type values and consumer surplus do not necessarily change at the same rate. It is just a measure of enlarged consumption, in terms of exchange value. In addition, there are other aspects of welfare, beyond consumption, like the possible gaps between the value attributed to goods and services by individuals according to their preferences, the objective effects of consuming certain goods and services, the existence of unemployment, distortions in income distribution, and other characteristics of a social nature.

It appears thus that the (use) value in monetary units of the (net) losses of environmental consumption services cannot be combined with the exchange value of conventional household consumption. The current monetary value, in terms of exchange value, of these services remains zero. What is measured, as changes in consumer surplus, is probably closer to a measure of the change in volume of environmental consumption services (and a corresponding measure of the change in an implicit price component), following an approach similar to that analysed in Vanoli (1995b, pp. 124–125). However, there remains an issue when comparing such a measure of the change in volume and other measures of volume based on exchange values or maintenance costs.

In conclusion, the welfare approach to environmental macro-accounting raises a number of issues which deserve further consideration.

Environmental accounting and the consumption of natural assets

When dealing with the losses of environmental services the SEEA in versions IV.3 (implicitly) and V.5 (explicitly) does not introduce any connections with the accounts of (non-produced) natural assets. No explanation for this disconnection is given. Is it that valuing the stock of non-market natural assets, following a welfare approach, is considered unrealistic?

In other versions of the SEEA, the approach is in terms of consumption of natural assets; it is based on a totally different methodology, namely maintenance costing. Maintenance costs are the costs 'that would have been incurred if the environment had been used in such a way as not to have affected its future use' (United Nations, 1993a, para. 50, see also para. 298). Maintenance costs are the costs imputed in addition to actual costs incurred. They are by definition hypothetical. The SEEA stresses the analogy with the concept of consumption of fixed capital in the case of produced assets.

Direct and indirect maintenance costs

Two main categories of costs can be distinguished:

(a) direct costs: prevention, protection or restoration costs can be termed 'direct' when corresponding to available techniques or products which could have actually been applied with available technologies. These costs would have permitted protecting the environment (totally or according to certain socially accepted standards) without technical changes in the nature of the output and/or the scale of economic activities;

(b) indirect costs: indirect costs are related to changes in the nature of output and/or the scale of economic activities. This is because available techniques or products do not permit fully protecting or restoring the environment to required levels. Restructuring the economic activities would become necessary.

Estimating maintenance costs can only be the result of modelling. It requires answering the question 'what would or could have happened if certain constraints had been imposed on economic activities?' With regard to modelling, it is difficult to understand the views expressed in the SEEA (para. 301). Arguing against the use of input–output models to associate the environmental costs with the final use of products rather than with the activities immediately responsible for causing the costs, the SEEA states: 'the accounting framework of the SEEA, however, deals by definition with the assessment of the economic-environmental interaction during the past accounting period. It is an information system that should avoid the inclusion of data that are the result of modelling. The analysis of environmental costs focuses, therefore, on the economic units immediately responsible for environmental impacts since those impacts can be readily identified, measured and attributed to the accounting period'. The impacts, perhaps, but the maintenance costs as they are described for instance in paragraph 307? Obviously maintenance costs cannot be observed; if they were, they would not be hypothetical (paras 50 and 54). They can be estimated only on the basis of assumptions. Even if hypothetical direct costs would have been sufficient to prevent the deterioration of natural assets, their estimation is not straightforward. Probably, there existed alternative solutions, the relations between technical solutions and their impacts were not perfectly known and indirect consequences had to be estimated.

When indirect maintenance costs are estimated, the use of modelling is evident. Actually, maintenance costing is similar to ex ante or simulation modelling discussed above. However, it is ex post modelling. Ex post modelling aiming at estimating hypothetical maintenance costs is more constrained than ex ante modelling: only available technologies can be taken into account, the economic and social policy is given, apart from imposing the relevant standards, no new relative prices are estimated; and no adaptation of economic behaviours both by producers and consumers is produced because it would necessitate reconstructing a totally different economy. Thus some interactions are taken into account, but others are not. This is partial modelling. By contrast, ex ante modelling with environmental constraints possesses many more possibilities as indicated above. On this basis, some authors (Aaheim and Nyborg, 1995) contrast the modelling approach with the accounting approach followed in the SEEA. The SEEA rather follows a truncated modelling approach.

Maintenance costs and consumption of fixed capital

A parallel is often drawn in the SEEA between the method of calculating the value of the depreciation of produced fixed assets and the concept of maintenance costs (see for instance para. 303). Even if the basic notion is the same, namely calculating the value of the consumption of assets, there seem to be significant differences between the cases of produced and non-produced assets. The values of produced fixed assets at the beginning and the end of the accounting period can be assessed in principle, though as discussed above expectations about the future pose problems for the determination of asset values. The value of non-market natural assets, however, is probably not measurable and definitely unknown. Thus we try calculating the value of the consumption of assets, when the value of the stock of these assets can probably not be calculated in monetary terms.

Morever, in the case of produced assets, there is a close relation between the consumption of fixed capital and the corresponding change in capital services (see above). The issue is more complex in the case of non-market natural assets, because disposal services and consumption services are supplied jointly by natural assets but move in different directions (Vanoli, 1995b, pp. 123–127). The maintenance costs measure the current monetary value of the disposal services provided by nature (when there is a degradation of natural assets). Contingent valuation (or other similar methods) could be thought of as another method for valuing the consumption of natural assets through the losses of consumption services due to environmental degradation. However contingent valuation estimates use value and not exchange value. There is no reason why *a priori* the two approaches should produce similar results (of course, a high level of individual WTP is a good indicator of the acceptability of difficult social choices).

At the macro-level, the maintenance costs approach seems more promising than the consumers' valuation for measuring the consumption of non-market natural assets, even if its basis in microeconomic theory is weak (some people may say nonexistent). However, the feasibility of calculating these costs on an annual basis is not obvious at the moment and must be tested. Resource limitations may lend priority to ex ante modelling. Such modelling may be more useful for calculating the costs of achieving a socially agreed pattern of sustainability and for analysing the constraints, ways and means, and the time dimension for reaching social objectives.

Inserting the consumption of non-market natural assets in an accounting framework

Supposing the annual maintenance costs have been calculated, the question is then how to record these costs in ex post economic and environmental accounting. This point has already been discussed at great length (see for instance Vanoli, 1995b, pp. 118–120). From an economic point of view, maintenance costs are costs which were avoided by economic production through the free use of non-market natural assets beyond their natural regeneration.[8]

The SEEA (versions IV.2, V.2 and V.6) charges maintenance costs as additional costs of production of industries and reduces correspondingly the net domestic product (NDP) when calculating the environmentally adjusted net domestic product (EDP); it does thus not change the valuation of output. The problem is that the value of output has been determined through market or market-like relations taking into consideration the actual costs of factors of production and individual consumers preferences or social preferences based on market or market-like prices. If disposal and other environmental services to production had to be paid for, the system of prices and quantities would have been significantly different. Thus, the SEEA's considering 'other things being equal' is not acceptable. Prices (exchange values) and quantities as they occurred during the accounting period, i.e. the disposal and other services to production, which are costly for nature and free to the economy, are a kind of surplus which benefited the economy as a whole.

Apart from the fact that it does not seem conceptually acceptable, the interpretation of the EDP thus obtained is not easy. If the objective is to apply to ex post accounting results similar to those calculated in the above-mentioned Dutch study (de Boer *et al.*, 1994), the result would be the calculation of the difference between actual NDP and something like the potential maintenance costs accumulated over time. Such accumulated potential costs would have to be allocated through a series of annual accounts (of course, it is not sure that a direct yearly calculation would give the same results). However calculated, annual EDPs cannot be considered as sustainable, because the potential maintenance costs are not actual costs and the natural assets may further deteriorate. Neither can EDP be considered as a measure of what NDP would really have been if tight constraints had been imposed on using the environment. The reason is that EDP calculation cannot take into account the adaptation of the economy to potential environmental policy and regulation (see for instance the reactions to the first oil crisis). Various aspects of the interpretation of annual EDP are discussed in Aaheim and Nyborg (1995).

Vanoli (1995b) proposed recording of the use of disposal services as a new item under final uses, leaving income unchanged (only as far as the consumption of non-market natural assets is concerned; see note 7 for market natural assets) and reducing saving. The capital account of the economy is rebalanced through a capital transfer received from nature.[9] Saving is the key aggregate. Such a recording would show explicitly that the economy consumes a part of nature, without disturbing the usual meaning of market-type flows. It can be interpreted as follows: the traditional figures in national accounts for final uses include only ultimately the value given to labour and (mostly produced) capital. Actually, the value of final uses should therefore also include the monetary value of the non-market natural assets consumed through their degradation. This global recording in the first instance does not prevent the breakdown of the disposal services consumed, by means of modelling, first according to their primary users and then according to their final users. This double analysis of the use of disposal services/maintenance costs caused is again in conflict with the surprising position against modelling taken by the SEEA in this respect.[10]

The reference to the concept of responsibility is probably the origin of this unexpected position of the SEEA Handbook. The authors of the SEEA perhaps see a contradiction between the polluter pays principle and the analysis of the maintenance (degradation) costs associated with final uses. The analysis of the direct and indirect costs, as the analysis of direct and indirect emissions of pollutants for instance, is part of an information system on the impact of economic activities on natural assets. It does not prejudge at what stages environmental policies will be implemented. Input–output analysis deals with techno-economic relations, not with relations of responsibility.

Referring to the concept of immediate responsibility seems to provide to the authors of the SEEA the justification of the reduction proposed in the value added of industries in proportion to the 'costs caused'. I do not myself advocate the reduction of value added in relation with the use of free disposal services. However, if somebody wants to do this what is the reason for reducing the value added of agriculture for the direct costs caused and not the value added of the food industry which uses a big part of agricultural output? We can see easily where the shoe pinches. If we look then at final consumption ultimately responsible for the consumption of most agricultural and food products, will we reduce the value of this final consumption? However, do we consume really a lower value of agricultural and food products? In fact, I am reluctant to use the concept of responsibility in the context of integrated economic and environmental accounting in lieu of the technical and economic interrelations. It has a moral connotation that is best avoided.

Perhaps the concept of responsibility is connected implicitly in the SEEA with the idea of negative externalities. The concept of externalities has not been much investigated in the context of macroeconomic accounting and needs further consideration. To the extent that responsibility is socially recognized and sanctioned by law or usage, it is very often insurable and dealt with in the SNA as transfers, not via negative output or additional intermediate consumption leading to a reduction in value added.

Conclusion

Modelling plays an important role in ex post environmental accounting. This is necessary because the valuation in money terms of many environmental phenomena is not done through the market mechanism. The term modelling in this context should be taken in a broad sense. It involves basing valuation methods on microeconomic theory. Generally environmental accounting in money terms relies to a significant degree on 'what if . . . ?' analyses.

The fact that modelling is used extensively does not deprive environmental accounting of its significance. However, the fact that the central framework of the national accounts also uses some modelling should not provide the justification for introducing any kind of modelling into the satellite accounts. Central national accounts are basically an observation system with some soft modelling. Satellite environmental accounts in money terms, with the objectives presented in the SEEA, seems largely a modelling construct, apart of course from the measurement of most actual environmental protection expenditures. The model perspective is a double one. On the one hand, additional flows are estimated, like costs caused at maintenance costs or repercussion costs (losses of environmental consumption services) at contingent values. On the other hand, these costs are combined with pre-existing flows in central accounting, sometimes redefined or re-routed, in order to give alternative pictures of certain economic relations and aggregates.

Such an approach requires a clear explanation of what type of flows and stocks are intended to be measured, what is not measurable, what is actually measured (or measurable), and what is the meaning of the aggregates that are proposed, both in current value and in volume. In this context, the feasibility and significance of applying, at the macro-level, methods whose justification is essentially based on the micro-level analysis deserves special scrutiny. When new flows are valued and introduced in an accounting framework, the model followed must be made explicit. 'Other things being equal' is not an acceptable answer without full justification, even more so when a high social significance is attached to the result of the exercise.

This chapter tried to illustrate some basic difficulties inherent in the interim version of the SEEA. The paper partly repeats, from a different angle, points already made in other papers. It also introduces some new ones, mainly the issue raised by the advocated combination of exchange values on the one hand and changes in consumer surplus on the other hand. This bears the question whether a measure in money units is always equivalent to a monetary value (an exchange value). The SEEA alludes to this issue when stating rightly (United Nations, 1993a, para. 58): 'the maintenance cost concept implies that uses of the environment that have no impacts on nature have a zero (monetary) value'. When comparing the concepts of market-priced capital and hypothetical maintenance costs, the SEEA also states (para. 55) 'if use does not affect capital, user cost will be zero independent of the value of such use for society. Neither concept has a direct welfare orientation'. By contrast, welfare measures based on the concept of consumer surplus and changes therein probably cannot be considered as measures of monetary value, even if they are expressed in money units.

Notes

1. We can sometimes observe expectations, which is a different issue.
2. See for instance the discussion in the 1993 SNA (Commission of the European Communities *et al.*, 1993 paras 6.188 to 6.200).
3. The following references in parentheses refer to paragraphs and tables of United Nations (1993a).
4. Of course the equivalent adjustment due to market valuation in version IV.3 (Table 4.8, row 16, columns 1, 2, 3) has no special significance. Rows 16 to 18 simply represent a bridge table.
5. For instance, the value attributed by consumers to natural assets and their services covers both use value – direct, indirect, option use value – and non-use value, the existence value; there are different measures of the surplus, etc.
6. For instance, if there is a river close to your house, you can swim twice a day if you like. However, you cannot choose swimming only once and have a chocolate bar in exchange for the second swim.
7. Beyond basic human needs, individuals are supposed to allocate freely their entire time between money earning activities and leisure (including housework). At equilibrium, the mar-

ginal values of market work and leisure coincide. The wage rates of individuals enjoying leisure are thus the basis for valuing leisure time. This valuation is obviously in terms of exchange value. The non-marginal leisure time has a greater value for individuals (more than the wage rate would be required for accepting to reduce their time of leisure) who benefit, in addition to the estimated exchange value, from something similar to a consumer surplus.

8. I leave aside the consumption of market natural assets. As the resources are marketed, they are valued in the system of market transactions. There is wide agreement nowadays, if not on the details at least on recording the consumption of such assets in the capital account and not just as other changes in volume of assets (as stipulated in the 1993 SNA).
9. Actually, the SEEA does not indicate how the accounts are closed. The rows reconciling EDP and NDP are simply bridge tables. On the issue of how to close the accounts, see Vanoli (1995a). Incidentally the authors of the SEEA probably misunderstood my proposal, stating (para. 367) that 'as an alternative . . . the direct deduction of the loss of those services [that is, the consumers services] from the final consumption of households has been proposed)'. I never proposed such a treatment.
10. I already referred to the last sentences of para. 301 of the SEEA. It is worth quoting now the first part of this paragraph: 'the concept of immediate responsibility is introduced for theoretical and statistical reasons. It is difficult to identify the economic activity that is ultimately responsible. Economic activities interact to a great extent and it is nearly impossible to trace the chain of economic dependencies and corresponding synergistic or antagonistic effects on the environment. It could be argued that the final demand for products is ultimately responsible for all stages of intermediate consumption and their environmental impacts. Following this argument, the environmental costs of different production activities should be associated, by using input–output models, with the final use of products. Such modelling is outlined in section D of chapter V'. Then comes the end of the paragraph contrasting modelling and the SEEA as an information system.

21

Alternative resource and environmental accounting approaches and their contribution to policy

HENRY M PESKIN

Most governments have yet to embrace resource and environmental accounting in spite of a long academic interest in the subject. While all the principal approaches have the potential of making some contribution to the policy process, proponents have failed to convince the policymaker. One reason is that the authors of different approaches interpret the meaning and functions of the concept of 'accounting' differently, a situation that can only lead to confusion. Another source of confusion is that the different approaches appear to be directed towards different objectives.

In order to promote clarity, this chapter reviews the principal approaches to resource and environmental accounting and assesses them especially in terms of their relevance for policy formation. The approaches differ in their formality. Many accounting efforts are mere collections of data or indices while other adhere more closely to formal, double-entry principles of modern accounting practice. Furthermore, several of the approaches are directed towards improved measures of economic and social performance (the 'scorekeeping' function of accounting) while others are intended primarily to generate information needed by policymakers in their efforts to manage the economy and the environment (the 'management' function of accounting).

The principal conclusions are that no single approach clearly dominates the others in terms of policy relevance. Furthermore, promotion of a single, standardized and 'best' approach, as recently suggested by the United Nations Statistics Division, may not be consistent with principles of efficient data management. A single, standard approach is only efficient if the relative costs and benefits of increasing the amount of accounting information beyond currently collected levels is the same for all countries. This situation is not likely since data collection costs and information needs are expected to differ, especially between industrialized and developing countries.

This chapter was prepared for the international conference *Applications of Environmental Accounting* sponsored by the Fondazione Eni Enrico Mattei and the State and Science Technological Commission of the People's Republic of China, 11–13 March, 1996, Beijing, China. Also, a contributed paper for the *International Symposium on Integrated Environmental and Economic Accounting in Theory and Practice* sponsored by the Economic Planning Agency of Japan, the United Nations University, and the International Association for Research in Income and Wealth, Tokyo, Japan, 5–8 March, 1996.

K. Uno and P. Bartelmus (eds.), Environmental Accounting in Theory and Practice, 375–394.

Introduction

It is now over 25 years since researchers and statisticians around the world began serious efforts in resource and environmental accounting. While interest in these efforts remains reasonably strong the interest still appears to be confined to those in academics or research or to those in the environmental movement who hope that such accounting will have a major role to play in preserving and protecting the natural environment. Official efforts in resource and environmental accounting on the part of governments remain quite sparse, being confined to tiny staff efforts in a handful of countries around the world. While the relative lack of support for these efforts could be explained by the same shortage of funds that plague other governmental programmes, there is also the possibility that proponents of resource and environmental accounting have failed to convince politicians and governmental leaders that these accounting efforts will yield any significant contributions to the policy process. Is there a failure in communication or is it the fact that resource and environmental accounting is only of academic interest and really provides little help to the policymaker?

As will be discussed below, addressing this question is complicated by the fact that the term resource and environmental accounting is open to several interpretations. Each interpretation, in turn, suggests different concepts, different data needs, and different implementation strategies. Furthermore, each interpretation appears to address different issues. Some of these are important for policy while others are primarily of theoretical, intellectual or historical interest.

What is meant by accounting?

One reason for some of the major dissimilarities between different versions of resource and environmental accounting is that not all practitioners use the word accounting in the same way. Some are clearly using the term to mean to take account of, or to keep track of. In this sense, an account is, loosely speaking, a report or similar body of descriptive information. Thus, one may praise Mrs. Jones for the good account she gave of her trip to Europe last summer. Similarly, the police officer may want an account of your time during the period when a crime was committed. However, there is another, more formal use of accounting. In this formal sense, an accounting is also a report or descriptive body of information but, more importantly, it is a structured body of information that describes a system, often at a single point in time, defined by the accounting period. A system, in turn, is a dynamic process that relates outputs to inputs. In a business, for example, the outputs are usually sales and the inputs are the costs of labour and materials necessary to support the sales. The role of the business accountant is to describe this system of production and sales. In order to ensure that the system is described consistently and completely, that is to ensure that inputs and outputs are in balance, the accounting profession has adopted a number of conventions and practices. Environmental and resource accounting in this more formal sense adopts many of the same conventions and practices.

It should be noted that the word accounting, whether interpreted formally or informally, can refer to particular objects (that is, a set of accounts or an accounting of something) or a process. Thus, it is not always clear whether a recommendation to undertake environmental accounting refers to a recommendation to produce a set of environmental accounts, to engage in the activity of accounting, or to do both. The activity of accounting does not always guarantee the production of a set of accounts or, at least, a set of accounts that would be universally accepted as valid and complete. Yet, in terms of the value of the activity for policy, the process of accounting can be quite valuable even if a final set of accounts never emerges. Thus, the distinction between accounting as a noun and accounting as a process is important in evaluating the benefits of the exercise. Indeed, it can be argued (see below) that if the principal purpose of resource and environmental accounting is only to produce a set of national economic accounts with theoretically 'better' numbers, the resulting 'improved' accounts may be far more of academic interest than of policy interest.

Functions of accounting: scorekeeping and management

Not only has discourse among practitioners of resource and environmental accounting been confused by a lack of unambiguous terminology, there does not appear to be general agreement on what these practitioners expect resource and environmental accounting to accomplish. I have distinguished between two principal functions of accounting: scorekeeping and management. By scorekeeping, I mean the function of maintaining a record of performance. In a business, this recording may be of profits, production, or sales. In formal national economic accounting, such as represented by the United Nations System of National Accounts (SNA), the scorekeeping function is fulfilled by aggregate measures such as gross domestic product (GDP) and net domestic product (NDP). Other national scorekeeping statistics, such as price levels, employment, and unemployment come from less formal accounting procedures – procedures that are more representative of accounting in the less formal sense of taking account of something. In fact, while formal accounting may assist in the generation of scorekeeping statistics (as is the case with national economic accounting), generally formal systems accounting is not necessary for this purpose.

Formal systems accounting, such as practised by the modern corporation, has a more significant role to play with respect to the management function of accounting. By the management function, I mean the production of statistics needed to support policy actions, either by a business, household, or by the government. The bottom line, the statement of profits and losses, provided by business accounting serves the scorekeeping function of letting the manager know how well he or she is doing; but it is the detailed set of consistent statistics on sales and expenses that guides the manager towards those policy decisions that, hopefully, will improve the score in the future. Indeed, score-keeping itself can be of little policy interest. One of the criticisms of the publication of improved net domestic product figures (improved because they account for the deterioration of natural resource stocks) is that the figures themselves give very little indication of what policy actions should have been different in the past and what policy actions are called for in the future (see below).

Historical impetus for resource and environmental accounting

While criticism of the national economic accounts as a measure of societal well-being has origins well before the second world war, the main academic and popular interest in resource and environmental accounting dates around the late 1960s or early 1970s. Both the popular press and the academic literature made three general criticisms of the conventional national economic accounts as a measure of social performance.

Of the three, the best-known criticism was that the common economic measures of economic performance, such as gross national product (GNP), net national product (NNP), GDP and NDP did not adequately reflect any degradation of the nation's environment. Indeed, the movement of these indicators is often perverse: GNP, GDP, and their net counterparts could increase in response to environmental degradation. Efforts to clean up an oil spill or health expenditures necessitated by poor air quality would tend to increase national production even though these actions could be no more than an attempt to maintain environmental quality at an acceptable, pre-pollution level.

Many economists join environmentalists in questioning the usefulness of these aggregate economic performance measures as indicators of social welfare. At best, as Hicks (1940) pointed out over 50 years ago, GNP could be an index of economic welfare. But economic welfare is not necessarily social welfare. Furthermore, if relative prices change over time and if some of the goods desired by society increase while other goods decrease, the GNP index could either increase or decrease depending on whether old or new prices are used in the calculation. Finally, at best, GNP can be interpreted as a linear approximation to some non-linear social welfare function. But as Arrow (1963) pointed out, there are no assurances that such a welfare function even exists, at least a function that consistently reflects the preferences of members of the society. It is no surprise, therefore, that economists are not especially impressed with the criticism that GNP overstates well-being because of its neglect of environmental degradation.

However, there are two other criticisms of the conventional national economic accounts and their neglect of the environment that should have more appeal to the professional economist. A second criticism is that the conventional accounts treat reproducible wealth and natural wealth inconsistently. Specifically, the stock of reproducible wealth is depreciated in order to calculate an income measure (NNP) that measures a nation's sustainable income – income that allows for replacement of losses in the capital stock. However, losses in the stock of natural resources are not similarly depreciated, suggesting that true net national product is overstated.

A third criticism is that the conventional national accounts are incomplete in that they neglect important inputs and outputs in the nation's production function: inputs and outputs that have economic significance but are neglected because they lack market-determined values and prices. An example of neglected inputs, which are of special interest to environmental policy makers, are the waste disposal services provided by the natural environment. These services are typically scarce (and, thus, have economic value) because they compete with other services also provided by the natural environment such as recreational and ecological services. Because all these environmental services are not bought and sold in conventional markets, they lack prices and therefore are neglected by conventional economic accounting.

For reasons that are not especially clear to me, of these two additional criticisms, the first has received most of the attention by those promoting resource and environmental accounting. Certainly, developing a truer measure of net or sustainable product is of important policy interest. However, without further information on environmental inputs and outputs in the national production function, it is very difficult to develop policy that achieves a socially acceptable balance between national economic objectives and environmental objectives or politically acceptable sustainable output goals. Given the interest in this issue, especially in developing countries, one would think that the third criticism would have received far more attention than it has up to now.

Responses to the criticisms: Alternative approaches

Responses to these criticisms of the conventional economic accounts have varied across countries. The differences in approach appear to reflect differences in the relative emphasis on the three criticisms and between the emphasis placed on the scorekeeping and management functions of accounting. For purposes of exposition, the various approaches can be grouped under four headings.

Pollution expenditure accounting

One of the earliest reactions to perceived weaknesses in the conventional economic accounts was to develop data series on pollution abatement and other environmental expenditures. Such data series have been maintained in the USA since 1972 (with, unfortunately, several breaks in the series) and are available in other countries. As these data refer to measured expenditures already incurred, either due to policy or to standard business and household practice, they should not be considered as additions to the conventional economic accounts but rather as a delineation or re-specification of information already accounted for. Merely identifying such data serves neither the score-keeping nor the management functions of accounting. Also, without a further accounting of the sources of the expenditures, mere assembly of expenditure data does not meet the criteria for formal, systems accounting either.

However, some would like to use such data for scorekeeping purposes. In particular, critics of the conventional accounts have argued that such environmental expenditures are inherently 'intermediate' and should, therefore, be deducted from final product indicators in order to generate an appropriately 'green' GDP. The idea of deducting environmental protection and similarly defensive expenditures from GDP has a long origin (e.g. Juster, 1973), but has never been adopted by national accountants, presumably because of difficulties in drawing the line between defensive and non-defensive outlays.[1] Such an adjustment has never been made in the United States or in any other country to my knowledge. The motivation, therefore, for the statistical series on pollution abatement expenditures appears not to be for better scorekeeping but rather for better management of the economy. Specifically, several investigators believe that such information may provide for an explanation for changes in measured productivity (e.g. Denison, 1979; Jorgenson and Wilcoxen, 1990). The argument is that these abatement (and other regulatory) expenditures divert resources from conventionally measured production.

While this hypothesis may be valid, the argument assumes that observed abatement expenditures are a good representation of the true opportunity costs of environmental protection. Yet, while expenditures may reflect true economic costs, costs and expenditures are two very different concepts. In fact, observed environmental protection expenditures probably underestimate true opportunity costs, primarily due to a failure of expenditures to adequately reflect process and product-mix responses to environmental protection needs on the part of business. How large is this underestimate? Data on this point are sparse. However, several years ago, during a detailed investigation of the costs of meeting water pollution regulations, a research team under my direction found that many plants met the US regulations without any expenditures on abatement equipment (Gianessi and Peskin, 1976). They simply ceased producing certain products, such as bright, highly coated writing papers, that were highly polluting. In these cases, abatement expenditures were zero, although costs in terms of lost profits on the forgone products were greater than zero. Furthermore, we found that in nearly all cases where there were purchases of end-of-pipe control equipment, changes in process were also necessary. Expenditures associated with these process changes often are not separable from ordinary production costs and are, therefore, missed in the pollution-abatement expenditure data.

On the other hand, abatement expenditure data can tend to overestimate true opportunity costs due to the fact that the reported expenditures include outlays for materials, the values of which may include pollution abatement expenditures of the sector producing the materials. Thus, there may be double counting. For this reason, the frequent practice of comparing pollution abatement expenditures with GDP is misleading since the latter covers only primary costs and is free from double counting. The expenditure data can be corrected using input–output techniques, but not easily (Nestor and Pasurka, 1995b).

Finally, it should be noted that reported environmental expenditures probably overestimate those pollution control costs that are engendered by regulations. Many pollution abatement activities are voluntary and are not in response to policy. Thus, a major component of the US pollution abatement expenditure series is expenditures associated with sewer hookups and septic tanks for newly constructed housing. Such sanitary practices have a history that far pre-dates the environmental movement. If expenditures associated with such conventional practices are not excluded from pollution abatement expenditure series, the use of such series to explain productivity changes can be very misleading.[2] How large might this overestimate be? In the investigation referred to above, we found that an average of about 20% of reported pollution abatement expenditures did not have origins in federal regulatory policy. In some sectors, nearly all reported expenditures pre-dated federal regulations.

Physical accounting

A second response to criticisms of the conventional economic accounts is to supplement these accounts with data bases containing physical information describing the natural environment and status of natural resources. The information can be arranged in a formal accounting system such as the material flow accounts as suggested, for example, by Ayres and Kneese (1969). As shown by Leontief (1970), the integration can be made quite tight by supplementing the conventional input–output matrix with rows and columns that trace the flows of environmental pollutants and that account for changes in the levels of environmental assets. This approach was implemented in the 1972 US input–output matrix as published by *Scientific American* magazine. Far more sophisticated versions of these input–output matrices have been generated by Duchin and her colleagues at New York University (Duchin and Lange, 1993). A very complete input–output matrix system, the National Accounting Matrix including Environmental Accounts (NAMEA), which fully integrates economic and physical environmental information, has also been developed by Keuning and his colleagues in the Netherlands (Keuning, 1995).

Similar physical accounting systems exist in Norway and in France although in both these cases the coverage in terms of assets covered and pollutants is not as complete as in the Netherlands.

To the extent that the accounting covers the processes that cause the changes in physical stock and environmental quality, and not just records the changes, a physical accounting meets the more formal definition of systems accounting. Those physical systems that are closely linked with the economic input–output matrix are clearly formal accounting systems in the sense that I have used this term. However, other physical accounting efforts also exist that do not meet the definition of formal systems accounting. These approaches attempt to take account of the environment by assembling large quantities of physical descriptive information such as indicators of air and water quality, species accounts, area of forest cover, etc. Typically, these informal accounting systems take the form of national state of the environment reports or of large physical environmental data bases such as the STRESS system in Canada and similarly large data bases maintained by several US governmental agencies (e.g. EPA's STORET, the USGS NASQAN system, and many others).

A physical accounting, whether formal or informal, can provide the inputs for the construction of various environmental indicators and thus be used for scorekeeping purposes. In addition, they are often used to support policy analyses and assessments. In particular, the formal matrix systems lend themselves to input–output and linear programming modelling and simulations. In Norway, for example, where this approach has been followed for many years, the intent is to generate information and analyses to support the nation's economic planning process. Of course, the more informal physical data systems available in many countries can be equally useful and are often essential for the development of complex environmental regulations.

While the concept of physical accounting is straightforward, and while such physical information has a central role to play in policy formulation, several factors complicate their use for policy purposes. In the first place, the choice of appropriate physical units of measure is not obvious. Typically, a decision must be made to choose a physical measure that is relevant for some environmental policy concern. A forest, for example, can be physically measured in terms of its acreage, the volume of its timber, the variety of its biota (as evidenced by the number of available species), the stock of non-timber resources such as firewood and grasses, etc. Which of these alternatives is chosen as the physical measure of the forest will depend on what are the relevant policy objectives: commercial timber management, assurance of firewood supply, adequate species diversity, etc. Second, there is the matter of incomparability of units. Unfortunately, numbers do not speak for themselves. While at times environmental priorities seem obvious, more often the policymaker needs some way of deciding whether limited budgets should be directed at one environmental problem or another or at one set of pollutants or another set. Physical information alone does not provide much help.

Related to this second issue are questions of coverage, detail and aggregation. In an effort to anticipate potential policy needs, the systems can become quite large and detailed. Yet, large data systems are almost useless for scorekeeping purposes and also for many modelling uses unless they can be collapsed through aggregation. With physical systems this aggregation is accomplished by finding some common physical unit of measure such as weight, volume, or energy content. One interesting solution to the aggregation problem is suggested by the Dutch NAMEA system. Dissimilar pollutants are converted to common units based on their contribution to environmental themes such as global warming or acid rain. Thus, individual pollutants can be measured in terms of their CO_2 equivalents, with respect to the first problem, or their contribution to acidity, with respect to the second. These aggregates can be further measured in terms of their contribution to environmental pressure by comparing the common unit aggregates with policy target levels. The procedures lead to an index which is larger the more emissions and the more the common unit aggregate exceed the policy target. Of course, the procedure assumes that the policy makers know what these targets are, *a priori.* While setting target levels may not be a problem in the Netherlands, it is in the USA. Policy makers often want to know the potential severity of the environmental problem, and often the value of this severity in dollar terms, before policy targets are set.

In summary, physical accounting systems appear to be most valuable for policy support when overall environmental objectives are already clear. The more formal systems can be used to provide data for linear programming, input–output and other simulation models. Solutions from these models, in turn, can yield shadow prices that can be used to provide monetary values for environmental assets and pollutants. When policy objectives are less clear, purely physical accounting may be less valuable. However, detailed physical information from both formal and informal accounting systems is an essential component of all rational environmental policy action.

'Green' indicators

A third approach to resource and environmental accounting, and perhaps the one with the longest history, is to construct a 'green' GDP or some other economic index to replace the conventional GDP and NDP. This work has proceeded along two parallel paths. First, there has been the effort to construct entirely new indicators of social well-being, usually by altering one or more of the components of the conventional aggregates (subtracting out pollution abatement expenditures would be an example) or by adding some new components (such as a factor measuring the negative effects of urbanization). The best known example of this approach is the Nordhaus–Tobin MEW (Measure of Economic Welfare) indicator (Nordhaus and Tobin, 1973). Similar indicator approaches have been developed by Japan (the NNW, Net National Welfare) (Economic Council, Japan, 1973) and, more recently, by Daly and Cobb (1989).

If these indicator approaches are considered accounting, then they provide a clear example of the use of the word accounting being interpreted to mean 'to take account of'. They also provide a clear example of accounting intended to support scorekeeping rather than management. Indeed, one criticism of these approaches is that while the various indexes may indicate that society is worse off than might be suggested by the conventional GDP, they give the policy maker little indication what to do about it. Even if it is fairly clear that the index is falling due to, say, a combination of increased pollution, more crime, and unemployment, there is no guidance on how scarce resources should be directed towards remedying these evils. Of course another problem is that the components of the index and their weights may appear politically arbitrary. While pollution is generally accepted as a bad thing, what about increases in urbanization or increases in defence? Increases in these social conditions can be a source of comfort to some and an evil to others. Even if there is general agreement that both are bad, it is not obvious how they should be weighted in an overall social performance index.

Because of a lack of rigour in their construction, the development of social performance indicators has not received much support in the policy community, especially from economists. However, it should be noted that one strong feature of these performance indicators is that in their development they draw on a wide range of social and economic information that affects social well-being. The integration of this information could serve an extremely valuable service to policy makers, who exist in a world of specialists who rarely talk the same language. Therefore, in spite of the problems referred to above, I, for one, would encourage continual efforts in the development of these indicators. These efforts would be a good example of where the action or process of accounting may be far more significant than the accounts (or, in this case, the index) themselves.

Another, perhaps more conservative, example of the 'green' indicators approach has been provided by Repetto and his colleagues at the World Resources Institute (WRI; Repetto *et al.*, 1989). The principal thrust of this effort is not to replace the conventional gross income aggregates. Rather the purpose is to modify the conventional measures of net product: NNP or NDP, defined as gross product less depreciation. Essentially, the idea is to depreciate natural assets such as forests, mineral stocks, fish stocks, and soils in order that reproducible capital and natural capital receive equal treatment in the computation of net income. The idea of treating natural assets the same as plant and equipment with respect to depreciation seems, at first, quite sensible and noncontroversial. But upon reflection, whether this is a good idea and, more importantly, whether it serves interests of policy will depend on what is meant by depreciation and net income. Not all definitions of depreciation and net income are mutually consistent or have the same implications for policy.

The idea of calculating net income, defined as gross income less depreciation, has its origins in a concern of Hicks that conventionally defined income as a flow of payments is inherently ambiguous. For example, if one is paid monthly, do we conclude that on dates between the monthly paychecks income is zero? To get around this absurdity, Hicks suggested that income be defined not in terms of payments but in terms of potential consumption: '... the amount [people] can consume without impoverishing themselves' (Hicks, 1946). Prevention of such self-impoverishment is assured if the individual maintains his or her wealth. Conversely, any loss in wealth should be deducted from gross income. The amount of money identified with the loss of wealth is depreciation.

While a less ambiguous concept of income than mere payments, Hicks recognized that this definition of income, which economists identify with Hicksian income or 'sustainable income', can also be criticized for a lack of precision since it is based on ex ante assumptions of future consumption and a related fuzziness regarding what it means to keep wealth intact. Does maintaining capital mean maintaining its original physical properties or does it mean maintaining its value? This issue is especially important when one is considering natural or environmental capital since there may be very little association between a natural asset's physical condition and its value. For the scorekeeping purposes of accounting, it can be shown that the choice between a physical and a value concept of depreciation depends on how capital is valued. Specifically, if the value of an asset is defined in terms of the gross incomes (net of costs but not of depreciation) it can generate over its lifetime, then the only definition of depreciation that is consistent with Hicksian income is the change in value of the asset. This change could occur with or without a change in the physical condition of the asset. This concept of depreciation is identified as economic depreciation to distinguish it from other concepts such as physical depreciation or depreciation as defined in the tax laws.[3]

To further clarify the difference between economic (value) depreciation and physical depreciation, suppose, for example, that the capital stock consisted of only machines and suppose further that the stock was depleted by one unit because a machine wore out. To calculate the economic depreciation, one would estimate the change in present value of the machines, defined as the discounted value of services associated with the original stock less the discounted value of the stock depleted by one machine. In contrast, physical depreciation would be calculated as the cost of replacing the machine and, thus, restoring the stock to its original condition. The distinction between these two depreciation concepts is not of any practical importance as long as the market for machines is essentially competitive. With competitive markets, the cost of the machine and its contribution to the present value of the stock of machines are brought into line through the competitive process. However, these equilibrating forces cannot function if capital markets are not very competitive or, as is the case with natural resource capital, the markets do not exist in the first place. In these non-market circumstances, there is no reason why replacement or restoration costs should have anything to do with the loss in value. Thus, while it may be easier (e.g. the calculation need not involve estimates of rates of return and lifetimes) and more convenient to measure the depreciation of the natural environment or of natural resources by the estimated costs of restoring the environment or the resources to some pre-existing condition, these costs need not in any way approximate true economic depreciation.

What are the policy consequences of these differences in calculation? Just how important is it to have a theoretically correct depreciation and a correct NDP? As an empirical matter, the distinction can be very significant indeed. When one calculates natural resource depreciation both ways, it turns out that there are orders of magnitude differences in value between estimated economic depreciation and depreciation based on replacement cost concepts. For example, for Colombian forests, depreciation based on the Repetto net rent approach, in essence depreciation based on restoration of lost rent due to timber extraction, was about 1.3×10^{10} pesos in 1995 while economic depreciation was estimated to be only 1.0×10^{6} pesos (Peskin and Michaels, 1991). Similar results were found in the Philippines for forests, fisheries, and a number of minerals. These may be extreme examples. Other depreciation approaches, such as those based on the El Serafy methods (El Serafy, 1989), provide results that are much closer to the economic depreciation estimates.

However, while there may be significant empirical differences between using the correct depreciation estimate and those that are based on replacement cost concepts, the policy implications of choosing one over the other may be of less importance. Of course, if the interest in NDP is to compare the well-being of one country to another at one point in time, it would be important to have a correct depreciation and, thus, a correct NDP. However, for many reasons such cross-country comparisons are extremely difficult to interpret even without the inclusion of non-market assets. Perhaps of more policy interest is the movement of NDP over time. Presumably, a falling NDP could be a sign of asset deterioration (including deterioration of the natural environment). Although it is possible for physical depreciation and economic depreciation to move in different directions over time, the empirical evidence seems to indicate that physical depreciation and true economic depreciation move in the same direction. Thus, for scorekeeping purposes, the differences between the two concepts may not be of much importance.

The policy management role of NDP and associated depreciation estimates seem less important than the scorekeeping role. The problem is that even if there were general agreement that particular NDP values, suitably adjusted for natural resource and environmental depreciation, were correct, the numbers themselves do not point in any particular policy direction. Thus, even if, for example, it were true that the Indonesians are overestimating their true income by not accounting for the depletion of their forests, petroleum and soils assets, it does not necessarily follow that Indonesia would have been better off by not depleting these assets in the first place. Depletion of these natural assets may have supported more than offsetting increases in other man-made assets and in human capital. Therefore, with respect to the management role of environmentally adjusted NDP estimates, the story is much the same as with the social accounting indicators such as MEW: the management role is quite limited.

Extensions of SNA-type systems

The fourth group of approaches builds upon the existing systems of national accounts. In this sense, they are quite conservative. However, because they do not focus on just one element of the conventional accounts, such as depreciation, but rather seek to cover all sectors that may interact with the environment, they are the most ambitious of all the four groups of approaches. In practice, these approaches require most of the same information used by the above approaches plus much more sector-specific information.

Examples are the United Nations SEEA (United Nations System for integrated Environmental and Economic Accounting) approach and the Environmental and Natural Resources Accounting Project (ENRAP) or Peskin framework. These two approaches have much in common. As noted, they both build on the conventional system of national economic accounts (and, thus, they are accounting in the more formal sense), they both require sector-specific information on the use of environmental assets, and they both are concerned with the management as well as the scorekeeping functions of accounting. Yet there are significant differences, especially in how they view the economic–environmental interaction, in their ties to economic theory, and their adherence to SNA concepts. With respect to these latter differences, SEEA appears much more concerned with consistency with the SNA and less concerned with consistency with economic theory. The ENRAP framework appears much more concerned with consistency with economic theory and less concerned with consistency with the SNA.

Perhaps one explanation for this difference in emphasis is a different view of the role of the accounts per se vs. the role of accounting. SEEA appears to emphasize the importance of attaining a defensible set of accounts. To meet this objective, a theoretically correct approach might be abandoned in favour of another method if the correct approach is felt to be impractical. Since those employing the ENRAP framework are more interested in developing the information and data systems that support the correct accounts rather than in the accounts themselves, there is no reason not to try the theoretically correct methods even if they prove unsuccessful in attaining a complete set of accounts. Thus, if willingness to pay to avoid environmental insult is the correct way to measure the value of pollution damage while calculating the cost of pollution controls the incorrect way, the ENRAP framework chooses the former over the latter, even though the latter is easier to measure. The gain in information from even an unsuccessful attempt to measure willingness to pay is felt to be of great value for policy purposes, especially since the ENRAP approach also measures cost of pollution controls anyway, although it uses the cost information for different purposes.

SEEA

In one sense, SEEA is more 'politically correct' than the ENRAP framework in that it attempts to encompass a wide variety of approaches, including the ENRAP framework. It accomplishes this trick by putting forth six versions of the system, some of which are subdivided into a number of sub-versions. This form of exposition makes it extremely difficult to criticize or even describe the system. If you find fault with something, the response can be: 'You are describing version IV.x. Don't worry, your concerns will be met in version V.y.' Therefore, the following attempts to cover features common to most of the versions, with the understanding that there may be at least one for which the criticisms are not relevant. In the spirit of the theme of this chapter, the focus will be on policy implications of these common features.

One notable feature of SEEA is its emphasis on costs. Not only does the system identify ex post costs (actually environmental protection expenditures) but also the ex ante costs of maintaining the environment at an essentially undisturbed and sustainable level. As discussed above, SEEA's identification of these prospective costs with environmental and natural resource depreciation is theoretically questionable. Their use to derive an environmentally adjusted GDP is equally objectionable, regardless of whether these costs are being interpreted as uncounted environmental depreciation or, alternatively, as a measure of environmental damage. In either case, the use of restoration or maintenance costs for these purposes does not yield a measure of income that is consistent with Hicksian or economically sustainable income.

Nevertheless, the cost and expenditure data can be extremely useful for policy purposes, especially if they can be compared with the benefits of attaining the levels of restoration underlying the cost estimates. As a practical policy matter, there may be concern over the sustainability of specific environmental assets such such as forests, agricultural land, water, etc. (as opposed to the sustainability of general economic income). As long as there are such environmental concerns and as long as there are policy actions to protect the environment (as opposed to the economy), maintenance and restoration cost information will be important. In addition, ex post expenditure data, if used with the understanding that these expenditures may not reflect true opportunity costs, can support analyses of the effect of environmental policy on conventional economic activity.

Not only is there a concern for costs but SEEA also displays a concern for the incidence of costs, a distinction between costs that are caused by economic units as opposed to costs that are borne by economic units but are caused by others. Unfortunately, SEEA's exposition on this distinction is not altogether clear since at times both costs caused and costs borne appear identified with environmental damage and, at other times, with actual expenditures to prevent this damage. Thus, the important distinction in the environmental management literature between environmental policy benefits and the costs to attain these benefits is clouded. Therefore, while the SEEA handbook compares total cost caused with total cost borne (Table 4.2), it is hard to interpret this comparison. However, the concern over incidence and, by extension, the distribution of costs among establishments and households, should be encouraged since policy makers working in a political environment may share this concern.

Another notable feature of SEEA is its reliance on an input–output framework for arraying information, whether in physical or in monetary terms. This use of input–output is reminiscent of the Dutch (NAMEA) system and is justified for similar reasons: an input–output framework clearly displays the interactions between environmental and conventional economic activities. However, there are problems with input–output that could have policy implications. One problem is that the input–output framework assumes that activities can be uniquely classified: for example, a sector can produce hammers or produce nails, but not both. Usually, when there is such 'jointness', the approach is to define a new sector, such as 'hardware', which receives the hammers and nails as inputs. However, this solution is less than satisfactory. One cannot tell whether the consumers of 'hardware' are consuming hammers or nails.

Unfortunately, 'jointness' is often a characteristic of environmental protection. Yes, there may be distinct sectors that produce equipment that is exclusively used for environmental protection. However, a sector often responds to pollution regulations by altering processes and product mix as well as by purchasing pollution abatement equipment from other sectors. Thus the product 'pollution abatement' is inherently a joint product produced with the primary output of the sector. The problem is further complicated by the fact that product mix changes and process changes are usually not reflected in clear expenditure numbers that can unambiguously be associated with the activity 'pollution abatement'. It is not clear how this problem is addressed in the SEEA framework. It would be unfortunate if the jointness problem were ignored since it can be the case that, with certain sectors, the majority of the costs associated with environmental protection are due to responses that do not require equipment purchases and are not reflected in expenditure records.

A third characteristic of SEEA is its close adherence to the SNA production boundaries. The close adherence to SNA concepts is an obvious strength since it will help assure consistency with the conventional accounts. Also, since the

SNA framework is widely adopted, the close adherence of SEEA with the SNA will help assure international comparability. However, this consistency comes at a price. In particular, presumably because coverage of the production activities of households is limited in the current revised SNA, SEEA also has a limited coverage of household production. Yet, some of this production, such as the production of non-marketed firewood, can have serious environmental consequences. My guess is that for those countries for which non-market household production can have a major environmental role, implementations of SEEA will go beyond the SNA production boundaries.

Of course, if countries do begin to tailor SEEA to their own needs, international comparability may be lost. How important is it to maintain such comparability? I feel the answer depends first, on the perceived relative importance of the scorekeeping and management roles of accounting, and second, on the cost of data development relative to the benefits. If it really is important to 'score' one nation's performance against another, perhaps to provide a rule-of-thumb for the dispersal of assistance funds, then obviously standardization is essential. However, in the absence of some international policy control mechanisms, such standardization is (or, at least, should be) irrelevant for national policy development. There is no reason why one country's choice of policy instruments to deal with some environmental problem should be influenced by the country's environmental performance relative to another country. Of course, I realize that in the real political world, policy actions are often justified based on what a neighbouring country is doing.

However, even if there are advantages to international comparability, it should be recognized that uniform standardization may be inconsistent with efficient data development. Presumably, efficient data development should strive for a balance between the marginal cost of data development and the marginal benefits. Yet, there is no reason why these marginal values should be the same for all countries. For some countries, the marginal cost of attaining a piece of SEEA-relevant data may be quite high and the benefits of this data element, quite low. The opposite may be the case in a neighbouring country. One would think, for example, that Nepal would want a full accounting of non-market household fuelwood production, given that non-market fuelwood constitutes about 75% of the nation's energy consumption. The importance of this sector is far less in many other countries. If so, why would it be in the interest of Nepal to adopt a system that contains the exact same pieces of information as, say, the USA.

The ENRAP framework

Like SEEA, the ENRAP framework is a formal accounting that integrates closely with the conventional economic accounts. Indeed, I have seen the system arranged as a large input–output matrix. In this form, it looks quite like SEEA. However, there are significant differences: there are additional production sectors not in the conventional SNA; there is a close adherence to neo-classical economic theory; there is explicit recognition that because of non-market assets, there may not be a balance between sector and aggregate inputs and outputs; there is less concern about consistency with the SNA; and there is more emphasis on the system as a guide to accounting as opposed to a guide to a particular set of accounts.

The two most significant sectors that are added to the conventional SNA are household production (primarily the production of non-market firewood) and nature. In the ENRAP framework, the natural environment or nature is viewed as a productive sector generating scarce, economic goods and services demanded both by households and by industry. Some of these goods are marketed, such as minerals, but most are non-marketed. The most significant non-marketed service demanded by industry is waste disposal, treated in the system as a productive input. Households also demand waste disposal services (for example, the disposal of automobile wastes or wastes from home heating and cooking) but they also demand more positive services of nature such as recreation, support of wildlife, and purely aesthetic enjoyment. These services are considered additions to the productive outputs of the economy.

Consumption of these services, even waste disposal services and the associated pollution, is not inherently bad in the ENRAP framework. What may be bad is that consumption of one service, e.g. waste disposal, may limit the consumption of another service, e.g. recreation, if, for example, the waste disposal generates pollutants that degrade air and water that support recreation activity. Thus, with the ENRAP system damaging pollutants are considered negative outputs.

Consistent with neo-classical economic theory, the services of nature are valued in terms of what members of society are willing to pay for them while damaging pollutants are valued in terms of what members of society are willing to pay to avoid them. ENRAP recognizes that such economic valuation, consistent with the principles of consumers' sovereignty, may undervalue the true social value of nature. Not all values are economic values. Furthermore, by relying on consumers' sovereignty, the interests of future generations are lost. However, the economic valuation was chosen simply because there appears to be no other alternative that is equally as democratic and non-arbitrary.

The underlying economic theory is neo-classical in that environmental service values reflect both supply (cost) and demand as opposed to classical economic theory, which posited that value depends solely on costs. The neo-classical economic model implicitly recognizes that in the absence of market equilibration, such as would be brought about by competition, there is no reason why costs should provide a unique measure of value. Thus, the theory would not support the idea that the costs of maintaining the environment necessarily equal true economic depreciation, an issue that was discussed earlier. Furthermore, the neo-classical theory suggests that if certain inputs and outputs are non-marketed, the sum of all inputs need not equal the sum of all outputs. Therefore, in the ENRAP system, in order to assure accounting balance, a residual is included (termed 'net environmental benefit'). The size of this term provides an approximation of how much the allocation of environmental services lacks an optimal economic balance. Indeed, one of the management goals of environmental policy is to allocate non-marketed environmental services in such a way as to make net environmental benefit as small as possible.

As with the conventional SNA and SEEA, the linkage between current and capital account is through investment (saving) and depreciation. The only difference between ENRAP and SEEA on this point is that ENRAP tries to estimate true economic depreciation for all natural assets including conventional natural resources such as minerals. In principle, the depreciation of minerals and other natural assets, such as water bodies and air sheds, is treated similarly. Depreciation is not equal to a restoration or maintenance number but rather is defined as the change in the discounted value of generated services over time.

It is recognized that implementation of the ENRAP concepts is extremely difficult and that full implementation may not be possible. Thus, all the empirical accounts that have been developed to date (in the USA and the Philippines) have had to rely on rather crude assumptions and approaches. Thus, for example, the willingness to pay for waste disposal services has been approximated by the costs that would have to be incurred by the user of the services were this use severely limited or denied. In practice, these costs are estimated as the costs necessary to attain environmental standards[4] and are thus comparable to the maintenance costs estimated in the SEEA system. The values of environmental damage and positive environmental services have been based on a number of environmental benefit estimates, some of which rely on travel-demand, hedonic, and contingent valuation methods.

While these methods are controversial and crude, the resulting estimates can still be informative for the policy process. Specifically, crude estimates of environmental damage, which can be linked to the costs of eliminating this damage, may be all the policymaker requires. Often it is clear that little will be gained by obtaining more refined estimates. Thus, the crude ENRAP estimates made in the Philippines have pointed the way towards a number of policy responses with respect to the role of industrial vs. household and non-point pollution and the importance of renewable vs. non-renewable assets as a source of environmental depreciation. It is unlikely that less crude estimates would have suggested different policy responses.

More importantly, even crude implementation requires data that are valuable in their own right for policy management. Specifically, ENRAP implementation has generated a coherent body of physical, cost, and benefits data that are essential for the development of rational environmental policy. These data have been assembled at a level of detail that will support policy both nationally and regionally. This aspect of ENRAP highlights the philosophy that accounting can be as important or more important than the accounts themselves. While it is true that the ENRAP accounts can provide scorekeeping numbers such as a GDP adjusted for non-marketed environmental services, pollution, and natural asset depreciation, the crudeness of the numbers may lead some to have doubts about the score's credibility. Yet, even if these account aggregate indices are rejected, the accounting process is nevertheless essential if there is to be rational environmental policy. The data needed to implement ENRAP are essentially the same data needed to support efficient environmental management.

Summary: policy contribution of environmental and resource accounting

When all the above approaches are assessed in terms of their potential contribution to the policy process, it is apparent that none of the systems really dominates the others. Even the more comprehensive ones such as SEEA or ENRAP fail to provide the full spectrum of physical environmental information provided by the more descriptive, informal physical accounting systems found in state of the environment reports. Certainly, no environmental protection agency could function very well without access to such physical information. Yet, given the intimate interconnection between the state of the environment, the state of the economy, and the general social and political condition of a country, it is equally clear that physical accounting systems are not enough either.

Each of the approaches has its own comparative advantages for the policy process. Cost and expenditure accounting is well suited for macro- and sectoral economic impact analyses; physical input–output accounting is ideal as a provider of inputs to policy simulation and other programming models; informal descriptive physical accounts provide the scientific support for regulatory policies; adjusted GDP and similar 'green' accounting indexes serve an important scorekeeping function; SEEA provides close integration with the SNA; and ENRAP is well suited to support benefit–cost assessments and rational policy targeting. One might conclude that the 'ideal' approach would be one that combined all of the others and, certainly, SEEA seems to be going in that direction. However, I feel that construction of an all-purpose environmental and resource accounting system would be a mistake since it neglects the fact that data development is not without real costs. The fact is that many countries do not need a SEEA or an ENRAP. Even if these systems were affordable in every country, the marginal benefits of the additional information generated may not nearly approximate their marginal costs.

Therefore, those of us who have authored and promulgated our own approaches should be careful with our missionary zeal. What we are doing may be the correct course in the countries for which we have gained our experience. It may be totally inappropriate elsewhere.

Notes

1. See Jaszi's criticism of Juster, in Juster (1973).
2. It should be noted that Denison, after talking with my colleagues, was careful to make the necessary adjustments in his study of the possible relationship between regulations and productivity changes. See Denison (1979).
3. The above definition of economic depreciation as the change in asset value is slightly different from the definition used by Hartwick and Hageman (1993). They define depreciation as the change in asset value under optimal use. This condition allows them to show an equivalence to hotelling rent but it is not especially useful in a national accounting context. The accountant must measure things as they are and not as they could be under optimal use. To follow Hartwick and Hageman, it would also be necessary to adjust all observed market prices since they all reflect the lack of optimality that exists in the real world.
4. This approach is reasonable as long as the regulations call for a near-zero level of residuals discharge. In practice, the costs estimated assume at least a 90% reduction in discharges.

22
Multisectoral policy modelling for environmental analysis

BERND MEYER AND GEORG EWERHART

Introduction

How would the economy perform if the targets of sustainable development were to be realized? This question can only be answered by the application of multisectoral macroeconometric models which depict the economic structures linked with the environment (Uno, 1995). The dynamic modelling approach described here is based on the input–output framework and describes variable consumption and production structures. The input–output framework provides necessary simulation properties such as accounting consistency as well as interindustry repercussions.

Modelling work based on a NAMEA data system (National Accounting Matrix including Environmental Accounts; De Haan *et al.*, 1994) uses a linear programming procedure which optimizes net domestic product with respect to environmental targets. Their results for the Netherlands show that sustainable development is only possible with a substantial decrease of net domestic product and final demand. They emphasize that the assumption of a constant technology and constant prices restricts the application of their model to cases where small changes are involved. We are, therefore, prevented from obtaining information about avoidance costs which result from a long-term ecological policy: this can only be reached by the application of a dynamic model (Radermacher, in press) which takes into account a change in technology and prices with reactions in resource allocation and the structure of final demand (van Dieren, 1995, Ch. 13).

The present chapter tries to answer the question 'what would be the avoidance costs for the West German economy, if the government would start now implementing ecological policy which reaches sustainability in the year 2005'? Our analysis is restricted to the problem of CO_2 pollution, and our computation therefore relates to 'partial' eco-domestic product.

Our instrument is the fully integrated 58 sector dynamic model PANTA RHEI. It can be characterized as the combination of an econometric input–

The authors would like to thank Andreas Bockermann and Christian Lutz for valuable assistance in running the model, and the Volkswagen-Stiftung for financial support.

K. Uno and P. Bartelmus (eds.), Environmental Accounting in Theory and Practice, 395–406.

output (I–O) model with an endogenized satellite account of physical CO_2 emission and energy consumption data. The name PANTA RHEI ('all things flow') goes back to Heraklit (Greek philosopher, 6th century B.C.) and has been chosen because the model generates structural change with price dependent input and investment coefficients as well as variable consumption structures. The five energy inputs of the German 58 sector I–O table (electric energy, coal, gas, crude oil and refined oil) are further subdivided into 29 different energy carriers. At this level of desegregation, energy consumption and CO_2 emissions are measured for each of the 58 industrial branches and for the private household sector.

In our dynamic simulations, the government sells tradable emission permits on a market, and the model calculates the market clearing price for the permits. The government reduces the supply of permits over time. The rising equilibrium price for pollution rights then transforms the structure of the economy toward a sustainable one. Our very preliminary results show that, in a scenario with compensation on labour costs for firms to be paid out of the revenue from the emission permits, the economy is able to absorb most of the increasing energy price by structural changes. If CO_2 pollution is reduced by 22.5% from 1996 to 2005, or 30% of the historic figure in 1990, the eco-domestic product will only be 3.1% lower than the GDP of the base forecast.

The model

PANTA RHEI is an ecologically extended version of the 58 sector econometric simulation and forecasting model INFORGE (Interindustry Forecasting Germany). Its performance is founded on the INFORUM philosophy (Almon, 1991), i.e. an econometric input–output model built bottom up and fully integrated. The bottom-up principle refers to the fact that each sector of the economy is modelled in great detail (INFORGE calculates about 150 variables for each of 58 sectors) and that the macroeconomic aggregates have to be actually calculated by explicitly aggregating sectoral figures calculated within the model. The fully integrated construction means that the model takes into account the I–O structure among all sectors of the economy, the complexity and simultaneity of income creation and distribution in the different sectors, its redistribution among the sectors, and its use for different goods and services. The sectors operate in the context of globalizing markets. In this way one succeeds to describe properly the role of each sector in the inter-industry relations, its role in the macroeconomic processes as well as its integration into international trade.

These conceptual advantages over other models allow consistent and powerful processing of sectoral and macroeconomic information. About 30,000 equations of INFORGE (of which about 7000 are behavioural equations) describe the inter-industry flows between the 58 branches plus their deliveries to personal consumption, government, equipment investment, construction, inventory investment, exports and prices, wages, output, imports, employment, labour compensation, profits, taxes, etc. for each sector as well as for the macroeconomy. In addition, the model describes the income redistribution in full detail. The model frequency is annual.

INFORGE describes the economy of West Germany, the former FRG. For East Germany, the former GDR, we can only offer a rather simple model of the circular income flow, which is linked to the I–O model for West Germany, due to lack of longer time series. As soon as the data situation improves we will apply the model structure described here to the whole of Germany.

INFORGE is a part of the INFORUM International System (Nyhus, 1991) which links 13 national I–O models on the sectoral level via export and import flows as well as the corresponding foreign trade prices. The consistent information obtained from this linked system, in comparison to isolated models, allows reliable analysis of the important contribution of exports for the performance of the German economy. The International System forecasts the economic development of Belgium, Germany, France, Great Britain, Italy, the Netherlands, Austria, Spain, USA, Canada, Mexico, Japan and South Korea in full sectoral desegregation. In the near future, models for China, Taiwan and Poland will be integrated into the system (Ma, 1995). Besides the goods market the INFORUM International System also represents the international financial markets, though in a less detailed way: American interest rates as an indicator for the international capital market condition have a weighty influence on German interest rates, with its repercussion on the German markets for goods and services.

The degree of endogenization is rather high. About 200 exogenous variables are primarily fiscal policy instruments, such as government purchases, tax rates, depreciation allowance rates and subsidies for the 58 sectors. Monetary policy exerts its influence through the discount rate, money supply, as well as reserve requirement rates. The variable relating to ecological policy is the supply of emission permits by the government on the market for CO_2 pollution rights. The supply of labour is an exogenous variable. From the variables representing the rest of the world, only the exchange rates between the 14 countries of the INFORUM International System are exogenous. All other variables indicating the development of the global markets that are necessary for the determination of German exports are endogenous.

Let us now turn to a more detailed description of intermediate demand, final demand, labour input and prices. The input coefficients are in general variables whose changes depend on relative prices and time trends. Material inputs are assumed to be technologically fixed at the deeper, non-observable levels of single, homogeneous commodities (Georgescu-Roegen, 1990). In input–output tables, we can only observe commodity groups. Their input coefficients are price dependent due to changes in the product mix of the sector involved. Business service inputs are actually substitutes for labour inputs in all industrial branches. Therefore, input coefficients of business services are dependent on the wage rate in the branch under consideration relative to the price of business services and on time trends. Energy inputs are modelled in two stages. In the first stage, we analyse the inputs of electric energy, gas, coal, crude oil and mineral oil. The input coefficient of crude oil is constant in chemistry and refined oil production. However, the inputs of gas, coal and refined oil are substitutes for each other in chemistry, steel production and electric power generation. In the remaining branches, electric power and refined oil inputs are estimated in relation to the invested capital stock: firms choose the energy intensity of their capital stock depending on relative prices and time trends. In the second stage, we disaggregate the inputs of electric energy, gas and mineral oil for each of the 58 sectors into 29 different energy carriers or fuel sources. At this stage we allow substitution in each product group (Bockermann, 1995).

Pollution by CO_2, SO_2 and NO_x is directly linked to the consumption of the 29 fuel sources in all 58 industrial branches and the private household sector. The market for pollution rights is simulated here for CO_2. Supply is exogenously given by the government and the price for pollution rights is adjusted to clear the market (Lutz, 1995). The clearing of the market for pollution rights is obtained in an iterative procedure which turns additionally to a solving routine for the whole system. The model starts in the first year of the simulation period with a given value for the price of the certificates. The model solution will contain an amount of pollution. If this figure is higher than supply, the price of pollution rights will be raised (in the case of supply surplus lowered) by an eta. A further iteration will be done with this new price. The procedure will be continued till the demand surplus converges to a certain accepted value.

The model describes the dual role of fixed investment. On the one hand it means a change in the capital stock of the investing industries, and on the other it represents a component of final demand. Our determination of durable investment is based upon the decision of the investing industry, which depends on the real interest rate, real profits and gross production of the industry. Using the investment bridge matrix which describes a supplying sector for each investing industry, we derive investment demand for the different sectors. Inventory investment is modelled by accelerator assumptions.

The capital stock of industries grows with investment in the past and is reduced by a constant rate of depreciation. Labour demand is given by definition, once the firms have made their decision on the capital intensity of labour input. We assume that this coefficient depends on the real wage rate of the industry in relation to its output price, the degree of capacity utilization and a technological time trend.

Behavioural equations for real personal consumption are estimated for 26 expenditure categories. Real disposable income of private households and the relative prices of the category, defined as the ratio between the category's price index and a macro-consumer price index, are the main explanatory variables.

For consumer durables, interest rates are included among the explanatory variables. Some special influences like the annual average temperature for heating expenses are also considered. Using a bridge matrix, we transform the 26 consumption categories into consumption demand for the 58 product groups.

Sectoral exports are linked to the INFORUM International System. The export of a sector is explained by the world import demand of the particular product and the relative price (i.e. the ratio of the export price of that sector and the particular world market price). Imports of a product group are divided into intermediate imports and finished product imports. The latter are determined by real disposable income and the relation between the domestic price of the product and its import price. Intermediate imports are observed for each of the 58 inputs in each sector. We define a matrix of import ratios, in which each element is variable and depends upon the relation of the domestic price and the particular import price. Multiplying the import shares with total intermediate inputs in that cell yields intermediate imports. Prices are estimated by the mark-up hypothesis: unit costs, competing import prices and the degree of capacity utilization explain the price of a product. Thus, the relation between price and unit costs is variable.

The model is far away from the neoclassical approach, which assumes optimization under conditions of competition using a well behaving substitutional technology. Such an approach to energy modelling is represented in econometric I–O models by Jorgenson and Wilcoxen (1990). Other energy models, such as the OECD GREEN model (Lee *et al.*, 1994) or the model of Whally and Wigle (1991) are general equilibrium models. They distinguish only a few sectors and do not use the full I–O information. This characterization also applies to the six energy models that were used in the simulation study of Dean and Hoeller (1992).

The neoclassical vision of a permanently realized equilibrium is wrong, especially if we look at disaggregated data. Firms do not have the information to determine the optimum, especially for long-term decisions. Many markets are far removed from the condition of perfect competition and the neoclassical production function is no appropriate instrument to describe the technology which relates to intermediate inputs. In contrast, INFORGE can be characterized as an evolutionary model, with firms following routines represented by mark-up pricing, for example. The production technology is implicitly described by the system of factor demand functions.

The scenario

Our method of analysis is to compare model runs with an active ecological policy with a status quo base forecast up to the year 2005. The main assumptions of the base forecast can be summarized as a unchanging behaviour of the government in tax policy and public consumption. The West German economy will grow, but with an average GDP growth rate of about 1.5% per year; the growth path will have a slightly weaker slope than in the past. This is caused by rising import shares.

The instrument of ecological policy of the government is the supply of pollution rights, which we assume is lowered from year to year. Beginning with relatively large steps in the first years, it follows a linear reduction path after 1999. With this schedule we compute four runs with a 15%, 20%, 25% and 30% reduction in 2005 against the 1990 CO_2 emission figure. We stress the relation to this historical figure because the German government has formulated the target to reduce CO_2 pollution in 2005 by 25% of the 1990 figure.

Every firm which produces or imports gas, coal or oil has to pay for the pollution costs. We go down to the level of the 29 different energy carriers. Their CO_2 emissions per ton are given technically: multiplying these emission coefficients by the amount of their energy input gives the emission of the particular energy input. If we multiply this figure by the market clearing price of the certificate market, we obtain the amount which the producer or importer has to pay to the government. We further assume that producers and importers of fossil energy will, in turn, raise their prices so that the buyers of fossil energy really have to bear the costs of pollution rights. The input share of nuclear energy in electric power generation is held constant during the simulations as is the amount of electric power imports.

We assume in our simulation scenario that the government uses the additional revenue obtained from selling the emission permit partly to finance social security. All the firms in the economy then receive compensation through making lower social security contributions. This means that firms with high energy inputs and low labour inputs will be faced with rising unit costs, whereas firms with low energy inputs and high labour inputs may reduce their unit costs. Thus, there will be a strong structural change which is driven by both rising energy costs and falling labour costs.

Simulation results for 2005

Introduction of the new policy instrument establishes the price for the emission permits. Figure 1 shows the time path of this price for the years 1996–2005 under the different reduction targets of our simulation. The higher the reduction target set by the government in a certain year, the higher the price. The model calculations show that the prescribed reduction of emissions leads to a slight reduction of consumption and production activities. The results are displayed in Figure 2 which shows the simulated macroeconomic avoidance–cost curve for CO_2 emissions. This is our very preliminary, model-based assessment of the trade off between CO_2 emissions and GDP for West Germany. The base forecast already produces a 7.5% reduction of CO_2 emissions against the 1990 figure (Figure 2), prolonging the historically observed declining CO_2/GDP relationship. The right end of the avoidance–cost curve

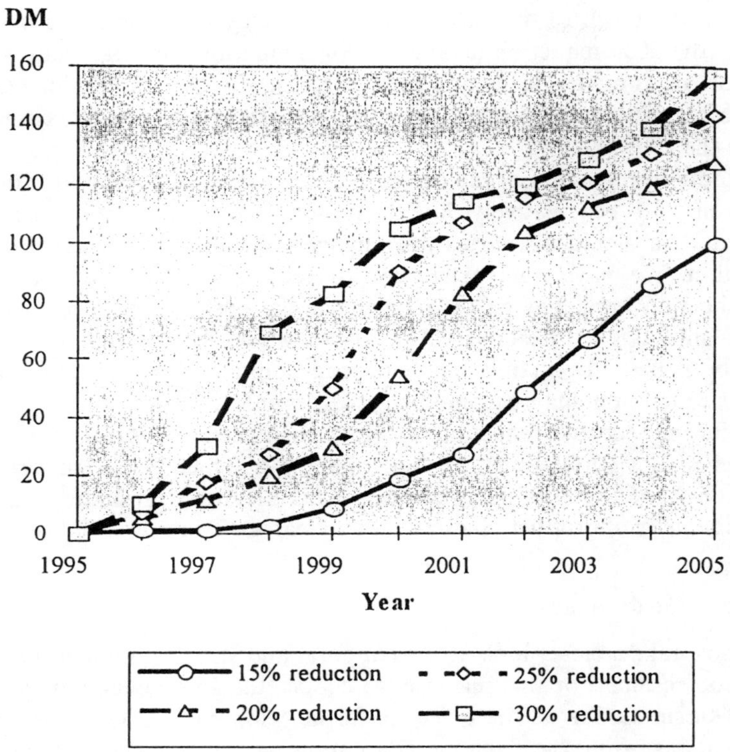

Figure 1. Development of the price for CO_2 emission permits under alternative reduction targets (simulation results, DM per ton CO_2 1996–2005). *Note*: reduction against 1990 figure.

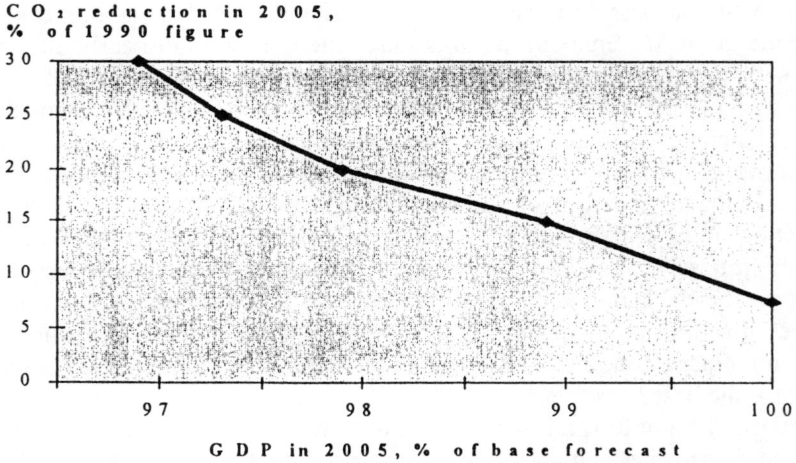

Figure 2. Effect of CO_2 reduction targets on GDP in 2005 (simulation results).

represents the result of the base forecast. The left end of the curve marks the result of the 30% reduction (against 1990) simulation run, signifying a 22.5% reduction in comparison to the base forecast. In this simulation, GDP is 96.9% of the base forecast figure, so in this case the avoidance costs amount to 3.1% of the 2005 GDP.

The induced reduction of the macro-economic consumption and production level seems to be relatively low. The simulated ecological policy leads, however, to a significant structural change towards a less energy intensive production. We have not yet studied the effects of CO_2 reduction targets above the 30% level. The results of such simulations cannot be predicted. But there might exist a threshold value, above which further reduction can no longer be realized by mainly changing consumption and input structures. A distinct reduction in macroeconomic activity will be necessary.

Despite the reported reduction in GDP, positive employment effects can be observed in our simulations, arising from the assumed relief concerning the labour costs of the producing sectors. This arrangement induces a higher labour intensity of production due to lower real wage rate.

Looking at the dynamics

Let us now take a closer look at the temporal and structural dimensions of the simulation results. For this purpose we choose the 25% reduction (compared with 1990) simulation as the object of further elaboration. The most remarkable point regarding the dynamic adaptation path to the ecological policy is that the evolution of the economy is not as steady and uniform as one would suspect. Figure 3 presents the time paths of the GDP components for the years 1996–2005. The strongest negative effect on GDP comes in the longer term from private consumption expenditures (PCE), which go down by 3.6% compared with the base forecast in 2005. Despite the relief of labour cost, the introduction of the emission permits raises the unit costs of nearly all sectors and with that the general price level. The assumed additional reduction for subsequent years makes this price effect permanent so that real personal disposable income and real PCE are restrained in comparison with the base forecast. Divergent from this long-term model property, we observe an increase in PCE for the first period of the simulation. This effect is attributable to a reduction of the real interest rate caused by the steadily rising price level. This interest rate effect is also the reason for the strong increase in construction investment, which moves up to a 5% increase in 2005. The same argument applies for the initial increase in equipment investment, but eventually the declining macroeconomic activity makes this component decrease in comparison with the base forecast.

Exports of goods and services follow the expected pattern. They are reduced due to the increase in domestic prices, while imports show a slightly positive effect despite the GDP reduction. The price elasticity of imports over-

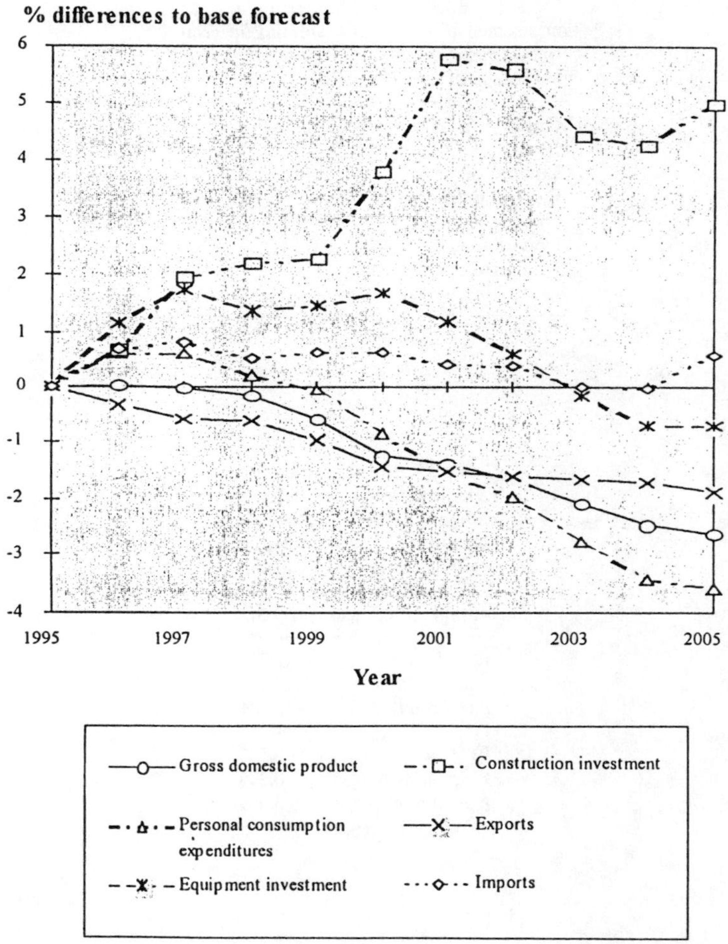

Figure 3. Effect of CO_2 reduction on the components of real GDP (17.5% simulation relative to the base forecast, 25% relative to 1990; percentage differences, 1996–2005).

compensates for the effect of lower GDP, and the impact of the simulated emission permits on import prices is not so strong as on domestic prices, so that imports become relatively cheaper.

In addition to the dynamic aspects of this ecological policy, significant structural change is observed. The price effect of the emission permits is accompanied by a distinct effect on relative prices. The structure of prices is changing according to the emission intensities of respective sectors: the higher the energy consumption and with that the emission of CO_2 per unit of output, the higher the increase in the unit costs of that sector caused by its emissions.

Table 1. Effect of a CO_2 reduction target on selected prices (17.5% simulation relative to the base forecast, 25% relative to 1990; percentage differences in 2005).

Crude petroleum, natural gas	+248.4%
Coal	+203.8%
Mineral oil	+146.3%
Electric power	+92.9%
Gas	+30.2%
Iron and steel	+9.8%
Non-ferrous metals	+4.8%
All product groups	+4.4%

Table 2. Effect of a CO_2 reduction target on selected categories of private consumption (17.5% simulation relative to the base forecast, 25% relative to 1990; percentage differences in 2005).

Gasoline	−18.7%
Coal	−13.1%
Automobiles	−12.3%
Gas	−11.3%
Fuel oil	−3.5%
Electric power	−0.2%
All categories	−3.6%

Table 3. Effect of a CO_2 reduction target on real gross production of selected sectors (17.5% simulation relative to base forecast, 25% relative to 1990; percentage differences in 2005).

Mineral oil	−46.7%
Electric power	−32.6%
Coal	−25.8%
Gas	−20.0%
Crude petroleum, natural gas	−8.0%
Iron and steel	−6.5%
Non-ferrous metals	−6.4%
Chemicals	−3.4%
Pulp, paper and paperboard	−1.9%
Building	+3.0%
Installation and building completion	+3.8%
All sectors	−2.4%

On the other hand, the cost reducing effect of the compensation for labour inputs is distributed relatively uniformly amongst the sectors.

On balance, the prices for primary energy are driven up drastically in comparison to the 4.4% macro-price increase in 2005 in the 25% CO_2 reduction case. Table 1 shows, for example, that the price of coal is about three times the price of the base forecast. Relative prices play a key role in determining the structure of final demand and production. The effect of a CO_2 reduction target on the structure of private consumption expenditures can thus be characterized as a substitution away from energy and related uses such as automobiles. Table 2 shows these results for selected categories of PCE. Comparing with the price changes reported in Table 1, one can conclude that the energy demand of private households is rather inelastic. The greater potential for an energy consumption reduction can be found in the business sectors.

If we finally take a look at the change of the output structure, we observe a shift into the same direction as the structure of private consumption. Table 3 reveals that all energy sectors face a severe reduction in the level of real gross output originating from lower intermediate and final demand. Sectors related to construction are the significant winners of this structural change induced by environmental policy.

Conclusions

Our preliminary results show that sustainable development may be possible without substantial losses in GDP, because of structural change induced by an ecological policy which raises the price of energy and, by assumption, reduces social security costs for firms. Of course such a process needs time. Our experience today with the simulation system is that a reduction of CO_2 emissions by more than 30% could overcharge the elasticity of the economic system. Furthermore it seems important to choose a time path with a continuous increase of the price for the emission permits.

Our analysis is only a first step on the way to analyse the path of sustainable development. Many variations of the scenario may be useful: for example an eco-tax approach is an alternative to our scenario with emission permits. This would influence the time path of the solutions. Another important point is the kind and the degree of compensation to be financed by the proceeds from ecological policy.

This chapter only discusses CO_2 emissions and, therefore, only relates to a 'partially adjusted' GDP. Further work has to be done for other greenhouse gases as well as for emissions which pollute water and soil. Emission reduction for other greenhouse gases may not restrict GDP more than CO_2, because their avoidance will be successfully achieved by the same mechanism as for CO_2. In addition, the implementation of end-of-pipe technologies can be analysed in the cases of NO_x, SO_2 and other greenhouse gases.

Air pollution and its economic implications are a global problem and therefore should be analysed by a global model. The INFORUM International System could be used for such a task. The paper at hand and a study with the Italian model of the INFORUM family called INTIMO (Bardazzi and Piacentino, 1993) on energy policy in Italy can be interpreted as pilot studies which demonstrate the analytical power of dynamic econometric input–output models in this field.

Part V

Identifying research priority

23
Identifying research priority

KIMIO UNO

Historical perspective

The first international exposure of the SEEA (System of integrated Environmental and Economic Accounting) occurred in 1991 when the IARIW (International Association for Research in Income and Wealth) Special Conference convened in Baden, Austria in order to discuss the manuscript of the framework being prepared by the United Nations. At the conference, the accounting framework was discussed by experts drawn from around the globe. Some of the papers from the conference were published in a proceedings volume (Franz and Stahmer, 1993). This publication was dubbed an *Interim Version*, making clear its tentative nature in the continued development of the framework. The manual became the basis of case studies in a host of countries, some assisted by the United Nations and the World Bank, some others in their own framework, results of which are reported in this volume.

The meeting in Tokyo in 1996 was intended as a follow up of the Baden meeting after 5 years of theoretical discussion, exchanging experience gained in implementing the United Nations SEEA. Other accounting frameworks with similar intentions exist, with some differences in their perspectives, including NAMEA which is being developed in the Netherlands (De Haan and Keuning, 1996), the ENRAP framework which was implemented in the Philippines (Peskin, 1991) and SEEDS (social, economic and environmental data set) which is an extended input–output framework designed to be the basis of environmental indicators, while at the same time incorporating the measures of quality of life (Uno, 1995). At this juncture, it would be useful to reflect upon what has been achieved and what is still to be expected. Whereas other chapters in this volume deal with individual issues relating to the SEEA, the focus here is on identifying a future research agenda.

The point of reference has to be the concept of sustainability which was the consensus reached at the United Nations Conference on Environment and Development, held in Rio de Janeiro, Brazil, in 1992. There, Agenda 21, an action plan for global sustainable development into the 21st century, was adopted. Being a document which was produced after world-wide consultation

K. Uno and P. Bartelmus (eds.), Environmental Accounting in Theory and Practice, 409–419.

and approved by more than 178 governments, Agenda 21 represents the recognition that humanity has reached a turning point in accommodating its aspirations within the confine of environmental capacity, and spells out the intention of the global community to search for a sustainable socioeconomic system.

The scope of the document is summarized in Table 1. 'While considerable data already exist, as the various sectoral chapters of Agenda 21 indicate, more and different types of data need to be collected, at the local, provincial, national and international levels, indicating the status and trends of the planet's ecosystem, natural resource, pollution and socio-economic variables. . . . Commonly used indicators such as the gross national product (GNP) and measurements of individual resource or pollution flows do not provide adequate indications of sustainability. Methods for assessing inter-

Table 1. AGENDA 21: Programme of Action

Social and economic dimensions:
International cooperation to accelerate sustainable development in developing countries and related domestic policies
Combating poverty
Changing consumption patterns
Domestic dynamics and sustainability
Protecting and promoting human health
Promoting sustainable human settlement development
Integrating environment and development in decision-making
Conservation and management of resources for development
Protection of the atmosphere
Integrated approach to the planning and management of land resources
Combating deforestation
Managing fragile ecosystems: combating desertification and drought
Managing fragile ecosystems: sustainable mountain development
Promoting sustainable agriculture and rural development
Conservation of biological diversity
Environmentally sound management of biotechnology
Protection of the oceans, seas, rational use and development of their living resources
Protection of the quality and supply of freshwater resources
Environmentally sound management of toxic chemicals
Environmentally sound management of hazardous wastes
Environmentally sound management of solid wastes and sewage-related issues
Safe and environmentally sound management of radioactive wastes
Strengthening the role of major groups
Means of implementation
Financial resources and mechanisms
Transfers of environmentally sound technology, cooperation and capacity-building
Science for sustainable development
Promoting education, public awareness and training
National mechanisms and international cooperation for capacity-building in developing countries
International institutional arrangements
International legal instruments and mechanisms
Information for decision-making

actions between different sectoral environmental, demographic, social and developmental parameters are not sufficiently developed or applied. Indicators of sustainable development need to be developed to provide solid bases for decision-making at all levels and to contribute to a self-regulating sustainability of integrated environment and development systems' (United Nations, 1993b, p. 284).

One direction of follow up is represented by the attempts at compiling indicators of sustainability. In 1995 the United Nations Commission on Sustainable Development (CSD) started a work programme aiming at collecting approximately 130 indicators of sustainable development (United Nations, 1996). The indicators are divided into four categories: social, economic, environmental and institutional. The environmental sphere covers water, land, other natural resources, atmosphere and waste. Subitems in each sphere are arranged so that they correspond to the chapters in Agenda 21. Within each sphere, indicators are divided into the ones representing driving force, state, and response. The framework will undergo empirical testing at national level at the next stage. 'Driving force' indicators relate to human activities, processes and patterns that affect sustainable development; state indicators refer to the current situation or level of environment and development; and response indicators represent policy measures and other social and/or technological decisions and reactions by economic agents to changes in the environmental quality.

Another important instance of indicator development is the compilation of empirical environmental data which has been undertaken by the OECD for its member countries since 1985 (Organisation for Economic Co-operation and Development, various years). The publication typically includes time-series data starting in 1980 or 1985 and cross-section international comparisons for the most recent year. The structure of the presentation is as follows, with a breakdown of individual spheres into a dozen or so indicators: the state of the environment (air, inland waters, land, forest, wildlife, waste); pressures on the environment (energy, transport, industry, agriculture); and managing the environment. These two frameworks include both physical and monetary variables. A similar survey of developing countries is currently being undertaken by the UNSD (United Nations Statistics Division), based on its own Framework for Indicators of Sustainable Development (FISD) (Bartelmus, 1994b).

Another direction of attempts to systematize environment-related data is the satellite analysis and accounts accompanying the SNA (System of National Accounts). 'The central framework of the SNA presents a number of characteristics which give it the advantage of an integrated accounting structure. It is exhaustive and consistent within the boundary of the economic activities it covers. . . . The counterpart of these benefits is that there are certain limitations as to what may be accommodated directly in the central framework.' 'Satellite accounts or systems generally stress the need to expand the analytical capacity of national accounting for selected areas of social concern in a flexible manner, without overburdening or disrupting the central system. . . . Typically satellite accounts or systems allow for:

(a) The provision of additional information on particular social concerns of a functional or cross-sector nature;
(b) The use of complementary or alternative concepts;
(c) Extended coverage of costs and benefits of human activities;
(d) Further analysis of data by means of relevant indicators and aggregates;
(e) Linkage of physical data sources and analysis to the monetary accounting system.'

'On the one hand, satellite accounts are linked with the central framework of national accounts . . . and through them to the main body of integrated economic statistics. On the other hand, as they are more specific to a given field or topic, they are also linked to the information system specific to this field or topic. They also call for better integration of monetary and physical data.' (Commission of the European Communities, *et al.*, 1993, p. 489).

It is in this context that a satellite system covering the interaction between environmental and economic spheres was proposed. Regarding the production boundary, the SNA treatment is that 'Economic production may be defined as an activity carried out under the control and responsibility of an institutional unit that uses inputs of labour, capital, and goods and services to produce outputs of goods and services. There must be an institutional unit that assumes responsibility for the process and owns any goods produced as outputs or entitled to be paid, or otherwise compensated, for the services provided' (Commission of the European Communities *et al.*, 1993, p. 123). 'In the SNA, only produced assets, including inventories, are explicitly taken into account in the calculation of net value added. The cost of their use is reflected in intermediate consumption and consumption of fixed capital. Non-produced natural assets – such as land, mineral resources and forests – are included in the SNA asset boundary insofar as they are under the effective control of institutional units. However, the cost of their use is not explicitly accounted for in production cost' (Commission of the European Communities, *et al.*, 1993, p. 508). 'The most important amendment introduced into environmental accounting as compared with the SNA is the extension of the asset boundary. In the SNA, natural assets are included only if they provide economic benefits to the owner, a characteristic that manifests itself through being controlled by an institutional unit. This often means explicit ownership, subject to government legislation in the case of natural forests, and/or availability of a market price. In the SEEA, the asset boundary is defined to be much wider. It includes in principle all assets; some may directly participate in production activities while others may be affected by environmental impacts of economic activities' (Commission of the European Communities, *et al.*, 1993, p. 513).

The publication of the SEEA framework corresponds to the inclusion of integrated environmental-economic satellite accounts in the 1993 SNA. There are different approaches of environmental and natural resource accounting, including physical accounting, monetary accounting, and net income approach. Reflecting this, the SEEA actually contains several different versions depending on the scope and valuation methods.

Version I is the starting point where the core of the SNA production account describing the supply and disposition of goods and services is preserved. Production accounts constitute the input-output framework. This is then extended to include flows of non-produced natural resources from the natural environment and flows of residuals back to the natural environment.

Version II is the disaggregation of environment-related activities in the SNA. The cases in question are environmental protection, environmental repercussions of production and consumption, as well as uses and stocks of non-produced natural assets.

Version III establishes linkage between monetary data and physical data on environment-related activities. Examples are incorporation of materials/energy balances and natural resource accounting in physical terms into the national accounting framework in value terms.

Version IV deals with imputed valuation of the economic use of the environment in order to reflect actual or potential deterioration of the natural assets. The imputation is done by either one of three valuation methods, including market valuation, maintenance cost valuation, and contingent valuation in order to deal with different aspects of environmental impact.

Version V introduces possible extensions of the SEEA such as the extension of the production boundary with regard to household activities, introduction of environmental services as an output of productive activities of the natural environment, and externalization of internal environmental protection activities.

Version VI is concerned with externalization of internal environmental protection activities. '... An additional gross output of those services, equal to the inputs, would have to be shown, since market prices of internal services do not exist. The externalized services would then be recorded as intermediate inputs of those industries ...' (United Nations, 1993a, p. 137).

The UN manual then goes on to describe environment-related input-output analysis very briefly, including analysis of physical flows of raw materials, produced goods, and residuals. This is in contrast to a draft of the 1993 publication, circulated among limited number of experts to collect their reactions, with a detailed discussion on the extension of input-output framework in order to accommodate environmental issues (United Nations, 1992).

Last but not least, a group of experts formed a forum under the initiative of the EUROSTAT in order to exchange views on the implementation and further development of environmental accounting. The group, called the London group after the first meeting in that city in 1994, produced a series of monographs through their annual meetings (London Group 1994, 1995, 1996). The 4th annual meeting took place in Ottawa in 1997.

Future research and development

Even when we assume that everything contained in the SEEA manual is realized, which is not the case to date, there remains a long list of items on the research agenda. The main issues in the further development of the SEEA might be summarized as follows.

Physical-monetary linkages and indicators

Physical measures should be pursued further, going beyond what is typically associated with the SEEA. One view holds that the accounting framework should focus on monetary estimates because of its ability to provide a system-wide overview (additivity, etc.) in terms of a common numeraire. An opposing view holds that physical estimates are essential in describing the characteristics of environmental quality and to shed light on sustainability. A more balanced view would be that both are essential in an environmental information system for the decision making. What is important then is the proper linkages between the physical and monetary spheres. One such case is land use accounting. Reports have been made on the linkages between land use accounting and environmental economic accounting. In addition to the usefulness of GIS (geographic information system) in itself, aggregated land use accounting in physical terms and its linkages to capital formation data in value terms will provide useful information. Transition matrices from various land uses to other, such as forests, agricultural, residential, and road and other social infrastructure, will provide clues to environmental impact of capital formation on land use (Uno, 1995, pp. 253–281). The SNA already includes land improvement in its asset accounts, which can be the linkage point between land use statistics and the national accounting. Automobile ownership statistics, distance travelled, and petroleum consumption are relevant information in establishing linkages between mode of transportation and environmental implications while being part of household consumption. The SNA data on the stock of consumer durables and consumption of fuels, both in value terms, and transportation statistics in physical terms can be linked in a satellite account. Among produced capital stock is included environment-related ones such as waste treatment facilities, water purification plants, and pollution prevention equipment, capacities of which are better described in physical terms. There have been attempts to provide functional classification to socio-economic activities including CEPA (classification of environmental protection activities) (UNECE 1992), COFOG (classification of the functions of government) (United Nations 1980), and COICOP (classification of individual consumption by purpose) (United Nations 1993). They can be useful in providing linkages between economy and environment if they are given empirical contents even on a limited scale which relates to environment.

Environmental indicators and accounting framework

Indeed, environmental indicators, developed by international organizations and individual countries, are mostly expressed in physical terms. Life cycle assessment also entails use of physical units. Material flow accounts enables us to trace the extraction, transformation, and disposal of industrial raw materials and the impact on the nature. We may say that physical measures are an indispensable part of environmental information. If efficiency of resource use is to be sought, it has to be measured in physical terms representing technological characteristics. One recent attempt to compile materials flows generated at home and (more importantly) abroad is available in (World Resource Institute, 1997) which compares the United States, Germany, and Japan in terms of Total Material Requirement (TMR), an indicator representing the material efficiency in each society.

We have seen the environmental concerns as described in Agenda 21. A set of environmental indicators have been selected by the UNCSD and the OECD, independent of each other, but employing a common scheme consisting of driving force (or pressure), state, and response (D-S-R or P-S-R). Indicators shed light on particular focal points and are effective in transmitting summary information to the general public and decision makers who are interested in knowing what the environment is going to be at the local and global levels. Expressed in measurement units in daily use, they are often effective in conveying a vivid message, albeit simple, compared to a complete accounting framework designed to capture the man–nature relationship in a consistent system, often expressed in terms of a common denominator, monetary valuation.

However, it is also true that indicator systems do not stand alone no matter how elaborate they are. For one thing, physical measures concerning various aspects of the environment are related to each other. Such interaction is difficult to establish unless one resorts to a more systematic description of the environment–economic activities. Indicators in this case represent a summary expression of the total information system aimed at easing communication. Second, different measurement units inherent in individual fields are not amenable for aggregation. This gap has to be filled by environmental and economic accounting in value terms. Third, policy measures are usually achieved through allocation of financial resources, which also has to be expressed in value terms.

International linkages

Sustainability is a global issue. Contrary to the rather elaborate formulation of the domestic part of the environmental accounts, less attention has been paid to describe international linkages of the accounts. There are pressing needs for an international or global framework encompassing environmental and economic spheres. The SEEA framework is a general scheme proposed by the

United Nations. Individual countries are implementing the parts of the framework to fit their own policy focus. The SEEA manual will facilitate development of a comparable framework in various countries in the near future, and may eventually facilitate international linkages of country frameworks.

Such development may take time. There are also problems inherent in international linkages such as the need to trace international trade and foreign direct investment. In the case of transborder flows of pollutants and wastes, the concept of environmental cost caused and cost borne may become relevant on an international scale. This points to the need to develop an international framework, without necessarily waiting for the completion of a country framework. The top-down and bottom-up approaches seem to be complementary in this case. The question is the feasibility of a global framework. Judged from the publication of time series input-output tables in standard industrial classification by OECD (Organisation for Economic Co-operation and Development, 1995), covering industrialized countries, and IDE (Institute of Developing Economies, 1992), covering Asian developing economies in more or less comparable format, it now seems possible to implement such a framework. International trade matrices, which can be compiled annually based on trade statistics, can be employed in order to establish linkages among countries through trade flows.

Inter-temporal comparisons

The trade-offs between the natural environment and the economy are a long-term phenomena. Many countries have tried to compile the SEEA, covering a lengthy period of time in order to capture the dynamism. Final demand has shifted from food to manufactured goods and then to services as the income level rose within a country. In the course of economic development, the industrial structure has shifted from agriculture to resource consuming heavy industries. Advanced economies are moving away from manufacturing and into private and business services, finance, retail trade, and information. Such structural change has its impact on resource requirements and the global environment.

The trial estimates of the SEEA currently made available quite often relates to one particular year. Even the core accounts of the SNA date back only 50 years. To ask for a time series of the SEEA to accommodate historical analysis in the environmental sphere is undoubtedly very demanding.

It is highly desirable to develop a scheme where long-term comparisons can be made in order to grasp dynamism running through economic development, technological change, resource requirements, environmental quality, and aspirations of the people. Economic development takes place in a country context. But it has happened sequentially in different parts of the world: first in Europe, then in the USA, followed by Japan. Other countries have come into the picture very quickly, including China. It may not be true that economic development will follow an identical path in different countries. Still, focusing on a single country experience would provide only a limited lesson if the structure of the economy undergoes change as a result of rising income and changing consumption patterns. Transplanting earlier experience into emerging participants in industrialization would be a valuable policy experiment on the future of the global community.

Technological change

The driving force of industrialization was energy. In this process, sources of primary energy has shifted from coal to crude oil and gas. Nuclear energy has been with us for nearly half a century, with increasing scepticism as to its future potentials. Solar and other new sources of energy are being tested. If technology remains constant, and if we aspire for higher levels of production, then environment and economy may be incompatible. This is precisely why the sustainability of our way of life has to be questioned. One possible route out of this dilemma is technological change. For a particular industrial sector, it may be possible to shift to resource-efficient technology. For the entire economy, it may be possible to move out of smoke stack industries and into high technology and services.

Compilation of a time-series SEEA is an interesting topic in order to capture the socio-economic performance over time in environmental spheres. This, however, is not an easy task due to rapid technological progress and quality change, a shift from retro-fitting of end-of-the-pipe technology to a new production process, and long lags from the time of environmental impact to the time when its harmful effect are felt. Thus, incorporation of technological factors in the accounting framework is a challenge to the statisticians.

In operational terms, production technology can be aptly described by input coefficients in the input–output framework which represent the proportion of different inputs into one particular activity. Energy is one particular input into various sectors of the economy. It is possible to look into the input structure and examine technological change. Research and development aimed at energy efficiency, alternative energy, recycling and pollution prevention also should be emphasized because they represent important policy variables in order to improve sustainability. Research and development should be a part of the environmental and economic accounting. The research and development expenditures and the expected flow of benefits to various industrial sectors and final demand can be formulated based on the input-output structure of the economy (Uno, 1991, 1995).

Quality of life and social dimensions

In the same vein, if we want to preserve our life style, and if the economy tries to accommodate ever-increasing demand by the expansion of supply capacity (particularly under unchanging technology), our way of life is on a collision course with environmental quality. In the economic sphere, we have not questioned the implications of huge income differentials in the global community. In some societies per capita income exceeds $30,000, while other societies struggle at $500 or less. If we rely on the market mechanism for the allocation of resources, including environmentally sensitive goods and services, and neglect the fact that the world is plagued by huge income disparities, advanced economies will dominate the global economy and, hence, the global environment.

The truth of the fact is that if low income economies succeed in catching up, the global environment cannot accommodate the enormous resource requirement. However, people in North America, Europe and Japan will not endorse policy measures directed at sustainability if this means forgoing the current level of the quality of life. Obtaining consensus is important, and it is especially true when we consider the amount of investment funds generated in industrially advanced economies. Thus, we are obliged to seek a way of life which is environmentally amenable and still preserves the current level of the quality of life.

The environmental burden can be broken down into four factors: population size, income per capita, resource requirement per unit of income, and environmental impact per unit of resource use. Maintaining and improving the quality of life while constraining resource use relates to the third item in this equation, while the technology factor mentioned above relates to the fourth item (Uno, 1995, pp. 9–15).

One important consequence of environmental degradation is its effect on human health, an aspect which is not covered by the current SEEA. Indeed, although there are empirical studies into dose-effect of particular factors on human health, they are sparse and disorganized to be incorporated in an accounting framework. This is a field where more intense communication is needed between natural sciences and social sciences.

From accounting to modelling

In order to develop a policy scenario for the betterment of people's well-being while safeguarding environmental quality, we need a scheme where various factors at play can be connected to each other in a systematic way. It is essential to be able to trace this interaction across national boundaries. For this, we need a simulation model. Sustainability of the global community can then be tested under various scenarios concerning economic performance, international division of labor, policy measures, technological development, and lifestyle. The model should contain all the factors mentioned above, including physical–monetary linkages, international linkages, intertemporal comparisons, technological change, and social dimensions. The model should also incorporate policy tools available to us. In the energy arena, the introduction of a carbon tax, measures aimed at modal shifts in transportation and energy efficiency in homes and offices, and incentives for new technology are the cases in question.

Toward Environmental Information System

The development of the SEEA, or indeed environmental-economic accounting in general, seems to be taking two directions. One is a theoretical refinement of the accounting framework toward more consistency which we have seen in the preceding chapters of this volume. This typically takes the form of a large matrix, with increasingly complex description of the man–nature relationship, measured in a common yardstick such as money terms or physical weights for aggregation and comparisons. We may call this intensive development, aiming at deepening the accounting practice for the sake of completeness of the system. Rigorous search for a more complete accounting practice, however, is sometimes accused of being monolithic. Another is an extensive development, often focusing on a particular segment of a complete system in a modular form and providing more detailed information in appropriate units of measurement. Freed from the need to reduce the information into a single measurement unit, the scope of the accounting can be expanded to a much larger sphere. The modules can be compared or linked according to the users' needs. Thus, we are seeking an integrated accounting framework where all relevant modules can be put in place in logical order on the one hand and, on the other, an environmental information system where all relevant information can be stored and accessed in a systematic way, allowing linkages, spatial disaggregation, intertemporal comparisons, and policy modelling.

References

Aaheim, A and Nyborg, K (1994). Green national product: good intention, poor device? In: *National Accounts and the Environment, Papers and Proceedings* from a Conference, London, UK, March 16–18, 1994. Ottawa: Statistics Canada.

Aaheim, A and Nyborg, K (1995). On the interpretation and applicability of a 'green national product'. *Review of Income and Wealth.* 41, 57–71.

Adams, Fiona, Werner, Wolfgang and Vale, Peter (1990). *The Land Issue in Namibia: An Inquiry.* Windhoek: University of Namibia, Namibia Institute for Social and Economic Research.

Adelman, M A, De Silva, Harindar and Koehn, Michael F (1991). User cost in oil production. *Resources and Energy.* 13, 217–240.

Adriaanse, A (1993). *Environmental Policy Performance Indicators.* The Hague: The Netherlands Ministry of Housing, Spatial Planning and the Environment.

Ahmad, Yusef, J, El Serafy, Salah and Lutz, Ernst eds. (1989). *Environmental Accounting for Sustainable Development.* A UNEP-World Bank Symposium. Washington, DC: The World Bank.

Alfieri, Alessandra and Bartelmus, Peter. *Unresolved Issues in Valuation.* Second Meeting of the London Group on Natural Resource and Environmental Accounting; Conference Papers. Washington, DC, 15–17 March 1995. Washington: Bureau of Economic Analysis, US Department of Commerce.

Alfsen, K H (1994). *Natural Resource Accounting and Analysis in Norway.* Documents 94/2. Oslo: Statistics Norway.

Alfsen, K H (1996). Macroeconomics and the environment, Norwegian experiences. In: Gandhi, V P ed. *Macroeconomics and the Environment.* Washington, DC.: IMF.

Alfsen, K H, Bye, T and Lorentsen, L (1987). *Natural Resource Accounting and Analysis, the Norwegian Experience 1978–1986.* Oslo: Central Bureau of Statistics.

Almon, Clopper (1991). The INFORUM Approach to Interindustry Modeling. *Economic Systems Research,* 3, 1–7.

Anderson, R J and Crocker, T (1971). Air pollution and residential property values. *Urban Studies.* 8, 171–180.

Andresen, Steinar, Skjæsetch, Jon Birger and Wettestad, Jørgen (1995). *Regime, the State and Society: Analyzing the Implementation of International Environmental Commitments.* WP-95-43. Laxenburg, Austria: International Institute for Applied systems Analysis.

Arrow, Kenneth J (1963). *Social Choice and Individual Values.* 2nd edn. Cowles Foundation for Research in Economics at Yale University Monograph 12. New Haven and London: Yale University Press. Originally published by John Wiley & Sons.

Arrow, K, Solow, R *et al.* (1993). *Report on the NOAH Panel on CV.* Washington, DC: Resources for the Future.

Auty, Richard and Warhurst, Alyson (1993). Sustainable development in mineral exporting economies. *Resources Policy.* March, pp. 14–29.

Ayres, Robert U and Kneese, Allen V (1969). Production, consumption and externalities. *American Economic Review.* LIX (3), 282–297.

Ayres, Robert U and Simonis, U E eds. (1994). *Industrial Metabolism. Restructuring for Sustainable Development.* Tokyo: United Nations University Press.

Baccini, Peter and Brunner, Paul H (1991). *Metabolism of the Anthroposphere.* Berlin: Springer-Verlag.

Balangue, T B (1990). *Dipterocarp Forest Resources Accounting.* NRAP Technical Report No. 1. Manila.

Bank of Korea (1993). *1990 Input Output Tables.* Seoul.

Bank of Korea (1994). *National Accounts.* Seoul.

Bardazzi, Rossella and Piacentino, Diego (1993). *The Effects of a Carbon Tax: Simulations for Italy in a Multisectoral Model.* Paper presented at the First Scientific Meeting of Environmental Economists. Pavia, March 1993.

Barnes, J I and de Jager, J L V (1995). *Economic and Financial Incentives for Wildlife Use on Private Land in Namibia and the Implications for Policy.* Research Discussion Paper No. 8.

Windhoek, Namibia: Directorate of Environmental Affairs, Ministry of Environment and Tourism.

Barr, C J, Bunce, R G H *et al.* (1993). *Countryside Survey 1990: Main Report.* Institute of Terrestrial Ecology and Institute of Freshwater Ecology. Report Prepared for the Department of the Environment. London.

Bartelmus, Peter (1992). Accounting for sustainable growth and development. *Structural Change and Economic Dynamics.* 3, 241–260.

Bartelmus, Peter (1994a). *Environment, Growth and Development, The Concepts and Strategies of Sustainability.* London and New York: Routledge.

Bartelmus, Peter (1994b). *Towards a Framework for Indicators of Sustainable Development.* Department for Economic and Social Information and Policy Analysis. Working Paper Series No. 7. New York: United Nations.

Bartelmus, Peter (1995). *Indicators of Sustainable Growth and Development – Linkage, Integration and Policy Use.* Paper submitted to the Scientific Workshop on Indicators of Sustainable Development, Wuppertal, 15–17 November 1995.

Bartelmus, Peter (1996). Environmental accounting: a framework for assessment and policy integration. In: Gandhi, V P ed. *Macroeconomics and the Environment.* Washington, DC: IMF.

Bartelmus, Peter and van Tongeren, J (1994). *Environmental Accounting: an Operational Perspective.* Department for Economic and Social Information and Policy Analysis. Working Paper Series No. 1. New York: United Nations.

Bartelmus, Peter and Tardos, A (1992). *Integrated Environmental and Economic Accounting: a Case Study for Thailand* (restricted).

Bartelmus, Peter, Lutz, E and Schweinfest, S (1992). *Integrated Environmental and Economic Accounting: A Case Study for Papua New Guinea.* Environment Working Paper No. 54. Washington, DC: The World Bank.

Behrensmeier, R and Bringezu, S (1995a) *Zur Methodik der volkswirtschaftlichen Material-Intensivität-Analyse: Ein quantitativer Vergleich des Umweltverbrauchs der bundesdeutschen Productionssektoren.* Wuppertal Papers Nr. 34.

Behrensmeier, R and Bringezu, S (1995b). *Zur Methodik der volkswirtschaftlichen Material-Intensivität-Analyse: Der bundesdeutsche Umweltverbrauch nach Bedarfsfeldern.* Wuppertal Papers Nr. 46.

Bezdek, Roger H (1993). Environment and economy: what's the bottom line? *Environment* 35, 7–11, 25–32.

Björsell, M (1994). Environmental expenditure in the Swedish manufacturing industries, an inventory of problems. In: Eurostat, ed. *Contributions of Member States and EFTA Countries to the SERIEE System 1992 and 1993.* Luxemborg: Eurostat.

Blomquist, G (1979). Value of life savings: implications of consumption activity. *Journal of Political Economy.* 87, 540–558.

Bockermann, A (1995). The modelling of energy inputs in the INFORGE model: the data and first specifications. In: Janoska, J J and Monaco, L S eds. *Papers from the Second INFORUM World Conference.* Osnabrück, Germany, September 1994. College Park, Maryland: International Economic Research Fund, Inc.

Born, Alice (1992). *Development of Natural Resource Accounts: Physical and Monetary Accounts for Crude Oil and Natural Gas Reserves in Alberta.* National Accounts and Environment Division Discussion Paper No. 11. Ottawa: Statistics Canada.

Born, Alice (1995). Valuation of Canada's mineral resources: a discussion of conceptual issues. In: *Second Meeting of the London Group on Natural Resource and Environmental Accounting, Conference Papers.* Washington, DC, 15–17 March, 1995. Washington, DC: Bureau of Economic Analysis, US Department of Commerce.

Born, Alice, Santiago, M and Dumas V (1995). Quantity and value of Canada's coal reserves. *Environmental Perspectives.* Catalogue 11-528, 55–64. Ottawa: Statistics Canada.

Bote, P (no publication date given) *Study on Timber Appraisal and Taxation Model for the Philippines.* Manila: Forest Management Bureau, DENR.

Bringezu, S (1993). Towards increasing resource productivity: how to measure the total material consumption of regional or national economies? *Fresenius Environmental Bulletin* 2, 437–442.

Bringezu, S ed. (1995). *Neue Ansätze der Umweltstatistik. Ein Wuppertaler Werkstattgespräch.* Basel, Berlin, Boston: Birkhäuser Verlag.

Bringezu, S and Schütz, H (1995). Wie mißt man die ökologische Zukunftsfähigkeit einer Volkswirtschaft? Ein Beitrag der Stoffstrombilanzierung am Beispiel der Bundesrepublik Deutschland. In: Bringezu, S ed. (1995).

Bringezu, S, Hinterberger, F and Schütz, H (1994). Integrating sustainability into the system of national accounts: the case of interregional material flows. *Intern. afcet Symposium, Models of Sustainable Development.* II, 669–680.

Bringezu, S, Schütz, H, Behrensmeier, R and Schmidt-Bleek, F (1995). *Indicating Environmental Pressure for National Economies and Industrial Sectors on the Basis of Integrated Economic and Environmental Accounting.* Paper presented at the SCOPE/BMU/UNEP/DGSII Workshop on Indicators of Sustainable Development. November 15–17, 1995. Wuppertal.

Bringezu, S *et al.* (1996). Material flow accounts indicating the environmental pressure of various sectors of the economy. Presented at the International symposium on Integrated Environmental and Economic Accounting in Theory and Practice, Tokyo, 5–8 March 1996.

Brookshire, David S, Thayer, Mark A, Schulze, William D and d'Arge, Ralph C (1982). Valuing public goods: a comparison of survey and hedonic approaches. *American Economic Review.* 72, 165–177.

Brown, C J ed. (1994). *Namibia's Green Plan.* Windhoek, Namibia: Ministry of Environment and Tourism.

Brown, L R *et al.* (various years). *State of the World.* London and New York: Norton.

Brown, Ronald H, O'Leary, Hazel and Browner, Carol (1993). *Environmental Technology Exports: Strategic Framework for US Leadership.* Report of the Interagency Environmental Technologies Exports Working Group. Washington, DC: US Department of Commerce.

Buenaventura, A, Ebarvia, M C and Mendez, A (1996). Environmental damages from water pollution: 1988 refinements and 1992 estimates. ENRAP III, Vol. V-5.

Bund, Misereor ed. (1996). *Sustainable Germany* (in German). Basel, Berlin, and Boston: Birkhäuser.

Bundestag Commission of Equity into the Protection of the Earth's Atmosphere, Germany, ed. (1994). *Schtz der grünen Erde – Klimaschutz durch umweltgerechte Landwirtschaft und Erhalt der Wälder.* Bonn.

Bundestag Commission of Equity into the Protection of Mankind and the Environment, Germany, ed. (1994). *Die Industriegesellschaft gestalten – Perspektiven für einen nachthaltigen Umgang mit Stoff- und Materialströmen.* Bonn.

Bureau of Agricultural Statistics, Philippines (1980–1993). *Philippine Fishery Statistics.* Manila: BAS.

Bureau of Agricultural Statistics, Philippines (1988–1993). *Survey of Wholesale Price for Selected Species of Philippine Fishery.* Manila: BAS.

Bureau of Economic Analysis (US) (1994). Accounting for mineral resources: issues and BEA's initial estimates. *Survey of Current Business,* April 1994.

Bureau of Fisheries and Aquatic Resources, Philippines (1951–1979). *Philippine Fishery Statistics.* Manila: BFAR.

Burns, M E (1989). *The Socio-Economic Impacts of Adelaide International Airport. An Independent Report.* Adelaide: Federal Airports Corporation.

Business Council for Sustainable Development (1995). *Achieving Eco-Efficiency in Business.* Report of the World Business Council for Sustainable Development. Second Antwerp Eco-Efficiency Workshop, March 14–15.

Carandang, A P *et al.* (1990). *Natural Pine Forest Resource Accounting.* Technical Report No. 3. NRAP-Phase 1. Manila.

Carson, C S and Landefeld, S (1994). Integrated economic and environmental satellite accounts. *Survey of Current Business.* 4, 33–49.

Carter, L A (1990). The Wildlife Survey of Skeleton Coast Park, Damaraland, and Kaokoland, Northwest Namibia, May/June 1990. Report to the Commission of the European Communities. Windhoek.

CBS (Centraal Bureau voor de Statistiek) (Statistics Netherlands) (1995). *Testing SERIEE: The Dutch Experience.* Eurostat meeting document SERIEE/95/11/REV., Luxembourg: Eurostat.

CBS (Statistics Netherlands) (1995). *National Accounts 1994.* The Hague: SDU-Publishers.

CBS (Statistics Netherlands) (1996). *Accounts and Indicators for the Economy and the Environment; the 1986–1992 NAMEAs* (in Dutch). Voorburg/Heerlen: Statistics Netherlands.

Central Statistical Office, Namibia (1995a). *Statistical Abstract 1995.* Windhoek, Namibia: National Planning Commission, Ministry of Finance.

Central Statistical Office, Namibia (1995b). *The Distribution of Economic Resources in the Population of Namibia.* Windhoek, Namibia: National Planning Commission, Ministry of Finance.

Clark, Colin W (1976). *Mathematical Bioeconomics: the Optimal Management of Renewable Resources.* New York: Wiley & Sons, Inc.

Cobb, C, Halstead, T and Rowe, J (1995). If the GDP is up, why is America down? *Atlantic Monthly.* October, 59–78.

Commission of the European Communities (CEC) (1994). Directions for the EU on Environmental Indicators and Green National Accounting. The Integration of Environmental and Economic Information Systems. COM(94)670. Brussels.

Commission of the European Communities (CEC), International Monetary Fund (IMF), Organisation for Economic Co-operation and Development (OECD), United Nations, and World Bank (1993). *System of National Accounts 1993.* United Nations Sales No. E.94.XVII.4, Document symbol ST/ESA/STAT/SER.F/2/Rev.4. Brussels/Luxembourg, New York, Paris, Washington, DC.

Commission of the European Communities (CEC) and US Environmental Protection Agency (1993). *Externalities of the Fuel Cycles: 'Externe Project'.* Working Documents 1, 2, 5, and 9. Brussels: Directorate-General XII, European Commission.

Conference of European Statisticians (1995). *Physical Environmental Accounting: Land Use*/Land Cover, *Nutrients and the Environment.* Orléans: Institut Français de l'Environnement (IFEN).

Cooper, A and Murray, R (1992). *Northern Ireland Countryside Survey.* Contract Report to DOENI. Coleraine: University of Ulster.

Costanza, Robert ed. (1991). *Ecological Economics: The Science and Management of Sustainability.* New York: Columbia University Press.

Dalvi, M Q (1988). *The Value of Life and Safety: A Search for a Consensus Estimate.* London: Department of Transport.

Daly, Herman E (1989). Toward a measure of sustainable social net national product. In: Ahmad, Y J, El Serafy, S and Lutz, E eds. *Environmental Accounting for Sustainable Development.* Washington, DC: The World Bank.

Daly, Herman E (1991a). From empty-world to full-world economics: recognizing an historical turning point in economic development. In: Goodland, R, Daly, H E, El Serafy, S and von Droste, B eds. *Environmentally Sustainable Economic Development: Building on Brundtland.* Paris: UNESCO.

Daly, Herman E (1991b). Sustainable development: from concepts and theory towards operational principles. In: Davis, K and Bernstein, M S eds. *Resources, Environment and Population: Present Knowledge, Future Options.* New York and Oxford: Oxford University Press.

Daly, Herman E (1992). Steady-state economics: concepts, questions, policies. *GAIA* 1, 333–338.

Daly, Herman E (1995). On Wilfred Beckerman's critique of sustainable development. *Environmental Values,* 4, 49–55.

Daly, Herman E and Cobb, John B Jr (1989). *For the Common Good: Redirecting the Economy Toward Community, the Environment, and a Sustainable Future.* Boston: Beacon Press.

Dasgupta, P (1994). Optimal versus sustainable development. In: Serageldin, I and Steer, A eds. *Valuing the Environment.* Proceedings of the First Annual International Conference on Environmentally Sustainable Development. Washington, DC: The World Bank.

Dasgupta, P and Mäler, K G (1991). *The Environment and Emerging Development Issues.* Beijer Reprint Series No. 1. Stockholm: Beijer.

Dean, Andrew and Hoeller, Peter (1992). *Costs of Reducing CO_2 Emissions: Evidence from Six Global Models.* OECD Economics Department Working Paper No. 122. Paris: OECD.

De Boo, Abraham (1995). Accounting for the costs of clean technologies and products. In: *Second Meeting of the London Group on Natural Resource and Environmental Accounting, Conference Papers.* Washington, DC: Bureau of Economic Analysis, US Department of Commerce.

De Boo, A J, Bosch, P R, Gorter, C N and Keuning, S J (1993). An environmental module and the complete system of national accounts. In: Franz, A and Stahmer, C eds. *Approaches to Environmental Accounting.* Proceedings of IARIW Conference on Environmental Accounting, 1991. Heidelberg: Physica-Verlag.

De Boer, B, de Haan, M and Voogt, M (1994). What would net domestic product have been in an environmentally sustainable economy? Preliminary views and results. In: *National Accounts and the Environment, Paper and Proceedings from a Conference,* London, England, March 16–18, 1994. Ottawa: Statistics Canada.

De Haan, M and Keuning, S J (1995). *Taking the Environment into Account: The Netherlands NAMEAs for 1989, 1990 and 1991*. National Accounts Occasional Papers, NA-074. Voorburg: Statistics Netherlands.

De Haan, M and Keuning, S J (1996). Taking the environment into account. The NAMEA approach. *Review of Income and Wealth*. 42, 131–148.

De Haan, M, Keuning, S J and Bosch, P R (1994). Integrating indicators in a national accounting matrix including environmental accounts (NAMEA); an application to the Netherlands. *National Accounts and the Environment, Paper and Proceedings from a Conference*, London, England, March 16–18, 1994. Ottawa: Statistics Canada.

Delos Angeles, M (1996). Depreciation of Philippine natural resources. ENRAP III, Vol. IV-7.

Denison, Edward F (1979). Pollution abatement programs: estimates of their effect upon output per unit of input, 1975–78. *Survey of Current Business*. 59, 58–59.

Denison, Edward F (1981). International transactions in measures of the nation's production. *Survey of Current Business*. 61, 17–28.

Department of the Environment, UK (1990). *This Common Inheritance: Britain's Environmental Strategy*. Command 1200. London: HMSO.

Department of the Environment, UK (1994a). *Biodiversity: The UK Action Plan*. Command 2428. London: HMSO.

Department of the Environment, UK (1994b). *Sustainable Development: The UK Strategy*. Command 2426. London: HMSO.

Department of the Environment, UK (1995). *A Living Countryside: Rural White Paper*. London: HMSO.

Department of Environment and Natural Resources (DENR), Republic of the Philippines *et al.* (1994). *The Philippine Environmental and Natural Resource Accounting Project (ENRAP II). Main Report*. Manila.

Department of Environment and Natural Resources (DENR), Republic of the Philippines *et al.* (1994). *The Philippine Environmental and Natural Resource Accounting Project (ENRAP II). Technical Appendices*. Manila.

Department of Environment and Natural Resources, Republic of the Philippines, *et al.* (1994). The Philippine Environment and Natural Resources Accounting Project (*ENRAP Phase II*), *International Workshop on the Contribution of Policy Environment and Natural Resources Accounting. Proceedings*. Manila.

Department of Water Affairs, Ministry of Agriculture, Water and Rural Development, Namibia (1991). *Perspective on Water Affairs*. Windhoek, Namibia: MAWRD.

Department of Water Affairs, Ministry of Agriculture, Water and Rural Development, Namibia (1993). *Central Area Water Master Plan*. Windhoek, Namibia: MAWRD.

Desaigues, B and Point, P (1993). *Economie du Patrimoine Naturel, la Valorisation des Bénéfices de Protection de l'Environment*. Paris: Economica.

Devall, Bill and Sessions, George (1985). *Deep Ecology: Living as if Nature Mattered*. Layton, Utah: Peregrine Smith.

Directorate of Veterinary Services (various years). *Veterinary Census*. Windhoek: Directorate of Veterinary Services, Ministry of Agriculture, Water and Rural Development.

Dixit, A, Hammond, P and Hoel, M (1980). On Hartwick's rule for regular Maximum paths of capital accumulation and resource depletion. *Review of Economic Studies*. XLVII, 551–556.

Dixon, J A, Scura, L F, Carpenter, R A and Sherman, P B (1994). *Economic Analysis of Environmental Impacts* (2nd edn.). London: Earthscan.

Dornbuscvh, Rudiger and Poterba, James M eds. (1992). *Global Warming: Economic Policy Responses*. Cambridge and London: MIT Press.

Duchin, Faye and Lange, Glenn-Marie (1993). *Development and the Environment in Indonesia: An Input–Output Analysis of Natural Resource Issues*. Final Report to the Canadian International Development Agency.

Duchin, Faye and Lange, Glenn-Marie (1994). *The Future of the Environment: Ecological Economics and Technological Change*. New York and Oxford: Oxford University Press.

Duchin, Faye and Lange, Glenn-Marie (1995). *Households and Eco-Restructuring: The Social Dimension of Ecological and Development Economics*. Institute for Economic Analysis, New York University.

Dunn, R and Harrison, A (1995). *Feasibility Study for a National Land Use Stock System*. Contract Report to DOE. Bristol: University of Bristol.

Economic Council, NNW Measurement Committee, Japan (1973). *Measuring Net National Welfare of Japan.* Tokyo.

Economic Planning Agency, Japan (1995). *Trial Estimates of Integrated Environmental and Economic Accounting: the Report of Research Study to Add the Environmental and Economic Accounting to the System of National Accounts.* English Summary.

Economic Planning Agency, Japan (1996). *Papers of International Symposium Integrated Environment and Economic Accounting in Theory and Practice.* Vol. 1 and Vol. 2. Tokyo.

Economist Intelligence Unit (1996). *Economic Intelligence Unit Report: Namibia, 4th Quarter 1995.* London: The Economist Intelligence Unit Limited.

ECOTEC (1996). *UK Environmental Protection Expenditure by Industry and Application of SERIEE in the United Kingdom.* Eurostat meeting document SERIEE/96/5. Luxembourg: Eurostat.

El Serafy, Salah (1989). The proper calculation of income from depletable natural resources. In: Ahmad, Y J, El Serafy, S and Lutz, E eds. *Environmental Accounting for Sustainable Development.* Washington, DC: The World Bank.

Environmental Agency, Japan (1992 and 1993). *Annual Report on Environment* (in Japanese). Tokyo: Ministry of Finance Printing Bureau.

Environmental Agency, Japan (1994). *Survey Concerning the Quantitative Analysis of Ecobusinesses* (in Japanese). Tokyo.

Environmental Agency, Japan (1996). *Report of Expert Workshop on Methodologies for Indicators of Sustainable Development.* 5–8 February 1996, Glen Cove, Long Island, New York. Tokyo.

Environmental Business Publishing, Inc. (1994). Foundation for the future. *Environmental Business Journal.* 7, 1–11.

Eurostat (1994). *SERIEE. The European System for the Collection of Economic Information on the Environment.* 1994 Version. Luxembourg: Office for Official Publication of the European Communities.

Eurostat (1994). *Environmental Protection Expenditure, Data Collection Methods in the Public Sector and Industry.* Luxembourg: Office for Official Publication of the European Communities.

Factor-10-Club (1994). *Carnoules Declaration.* Carnoules, France.

Farber, Kitt and Rutledge, Gary (1989). *Pollution Abatement and Control Expenditures: Methods and Sources for Current-Dollar Estimates.* Mimeo.

Färe, Rolf, Grosskopf, Shawna, Knox Lovell, C A and Pasurka, Carl (1989). Multilateral productivity comparisons when some outputs are undesirable: a nonparametric approach. *Review of Economics and Statistics.* 71, 90–98.

Färe, Rolf, Grosskopf, Shawna, Knox Lovell, C A and Yaisawarng, S (1993). Derivation of shadow prices for undesirable outputs: a distance function approach. *Review of Economics and Statistics,* 75, 374–380.

Faucheux, S, Froger, G and O'Connor, M (1994). *The Costs of Achieving Sustainability: the Differences between 'Environmentally Corrected National Accounts' and 'Sustainable National Income' as Information for Sustainability Policy.* University of York: Discussion Papers in Environmental Economics and Environmental Management.

Faucheux, S and Froger, G (1994). Le 'revenue national soutenable' est-il un indicateur de soutenabilite? *Revue Française d'Economie.* 9, 3–37.

Faucheux, S and O'Connor, M (1995). *Capital Theory and the Extensions to Natural Capital.* Paper prepared for the Final Project Report on the EC DG-XII Research Study Application of Non-Monetary Procedure of Economic Valuation for Managing a Sustainable Development. Paris: Centre Economie Espace Environment.

Faucheux, S and Noël, J-F (1995). *Economie des Resources Naturelles et de l'Environment.* Paris. Armand Colin.

Federal Statistical Office (Germany) (1995). *Umwelt: Umweltökonomische Gesamtrechnungen, Material und Energieflußrechnungen.* Fachserie 119, Reihe 5, Wiesbaden.

Flassbeck, H and Maier-Rigaud, G (1982). *Zur Diskriminierung des Umweltschutzes in der ökonomischen Analyse.* Umwelt und Wirtschaft 88. Tübingen: Mohr.

Food and Agricultural Organization of the United Nations (FAO), United Nations Fund for Population Activities (UNFPA), and International Institute for Applied Systems Analysis (IIASA) (1982). *Potential Population Supporting Capabilities of Lands in the Developing World.* Rome: FAO.

Forest Management Bureau, Philippines (1990). *Master Plan for Forestry Development.* Manila:

FMB, DENR.

Forest Management Bureau, Philippines (1990–1994). *Philippine Forestry Statistics.* Manila: FMB, DENR.

Franz, Alfred and Stahmer, Carsten eds. (1993). *Approaches to Environmental Accounting.* Heidelberg: Physica-Verlag.

Fuller, R and Groom, G (1994). *CORINE Land Cover Map: Pilot Study.* Contract Report to DOE. London: Department of the Environment.

Galbraith, J K (1986). *The New Industrial State* (4th edn.) New York: Mentor.

Georgescu-Roegen, N (1990). Production process and dynamic economies. In: Baranzini, M and Scazzieri, R eds. *The Economic Theory of Structure and Change.* Cambridge: Cambridge University Press.

Gerking, S and Stanley, L R (1986). An economic analysis of air pollution and health: the case of St. Louis. *Review of Economics and Statistics.* 68, 185–199.

Ghana Statistical Service (1996). *Integrated Environmental and Economic Accounts for Ghana 1991–1993.* Report of a joint UNEP, UNSD, Ghana Statistical Service project.

Gianessi, Leonard P and Peskin, Henry M (1976). *The Costs to Industries of Meeting the 1977 Provisions of the Water Pollution Control Act Amendments of 1972.* A Report to the US Environmental Protection Agency.

Gore, A (1992). *Earth in the Balance: Ecology and the Human Spirit.* New York: Houghton Mifflin Company.

Haines-Young, R, Watkins, C, Bruce, R G H and Hallam, C (1996). *Environmental Accounts for Land Cover.* London: Department of the Environment.

Hamilton, Kirk (1994). Green adjustments to GDP. *Resources Policy.* 20, 155–168.

Hamilton, K (1996). Pollution and pollution abatement in the national accounts. *Review of Income and Wealth.* 42, 13–33.

Hamilton, K and Atkinson, G (1995). Valuing air pollution in the national accounts. In: *Second Meeting of the London Group on Natural Resource and Environmental Accounting, Conference Papers.* Washington, DC: Bureau of Economic Analysis, US Department of Commerce.

Hamilton, K and Atkinson, G (1996). Air pollution and green accounts. *Energy Policy.* 24, 675–684.

Haneman, M W (1991). Review of a contingent valuation survey in Kakkadu conservation zone. Cited in Imber, D, Stevenson, G and Wilks, L (1991). *A Contingent Valuation Survey of the Kakkadu Conservation Zone.* Research Paper No. 3. Australia: Resource Assessment Commission. pp. 187–192.

Harrison, A (1989). Environmental issues and the SNA. *Review of Income and Wealth.* 4, 377–388.

Hartwick, J M (1977). Intergenerational equity and the investing of rents from exhaustible resources. *American Economic Review.* 67, 972–974.

Hartwick, J M (1990). Natural resources, national accounting and economic depreciation. *Journal of Public Economics.* 43, 291–304.

Hartwick, J M (1992). Deforestation and national accounting. *Environmental and Resource Economics.* 2, 513–521.

Hartwick, J M (1994). Untitled. Paper presented at the Workshop on Environmental Economics and Policy Analysis, organized by the Harvard Institute for International Development, Cambridge, MA, 27 June–29 July 1994.

Hartwick, John and Hageman, Anja (1993). Economic depreciation of mineral stocks and the contribution of El Serafy. In: Lutz, Ernst ed. *Toward Improved Accounting for the Environment.* Washington, DC: The World Bank.

Hartwick, J M and Lindsey, R (1989). *NNP and Economic Depreciation of Exhaustible Resource Stocks.* Department of Economics Discussion Paper No. 741. Queens University.

Hayami, H (1992). *Estimates of Air-Polluting Substances from Mobile Sources* (in Japanese). Keio Economic Observatory. Tokyo: Keio University.

Hicks, J R (1940). The valuation of the social income. *Economica.* VII, 105–124.

Hicks, J R (1946). *Value and Capital.* 2nd edn. Oxford: Oxford University Press.

Hill, P and Harrison, A (1994). Accounting subsoil assets in the 1993 SNA. In: *National Accounts and the Environment, Paper and Proceedings from a Conference*, London, England, March 16–18, 1994. Ottawa: Statistics Canada.

Hill, P and Harrison, A (1995). Accounting for depletion in the 1993 SNA. In: *Second Meeting of the London Group on Natural Resource and Environmental Accounting, Conference Papers.*

Washington, DC: Bureau of Economic Analysis, US Department of Commerce.

Hinterberger, F, Kranendonk, S, Welfens, M J and Schmidt-Bleek, F (1994). *Increasing Resource Productivity through Eco-efficient Services.* Wuppertal Papers Nr.13.

Hueting, R (1989). Correcting national income for environmental losses: towards a practical solution. In: Ahmad, Y J, El Serafy, S and Lutz, E eds. *Environmental Accounting for Sustainable Development.* Washington, DC: The World Bank.

Hueting, R and Bosch, P (1994). Sustainable national income in the Netherlands: the calculation of environmental losses in monetary terms. In: *National Accounts and the Environment, Paper and Proceedings from a Conference,* London, England, March 16–18, 1994. Ottawa: Statistics Canada.

Hueting, R, Bosch, P and De Boer, B (1991). *Methodology for the Calculation of Sustainable National Income.* Voorburg, Netherlands: Central Bureau of Statistics.

Hufschmidt, M M, James, D E, Meister, A D, Bower, B T and Dixon, J A (1983). *Environment, Natural Systems, and Development: An Economic Valuation Guide.* Baltimore: Johns Hopkins University Press.

IFEN (Institut Français de l'Environnement) (1996). *SERIEE's Environmental Protection Expenditure Account.* Eurostat meeting document SERIEE/95/9. Luxembourg: Eurostat.

Imber, D, Stevenson, G and Wilks, L (1991). *A Contingent Valuation Survey of the Kakkadu Conservation Zone.* Research Paper No. 3. Australia: Resource Assessment Commission.

INFRAS (1996). *SERIEE – Tables for Switzerland.* Eurostat meeting document SERIEE/96/2. Luxembourg: Eurostat.

Institute of Developing Economies, Japan (1992). *Asian International Input–Output Table 1985.* Tokyo.

Institute of Energy Economics, Japan (1993). *Statistical Abstract of Energy and the Economy '93* (in Japanese). Tokyo: Energy Saving Center.

Institute of Developing Economies, Japan (1995). *International Trade Matrix for Asia-Pacific Region by Industrial Group, 1975–1992.* Vols. I, II, and III. I.D.E. Statistical Data Series No. 68. Tokyo.

Jacobs, Michael (1994). The limits to neoclassicism: towards an institutional environmental economics. In: Redclift, M and Benton, T eds. *Social Theory and the Global Environment.* London and New York: Routledge.

Japan Association of Industrial Machinery Manufacturers (various years). *Production of Environment Equipment* (in Japanese). Tokyo.

Japan Association of Industrial Machinery Manufacturers (1993). *Guidebook of Environmental Equipment in Japan* (in Japanese). Tokyo: Ryoin Press.

Japan Society of Global Environment and Economy (1990). *Experience of Public Nuisance in Japan* (in Japanese). Tokyo: Goudo Press.

Jaszi, G (1971). An economic accountant's ledger. *Survey of Current Business.* 51, 183–227.

Jensen, Peter Rømose and Møllgaard, Elisabeth (1995). *On the Measurement of a Welfare Indicator for Denmark 1970–1990.* Copenhagen: The Rockwool Foundation Research Unit.

Jesinghaus, J (1994). EXTASY: towards an environmental index with the help of expert surveys. In: *National Accounts and the Environment, Papers and Proceedings from a Conference,* London, England, March 16–18, 1994. Ottawa: Statistics Canada.

Johansson, Per-Olov (1987). *The Economic Theory and Measurement of Environmental Benefits.* Cambridge: Cambridge University Press.

Jorgenson, Dale W and Wilcoxen, Peter J (1990). Environmental regulation and US economic growth. *Rand Journal of Economics,* 21, 314–340.

Juster, F Thomas (1973). A framework for the measurement of economic and social performance. In: Moss, Milton ed. *The Measurement of Economic and Social Performance.* Studies in Income and Wealth. Vol. 38. New York: Columbia University Press for the National Bureau of Economic Research.

Keuning, Steven J (1993). An information system for environmental indicators in relation to the national accounts. In: De Vries, W F M, den Bakker, G P, Gircour, M B G, Keuning, S J and Lenson, A eds. *The Value Added of National Accounting.* Voorburg, Heerlen: Netherlands Central Bureau of Statistics.

Keuning, S J (1995). An information system for environmental indicators in relation to the national accounts. In: de Vries, W F M, den Bakker, G P, Gircour, M B G *et al.* eds. *The Value Added of National Accounting.* Netherlands, Central Bureau of Statistics.

Keuning, S J (1996a). The NAMEA experience, an interim evaluation of the Netherlands integrated accounts and indicators for the environment and the economy. Presented at the International symposium on Integrated Environmental and Economic Accounting in Theory and Practice, Tokyo, 5–8 March 1996.

Keuning, S J (1996b). *Accounting for Economic Development and Social Change.* Amsterdam, IOS Publishers.

Keuning, Steven J and de Gijt, J (1992). *A National Accounting Matrix for the Netherlands.* National Accounts Occasional Paper Series No. 59. Voorburg, Heerlen: Statistical Netherlands.

Keuning, S J and Timmerman, J G (1995). An information system for economic environmental and social statistics: integrating environmental data into the SESAME. In: *Second Meeting of the London Group on Natural Resource and Environmental Accounting, Conference Papers.* Washington, DC, Bureau of Economic Analysis, US Department of Commerce.

Knetsch, J (1993). *Environmental Valuation: Some Practical Problems of Wrong Questions and Misleading Answers.* Occasional Publication No. 5. Resource Assessment Commission.

Konijn, P, De Boer, S and van Dalen, J (1995). *Material Flows, Energy Use and the Structure of the Economy.* National Accounts Occasional Paper Series No. 77. Voorburg, Heerlen: Statistics Netherlands.

Korea Energy Economics Institute (1987, 1990, 1992). *Report on Energy Census.* Seoul.

Korean Environmental Technology Research Institute (KETRI) *et al.* (in prep.). *Report on the Pilot Compilation of Environmental-Economic Accounts for the Republic of Korea.*

Krupnic, A J, Harrington, W and Ostro, B (1990). Ambient ozone and acute health effects: evidence from daily data. *Journal of Environmental Economics and Management.* 18, 1–18.

Kuhn, M, Radermacher, W and Stahmer, C (1994). Umweltökonomische Trends 1960 bis 1990. *Wirtschaft und Statistik.* 8, 658–677.

Kunimatsu, T and Muraoka, K (1990). *Model Analysis of River Pollution* (in Japanese). Tokyo: Gihoudo Press.

Kuznets, Simon (1946). *National Income: A Summary of Findings.* New York: National Bureau of Economic Research.

Landefeld, J S and Hines, J R (1985). National accounting for non-renewable natural resources in the mining industries. *Review of Income and Wealth.* 31, 1–20.

Landefeld, J Steven, Carson, Carol S, *et al.* (1994). Integrated economic and environmental satellite accounts. *Survey of Current Business.* 74, 33–49.

Landefeld, J Steven, Carson, Carol S, *et al.* (1994). Accounting for mineral resources: issues and BEA's initial estimates. *Survey of Current Business.* 74, 50–72.

Landefeld, S K and Seskin, E P (1982). The economic value of life: linking theory to practice. *American Journal of Public Health,* 72, 555–566.

Lange, G (1996). *Construction of Accounts for Water in the Natural Resource Accounts for Namibia.* Report to USAID and Directorate of Environmental Affairs, Windhoek, Namibia: Namibia Ministry of Environment and Tourism.

Lave, L B and Seskin, E P (1977). *Air Pollution and Human Health.* Baltimore: Johns Hopkins University Press.

Lee, H, Oliveira-Martins, J and van der Mensbrugghe, D (1994). *The OECD GREEN Model: An Updated Overview.* Technical Paper No. 97. Paris: OECD Development Center.

Leipert, C (1989). National income and economic growth: the conceptual side of defensive expenditures. *Journal of Economic Issues.* 23, 843–856.

Leipert, C (1995). *Environmentally Defensive Expenditure and Their Role in an Environmentally Adjusted NDP Figure.* Mimeo. Berlin.

Leontief, Wassily (1970). Environmental repercussions and the economic structure: an input–output approach. *Review of Economics and Statistics.* 52, 262–271.

Lerner, A P (1963). Consumer's surplus and micro-macro. *Journal of Political Economy.* 61, 76–81.

Liedtke, D, Manstein, C and Merten, T (1994). *MIPS, Resource management and sustainable development.* ASM Intern. Conf. on the Recycling of Metals. October 1994. Amsterdam.

London Group (1994). *National Accounts and the Environment.* Papers and Proceedings from a Conference, London, England, March 16–18, 1994. Ottawa: Statistics Canada.

London Group (1995). *Second Meeting of the London Group on Natural Resource and Environmental Accounting.* Conference Papers. Washington, DC: US Department of Commerce, Bureau of Economic Analysis.

London Group (1996). *Third Meeting of the London Group on Natural Resource and Environ-*

mental Accounting. Proceedings Volume. Stockholm: Statistics Sweden.

Looß, A and Katz, C (1993). *Abfallvermeidung: Strategien, Instrumente und Bewertungskristerien.* TAB-Arbeitsbericht Nr.16. Bonn: Office for the Assessment of the Consequences of Technology at the German Bundestag (TAB).

Lutz, C (1995). Modeling air pollution in Germany: a first approach. In: Janoska, J J and Monaco, L S eds. *Papers from the Second INFORUM World Conference*. Osnabrück, Germany, September 1994. College Park, Maryland.

Lutz, Ernst (1993). *Toward Improved Accounting for the Environment*. An UNSTAT–World Bank Symposium. Washington, DC: The World Bank.

Ma, Q (1995). The road to a bilateral trade model. In: Janoska, J J and Monaco, L S eds. *Papers from the Second INFORUM World Conference*. Osnabrück, Germany, September 1994. College Park, Maryland.

Mäler, K-G (1991). *National Accounts and Environmental Resources*. Beijer Reprint Series No. 4. Stockholm: Beijer.

Management and Coordination Agency, Japan (1985 and 1990). *Input–Output Tables* (in Japanese). Tokyo: National Federation of Statistical Association.

Mayer, Helmut and Stahmer, Carsten (1989). *Energy Consumption and Sulphur Dioxide Emissions in the Federal Republic of Germany in 1980 and 1986*. Mimeo.

Meadows, D H, Meadows, D L and Randers, J (1992). *Beyond the Limits*. Post Mills, VT, USA: Chelsa Green Publishing.

Meyer, B and Ewerhart, G (1996). Modelling towards eco-domestic product. Presented at the International symposium on Integrated Environmental and Economic Accounting in Theory and Practice, Tokyo, 5–8 March 1996.

Mikesell, Raymond (1994). Sustainable development and mineral resources. *Resources Policy*. 20, 83–86.

Miller, M H and Upton, C W (1985). A test of the Hotelling valuation principle. *Journal of Political Economy*. 93, 1–25.

Mines and Geo-Sciences Bureau, Philippines (1987–1992). *Philippine Metallic Mineral Reserves*. Manila: MGB.

Mines and Geo-Sciences Bureau, Philippines (1987–1992). *Integrated Annual Report*. Manila: MGB.

Ministry of Agriculture, Forestry, and Fisheries, Japan (1978 and 1993). *Statistics on National Forest Production* (in Japanese). Tokyo.

Ministry of Agriculture, Forestry, and Fisheries, Japan (1994). *Statistical Abstract of Forestry* (in Japanese). Tokyo: Rinya Koseikai.

Ministry of Agriculture, Forestry, and Fisheries, Korea (various years). *Statistical Yearbook of Agriculture, Forestry and Fisheries*. Seoul.

Ministry of Construction, Japan (1994). *Comprehensive Program for the Arrangement of Regional Sewerage System: Guide and Commentary* (in Japanese). Tokyo: Nihon Gesuido Kyokai.

Ministry of Environment, Korea (1989–1994). *Korea Environmental Yearbook*. Seoul.

Ministry of Environment, Korea (1986–1992, 1994). *The Annual Survey Report on the Industrial Waste Water Discharge*. Seoul.

Ministry of Environment, Korea (1992). *Survey on the Environmental Protection Expenditures by Manufacturers*.

Ministry of Environment, Korea (1996). *Environmental Vision in the 21st Century*. Seoul.

Ministry of Finance, Korea (1985–1992). *Government Revenue and Expenditures*.

Ministry of Health and Welfare, Japan (various years). *Waste Disposal in Japan* (in Japanese). Tokyo: Zenkoku Toshi Seiso Kaigi.

Ministry of International Trade and Industry, Japan (1990 and 1993). *Energy Balance Tables in Japan* (in Japanese). Tokyo: Tsusho Sangyo Kenkyusha.

Ministry of International Trade and Industry, Japan (1985 and 1990). *Census of Manufacturers* (in Japanese). Tokyo: Ministry of Finance Printing Bureau.

Ministry of International Trade and Industry, Japan (1993). *Survey of Investment in Environmental Pollution Prevention Equipment* (in Japanese). Tokyo.

Ministry of International Trade and Industry, Japan (1994). *Statistical Abstract of Mining* (in Japanese). Tokyo: Research Institute of International Trade and Industry Publication Department.

Ministry of International Trade and Industry, Japan (1994). *Yearbook of Minerals and Non-ferrous*

Metals Statistics (in Japanese). Tokyo: International Trade and Industry Statistics Association.

Mishan, E J (1982). *Cost Benefit Analysis: An Informal Introduction.* 3rd edn. London: George Allen and Unwin.

Møllgaard, Elisabeth (1995). Valuation of pollution in Denmark by the avoidance cost method. In: *Second Meeting of the London Group on Natural Resource and Environmental Accounting, Conference Papers.* Washington, DC: Bureau of Economic Analysis, US Department of Commerce.

Munashinghe, M (1993). *Environmental Economics and Sustainable Development.* World Bank Environment Paper No. 3. Washington, DC: The World Bank.

Muraleedharan, T R, Hadkar, N, Sharma, S, Ashish, D, Geetha, S and Babu, P G (1995). *Are People in Developing Countries Willing to Pay for Natural Resources Preservation: Evidence from a Contingent Valuation of the Borivali National Park.* Bombay.

Musu, Ignazio and Siniscalco, Domenico eds. *National Accounts and the Environment.* Dordrecht: Kluwer Academic Publishers.

Naess, A (1976). The Shallow and the Deep, Long-Range Ecology Movement, A Summary. *Inquiry.* 16, 95–100.

National Institute of Economic Research and Statistics Sweden (1994). *SWEEA, Swedish Economic and Environmental Accounts.* Preliminary Edition.

National Land Agency, Japan (1990 and 1993). *Statistical Abstract of Land* (in Japanese). Tokyo: Taisei Press.

National Planning Commission, Namibia (1995). *First National Development Plan.* Windhoek: Government of Namibia.

National Statistical Coordination Board, Philippines (1988–1992). *Factor Share for Mineral Sector.* Manila: NSCB.

National Statistical Office, Korea (1987–1994). *The Report on Mining and Manufacturing Survey.* Seoul.

National Statistical Office, Korea (1986–1993). *The Report on the Construction Work Survey.* Seoul.

National Statistical Office, Korea (1989). *1987 National Wealth Survey of Korea.* Vol. I–V. Seoul.

National Statistical Office, Korea (1989–1993). *The Annual Report on the Survey of Services.* Seoul.

National Statistics Office, Philippines (1989). *Annual Survey of Establishments.* Manila: NSO.

Nelson, J P (1980). Airport and property values: a survey of recent evidence. *Journal of Transport and Policy.* 16, 117–130.

Nestor, Deborah Vaughn and Pasurka, Jr, Carl A (1994). *Productivity Measurement with Undesirable Outputs: A Distance Function Approach.* Washington, DC: Mimeo.

Nestor, Deborah Vaughn and Pasurka, Jr, Carl A (1995a). Alternative specifications for environmental control costs in a general equilibrium framework. *Economics Letters.* 48, 273–280.

Nestor, Deborah Vaughn and Pasurka, Jr, Carl A (1995b). Environmental–economic accounting and indicators of the economic importance of environmental protection activities. *Review of Income and Wealth.* 41, 265–287.

Nestor, Deborah Vaughn and Pasurka, Jr, Carl A (1995c). Welfare effects of environmental control: the German example. *Economic Modelling.* 12, 53–59.

Netherlands Ministry of Housing, Spatial Planning and the Environment (1989). *National Environmental Policy Plan* (in Dutch). The Hague.

Netherlands Ministry of Housing, Spatial Planning and the Environment (1993). *Second National Environmental Policy Plan* (in Dutch). The Hague.

Nordhaus, William, and Tobin, James (1973). Is growth obsolete? In: Moss, Milton ed. *The Measurement of Economic and Social Performance.* Studies in Income and Wealth, Vol. 38. New York: Columbia University Press for the National Bureau of Economic Research.

Nyhus, D (1991). The INFORUM international system. *Economic Systems Research.* 3, 55–64.

Office of the Prime Minister, Government of Namibia (1991). *National Conference on Land Reform and the Land Question.* Proceedings of a Conference, 25 June–1 July, 1991. Windhoek, Namibia: Namibian Economic Policy Research Unit on behalf of the Office of the Prime Minister.

Organisation for Economic Co-operation and Development (OECD) (1973). *List of Social Concerns Common to Most OECD Countries.* Paris: OECD.

Organisation for Economic Co-operation and Development (OECD) (1982). *Economic and Ecological Interdependence.* Paris: OECD.

Organisation for Economic Co-operation and Development (OECD) (1988). *Transport and the*

Environment. Paris: OECD.
Organisation for Economic Co-operation and Development (OECD) (1989). *Environmental Policy Benefits: Monetary Valuation.* Paris: OECD.
Organisation for Economic Co-operation and Development (1992). *The OECD Environment Industry: Situation, Prospects, and Government Policies.* Paris: OECD.
Organisation for Economic Co-operation and Development (OECD) (1995). *The OECD Input–Output Database.* Paris: OECD.
Organisation for Economic Co-operation and Development (OECD) (1996). *Pollution Abatement and Control Expenditure in OECD Countries.* Environmental Monograph OCDE/GD(96). Paris: OECD.
Organisation for Economic Co-operation and Development (OECD) (1996b). *The Environment Industry, The Washington Meeting.* Paris: OECD.
Organisation for Economic Co-operation and Development (OECD) (various years). *OECD Environmental Data: Compendium.* Paris: OECD.
Ostro, B D (1983). The effects of air pollution on work loss and morbidity. *Journal of Environmental Economics and Management.* 10, 371–382.
Ostro, B D (1987). Air pollution and morbidity revisited: a specification test. *Journal of Environmental Economics and Management.* 14, 87–98.
Pabuayon, I M (1990). *Natural Resources Accounting: Rattan.* NRAP Technical Report No. 5. Manila.
Padilla, J E and de Guzman, F (1994). *Fishery Resources Accounting in the Philippines: Applications to the Small-Pelagic Fishery.* Technical Report No. 6. Manila: ENRAP.
Parikh, K, Parikh, J, Muraleedharan, T R and Hadker, N (1996). *Valuing Health Damage Costs due to Air Pollution in Bombay.* Report for Metropolitan Environmental Improvement Programme of the World Bank. IGIDR, Mumbai, India.
Parker, Larry, Blodgett, John and Aguirre, John (1993). *The Pollution Control Industry: An Emerging Economic Presence.* The Library of Congress, 93-811 ENR. Washington, DC: Congressional Research Service.
Pearce, David W and Turner, R Kerry (1990). *Economics of Natural Resources and the Environment.* Hemel Hempstead: Harvester Wheatsheaf.
Pearce, D W (1994). Valuing the environment: past practice, future prospect. In: Serageldin, I and Steer, A eds. *Valuing the Environment. Environmentally Sustainable Development Proceedings.* Series No. 2. Washington, DC: The World Bank.
Pearce, D W and Atkinson, G (1993). Capital theory and the measurement of sustainable development: an indicator of weak sustainability. *Ecological Economics.* 8, 103–108.
Pearce, D W and Ulph, D (1995). *A Social Discount Rate for the United Kingdom.* University College London: Centre for Social and Economic Research on the Global Environment, Working Paper 95-01.
Pearce, David, Markandya, Anil and Barbier, Edward B (1989). *Blueprint for a Green Economy.* London: Earthscan Publications.
Peskin, Henry M and Lutz, Ernst (1990). *A Survey of Resource and Environmental Accounting in Industrialized Countries.* Environment Working Paper No. 37. Washington, DC: The World Bank Environment Department.
Peskin, Henry M (1991). Alternative environmental and resource accounting approaches. In: Costanza, Robert ed. *Ecological Economics: The Science and Management of Sustainability.* New York: Columbia University Press.
Peskin, Henry M and Gregory Michaels, R (1991). *Environmental and Natural Resource Accounting in Colombia.* Report to the World Bank.
The Philippine Lumberman (1987). *RP-German Forestry Project.* Manila.
Powell, M (1996). Integrated environmental and economic accounts for Ghana. In: *Integrated Environmental and Economic Accounting in Theory and Practice*, Papers of International Symposium (Tokyo, 5–8 March 1996). Tokyo: Economic Planning Agency.
Puolamaa, Maila, Kaplas, Marianne and Renikainen, Tapro (1996). *Index of Environmental Friendliness, A Methodological Study.* Helsinki: Statistics Finland.
Pyo, H K *et al.* (1993). *Industrial Growth Factor Analysis and Productivity Estimation in Korea.* Seoul: Korea Economic Research Institute.
Radermacher, W (1995). General aspects of the maintenance cost approach in the EU-project on methodological problems in the construction of an adjusted income figure. In: *Second Meeting*

of the London Group on Natural Resource and Environmental Accounting, Conference Papers. Washington, DC: Bureau of Economic Analysis, US Department of Commerce.

Radermacher, W (in press). *Remote Sensing for Physical Environmental Accounting – The Contribution of CORINE Land Cover.* Esquilino Seminar. Rome, Italy.

Radermacher, W and Stahmer, C (1994). Environmental accounting and national accounts in Germany. In: *National Accounts and the Environment, Paper and Proceedings from a Conference,* London, England, March 16–18, 1994. Ottawa: Statistics Canada.

Radermacher, W and Stahmer, C (1996). *German Material and Energy Flow Information System: Concepts, Available Data and Applications.* Presented at the International Symposium on Integrated Environmental and Economic Accounting in Theory and Practice, Tokyo, 5–8 March 1996.

Repetto, Robert, Magrath, William, Wells, Michael, Beer, Christine and Rossini, Fabrizio (1989). *Wasting Assets: Natural Resources in the National Income Accounts.* Washington, DC: World Resources Institute.

Rezeler, J (1993). *Economic Costs of Environmental Degradation in Bombay Metropolitan Region (BMR).* Presented at the Workshop on Economic Valuation of Environmental Degradation, IGIDR, November 5–6, 1993, Bombay.

Rodwell, T C, Tagg, J and Grobler M. (1995). Wildlife Resources in the Caprivi, Namibia: The Results of an Aerial Census in 1994 and Comparison with Pat Surveys. Research Discussion Paper No. 9. Namibia: Directorate of Environmental Affairs, Ministry of Environment and Tourism.

Rosen, S (1974). Hedonic prices and implicit markets: product differentiation in pure competition. *Journal of Political Economy.* 82, 34–56.

Roy, G D (1977). *Anatomy of Depreciation.* Calcutta: World Press Private.

Samuelson, Paul A and Nordhaus, William D (1992). *Economics.* 14th edn. New York: McGraw-Hill.

Schäfer, D and Stahmer, C (1989). Input–output model for the macroeconomic analysis of environmental protection activities. *Economic Systems Research.* 1, 203–228.

Schäfer, D (1995a). *Indicators of the Quality of the Environment – Concepts of a Pilot Project in Germany* Contribution to the Joint ECE/Eurostat Work Session on Methodological Issues of Environment Statistics. Mèze, France, September 1995.

Schäfer, D (1995b). *Effects of Environmental Protection Measures on Gross Domestic Product.* Eurostat meeting document SERIEE/95/16. Luxembourg: Eurostat.

Schmmidt-Bleek, F (1994). *Wieviel Umwelt braucht der Mensch*? Basel, Berlin, and Boston: Birkhäuser.

Schnare, A B (1976). Racial and ethnic price differential in urban housing market. *Urban Studies.* 13, 107–120.

Schrecker, Ted (1995). Commonplaces and heresies about the human–ecosystem interface. In: Hedge, Tony *et al.* eds. (1995). *Pathway to Sustainability: Assessing Our Progress.* National Round Table on the Environment and the Economy. Ottawa.

Schütz, H and Bringezu, S (1993). Major material flows in Germany. *Fresenius Environmental Bulletin.* 2, 443–448.

Serageldin, I and Steer, A (1994). Epilogue: expanding the capital stock. In: Serageldin, I and Steer, A eds. *Making Development Sustainable – From Concepts to Action.* Environmentally Sustainable Development Occasional Paper Series No. 2. Washington, DC: The World Bank.

Shin, E, Gregory, R, Hufschmidt, M, Lee Y, Nickum, J E and Umetsu, C (1993). *Valuing the Economic Impacts of Environmental Problems, Asian cities.* Washington DC: The World Bank.

Silvestre, G and Pauly D (1987). Estimate of yield and economic rent from Philippine demersal stocks (1964–1984) using vessel horsepower as an index of fishing effort. *UPV Fishery Journal.* 1(2)/2(1-2)/3(1-2), 11–24.

Smith, Philip (1994). The Canadian national accounts environmental component: a status report. *National Income and Expenditure Accounts, Annual Estimates, 1982–1993.* Catalogue No. 13-201. Ottawa: Statistics Canada.

Smith, Roger S (1992). Income growth, government spending and wasting assets – Alberta's oil and gas. *Canadian Public Policy.* 18, 387–412.

Solow, R (1974). Integrated equity and exhaustible resources. *Review of Economic Studies,* Symposium 29–46.

Solow, R (1992). *An Almost Practical Step Toward Sustainability.* Washington, DC: Resources for

the Future. Also (1993) *Resources Policy*. September, 162–172.
Solórzano, R *et al.* (1991). *Accounts Overdue: Natural Resource Depletion in Costa Rica.* San José, Costa Rica: Tropical Science Center, and Washington DC: World Resources Institute.
Stahmer, C (1995a). Estimating sustainable national income. In: Van Dieren, Wouter ed. *Taking Nature into Account.* Report to the Club of Rome. New York: Copernicus.
Stahmer, C (1995b). Environment adjustment of the system of national accounts. In: Van Dieren, Wouter ed. *Taking Nature into Account.* Report to the Club of Rome. New York: Copernicus.
Statistics Finland (1996). *Testing SERIEE's Environmental Protection Expenditure Account in Finland.* Eurostat meeting document SERIEE/96/3. Luxembourg: Eurostat.
Statistics Sweden (1996). *Testing SERIEE's Environmental Protection Expenditure Account in Sweden.* Eurostat meeting document SERIEE/96/4. Luxembourg: Eurostat.
Statistisches Bundesamt, Germany (1994). *Input–Output-Tabellen 1986, 1988, 1990.* Fachserie 18, Reihe 2.
Statistisches Bundesamt, Germany (1995). *Umwelt: Umweltökonomische Gesemtrechnungen – Material und Energieflurechnungen.* Fachserie 19, Reihe 5. Wiesbaden.
Statistisches Bundesamt, Germany (1996). *Implementation of SERIEE's Environmental Protection Expenditure Account in Germany.* Eurostat meeting document SERIEE/95/10 REV.1. Luxembourg: Eurostat.
Steuer, A (1992). Stoffstrombilanz Östereich 1988. In: Schrift, ed: *Sozial Ökologie des iff.* Band 26. Wien.
Steuer, A (1995). *Issues Arising from the Implementation of the SERIEE's EPEA.* Eurostat meeting document SERIEE/95/13. Luxembourg: Eurostat.
Streeting, M C (1990). *A Survey of Hedonic Price Technique.* Research Paper No. 1, Australia: Resource Assessment Commission.
Theys, J (1989). Environmental accounting in development policy: the French experience. In: Ahmad, Y J, El Serafy, S and Lutz, E, eds. *Environmental Accounting for Sustainable Development.* Washington: The World Bank.
Tietenberg, Tom (1992). *Environmental and Natural Resources Economics.* New York: Harper Collins Publishers.
Timmerman, J G and van de Ven, P (1994). *A SAM for the Netherlands, Concepts and Results.* National Accounts Occasional Paper Series No. 68. Voorburg.
Tolba, Mostafa K. *Saving Our Planet.* See United Nations Environment Programme.
Ulgiati, S, Odum, H T and Bastianoni, S (1994). Energy use, environmental loading and sustainability – an energy analysis of Italy. *Ecological Modelling.* 73, 215–268.
United Nations (1968). *A System of National Accounts.* Sales no. E.69.XVII.3.
United Nations (1975). *Toward a System of Social and Demographic Statistics.* Sales No. E.74.XVII.8. New York.
United Nations (1980). *Classification of the Functions of Government.* Statistical Papers, Series M, No. 70. Sales No. E.80.XVII..17. New York.
United Nations (1984). *A Framework for the Development of Environment Statistics.* Sales No. E.84.XVII.12. New York.
United Nations (1990). *International Standard Classification of All Economic Activities.* Sales No. E.90.XVII.11. New York.
United Nations (1991). *Concepts and Methods of Environment Statistics: Statistics of the Natural Environment – A Technical Report.* Sales No. E.91.XVII.18. New York.
United Nations (1992). *Handbook on National Accounting, Integrated Environmental and Economic Accounting. Interim Version (Draft).* New York.
United Nations (1993a). *Integrated Environmental and Economic Accounting.* Sales No. E.93.XVII. 12 New York: United Nations Statistical Division.
United Nations (1993b). *Report of the United Nations Conference on Environment and Development, Rio de Janeiro, 3–24 June 1992. Vol. 1. Resolutions Adopted by the Conference.* (Sales No. E.93.I.8).
United Nations (1993b). *AGENDA 21: The United Nations Programme of Action from Rio.* Sales No. E.93.I.11. New York: United Nations Department of Public Information.
United Nations (1995). *Work Programme on Indicators of Sustainable Development of the Commission on Sustainable Development.* Prepared for the Department of Policy Coordination and Sustainable Development. (Internal paper).
United Nations (1996). *Indicators of Sustainable Development: Methodology Sheets.* Prepared for

the Division for Sustainable Development for the Commission on Sustainable Development, Fourth Session. 18 April–3 May 1996, New York.

United Nations Development Programme (1991). *Human Development Report.* New York and Oxford: Oxford University Press.

United Nations Environment Programme (1992). *Saving our Planet – Challenges and Hopes.* Nairobi: UNEP.

Uno, Kimio (1989). *Measurement of Services in an Input-Output Framework.* Amsterdam: Elsevier Science Publishers. B.V.

Uno, Kimio (1991). *Technology, Investment, and Trade.* New York: Elsevier Science Publishing Co.

Uno, Kimio (1995). *Environmental Options: Accounting for Sustainability.* Dordrecht, Boston, and London: Kluwer Academic Publishers.

US Department of Commerce, Bureau of Economic Analysis (1984). The input–output structure of the US economy, 1977. *Survey of Current Business.* 64, 42–84.

US Department of Commerce, Bureau of Economic Analysis (1991). Benchmark input–output accounts for the US economy, 1982. *Survey of Current Business.* 71, 30–71.

US Department of Commerce, Bureau of Economic Analysis (1994). Accounting for mineral resources: issues and BEA's initial estimates. *Survey of Current Business.* April, 50–72.

US Department of Energy (various issues). *Manufacturing Energy Consumption Survey.*

US Environmental Protection Agency (1990). *Environmental Investments: The Cost of a Clean Environment.* Report of the Administrator of the Environmental Protection Agency to the Congress of the United States. EPA 230-11-90-083. Washington, DC.

US Environmental Protection Agency, Office of Policy, Planning and Evaluation (1993). *International Trade in Environmental Protection Equipment: An Assessment of Existing Data.* EPA 230-R-93-006. Washington, DC.

US Environmental Protection Agency, Office of Policy, Planning and Evaluation (1995a). *The US Environmental Protection Industry: A Proposed Framework for Assessment.* EPA 230-R-95-011. Washington, DC.

US Environmental Protection Agency, Office of Policy, Planning and Evaluation (1995b). *The US Environmental Protection Industry: The Technical Document.* EPA 230-R-95-012. Washington, DC.

US Environmental Protection Agency (various issues). *Toxic Release Inventory.* Washington, DC.

US Water Resources Council (1979). *Procedures for Evaluation of National Economic Development (NED), Benefits and Costs in Water Resources Planning (Level C).* In Final Rule Federal Register, Washington, DC: Government Printing Office. 14th December.

Van Dieren, Wouter ed. (1995). *Taking Nature into Account.* New York: Springer-Verlag, New York, Inc.

Vanoli, André (1995a). *Comments on the SEEA.* Paper prepared for the Workshop on Environmental and Natural Resource Accounting, United Nations Environment Programme, Hungarian Central Statistical Office, Budapest, March 27–30, 1995.

Vanoli, André (1995b). Reflections on environmental accounting issues. *Review of Income and Wealth.* 41, 113–137.

Van Tongeren, J *et al.* (1991) *Integrated Environmental and Economic Accounting: A Case Study for Mexico.* Environment Working Paper No. 50. Washington, DC: The World Bank.

Van Tongeren, Jan, Kim, Seung-Woo and Alfieri, Alessandra (1994). *Design of Accounting Framework for SEEA in Korea.* Proceedings of International Conference on Green GNP and Sustainable Development, Seoul, 16–17 June 1994 organized by Korea Resource Economics Association.

Van Tongeren, Jan W and Becker, Bernd (1995). *Integrated Satellite Accounting, Socio-Economic Concerns and Modelling.* Draft to be published in the Review of Income and Wealth.

Victor, Peter (1990). Indicators of sustainable development: some lessons from capital theory. In: *Indicators of Ecologically Sustainable Development: Delineation Studies in Economics, Ecological and Decision Theory.* Ottawa.

Vitousek, P M, Ehrlich, P R, Ehrlich, A H and Matson, P A (1986). Human appropriation of the products of photosynthesis. *Bioscience.* 36, 368–373.

Von Below, M A (1993). Sustainable mining development hampered by low mineral prices. *Resources Policy.* September, 177–181.

Vu, V and van Tongeren, J (1995). An analytical approach to the calculation of 'green GDP'. In:

Second meeting of the London Group on Natural Resources and Environmental Accounting, Conference Papers. Washington.

Weber, J-L (1996). *A General Model for Land Cover and Land Use Accounting.* Paper presented at the IARIW Special Conference on Integrated Environmental and Economic Accounting in Theory and Practice. Tokyo.

Weitzman, M L (1976). On the welfare significance of national product in a dynamic economy. *Quarterly Journal of Economics.* 90, 156–162.

Weizsäcker, E U von, Lovins, A B and Lovins, H (1995). *Faktor Vier.* München: Droemer Knaur.

Weterings, R A P M and Opschoor, J B (1992). *The Ecocapacity as a Challenge to Technological Development.* Advisory Council for Research on Nature and Environment. Rijswijk.

Whalley, J and Wigle, R (1991). The international incidence of carbon taxes. In: Dornbusch, R and Poterba, J M eds. *Global Warming: Economic Policy Responses.* Cambridge, Massachusetts, and London: MIT Press.

Wieand, K F (1973). Air pollution and property values: A study of the St. Louis area. *Journal of Regional Science.* 13, 91–95.

Wier, Mette (1993). *A Decomposition of Changes in Emissions Through 25 Years: An Analysis of Emissions of Nutrients, Organic Matter, and Emissions from Fossil Fuels.* Tenth International conference on Input–Output Techniques, Sevilla, Spain, March 29–April 2.

Wilks, L C (1990). *A Survey of Contingent Valuation Method.* Research Paper No. 2. Australia: Resource Assessment Commission.

Wolters, S, ed. (1994). *Proceedings of Namibia's National Workshop to Combat Desertification.* Windhoek, Namibia: Desert Research Foundation of Namibia.

World Bank (1992). *Namibia: Poverty Alleviation with Sustainable Growth.* Washington, DC: World Bank.

World Bank (1995). *Monitoring Environmental Progress: A Report on Work in Progress.* Washington, DC: The World Bank.

World Commission on Environment and Development (1987). *Our Common Future.* Oxford: Oxford University Press.

World Resources Institute (1991). *Accounts Overdue: Natural Resource Depreciation in Costa Rica.* See R Solórzano, *et al.*

World Resources Institute, Wuppertal Institute, Netherlands Ministry of Housing and Spatial Planning and Environment. (1997). *Resource Flows: The Material Basis of Industrial Economics.* Washington, D.C.

Wyatt, B, Greatorex-Davies, J, *et al.* (1994). *Comparison of Land Cover Definitions.* London: Department of the Environment.

Yoshioka, K, Hayami, H, Suga, M, Toyooka, Y and Ikeda, A (1992). *Input–Output Tables for Analysis of Environment* (in Japanese). Tokyo: Keio Economic Observatory.

Notes on Contributors

Alessandra Alfieri

Alessandra Alfieri is a Statistician in the Environment Statistics Section of the United Nations Statistics Division. She is currently working on her PhD dissertation in Statistics/Econometrics at the University of Iowa. She worked in EUROSTAT and The World Bank. Her research interests include integrated environmental and economic accounting and econometric modelling. She has been involved in the preparation of a training manual on the implementation of the SEEA and in country projects in the Republic of Korea, Ghana and the Philippines.

Marian S. delos Angeles

Marian S. delos Angeles earned varied training which prepared her for leadership of the Philippines Environmental and Natural Resources Accounting Project (ENRAP) since 1991, namely: BS in Mathematics (magna cum laude), University of San Carlos, Cebu City; MA and PhD (Economics) from the University of the Philippines, Quezon City, and courses in natural resource and environmental economics at the University of Washington, Seattle, USA, through a UNDP post-doctorate fellowship grant in 1986–87.

Known as a policy-oriented researcher, Marian has been conducting studies on natural resources since 1975 while at the Development Academy of the Philippines. She has pursued various studies on natural resource management issues at the Philippine Institute for Development Studies, the Philippines' foremost policy research institution, where she has been based since 1983. Marian's researches have expanded to include environmental issues on air and water pollution, particularly through ENRAP and collaboration with the International Resources Group, Ltd, of Washington, DC, Henry Peskin of Edgevale Associates and the Philippines' foremost environmental economists at Resources, Environment and Economics Center for Studies, Inc. (REECS). Marian co-founded REECS in 1992 whose board she chairs and through which local research capacity for environmental economics analysis is being developed.

Peter Bartelmus

Peter Bartelmus is the Chief of the Environment, Energy and Industry Statistics Branch of the United Nations Statistics Division. He holds a PhD in economics from the University of Heidelberg. His research interest is the interface between statistics and policy analysis, as evidenced in his latest book on

Environment, Growth and Development – the Concepts and Strategies of Sustainability. As co-author of the United Nations System of integrated Environmental and Economic Accounting (SEEA) he has been instrumental in the implementation of the SEEA in country projects. He is currently involved in the revision of the SEEA and the development of environmental and sustainable development indicators.

Alice Born

Alice Born has been with the National Accounts and Environment Division of Statistics Canada since August, 1990. As a research economist, Ms. Born has conducted extensive research in the area of natural resource valuation and development of natural resource accounts. She has published numerous papers mainly focusing on monetary and physical accounts of Canada's energy and metal resources. Currently, she is completing the monetary value of Canada's mineral resources which will be included in the National Balance Sheet Accounts in 1997, as well as indicators of sustainable development. She studied Geology and has worked as an exploration geologist in many areas of Canada. In 1990, she completed an MBA.

Stefan Bringezu

Stefan Bringezu, born in 1958, was trained in biology. His PhD work was field research on mechanisms regulating the flow of materials and energy through natural ecosystems. From 1987 to 1992 he was deputy leader of the Department of Biocidal Substances within the Chemicals Control Division of the German Federal Environmental Agency. Since 1992 he has been Senior Researcher at the Wuppertal Institute in the Department of Material Flows and Structural Change. His main subjects are the development of indicators for sustainability on different levels (local, regional, national), the analysis of the industrial metabolism, and the development of practical instruments for an integrated economic and environmental accounting. He is lecturer at the Technical University of Berlin and the University of Dortmund. He edited for example *New Approaches to Environmental Statistics* (in German), and *Regional and National Material Flow Accounting* (Birkhäuser Verlag, Basel).

Mark de Haan

Mark de Haan is employed at the National Accounts Department of Statistics Netherlands. He is responsible for the implementation and further development of the National Accounts Matrix including Environmental Accounts (NAMEA). He obtained his Masters degree in economics at Erasmus University Rotterdam. He has published various articles on environmental accounting issues, including the incorporation of environmental issues into macroeconomic models.

Estrella V Domingo

Currently Acting Director, National Statistical Coordination Board (NSCB), the agency tasked with compiling the Philippine System of National Accounts, Ms. Domingo is in the forefront of the development of environment accounting and statistics in the Philippines. She is the Project Manager of the Sub-Programme Component on Environment and Natural Resource Accounts (ENRA) of the Department of Environment and Natural Resources (DENR) Project on Integrated Environment Management for Sustainable Development, a UNDP Project. She also served as a Consultant for the ENTRAP, a DENR Project on the institutionalization of ENRA using the Peskin's Model. A certified Public Accountant, she has had extensive academic and applied knowledge in socioeconomic accounting and indicator system from the Philippines and abroad.

Georg Ewerhart

Georg Ewerhart was born in 1958 and graduated as a Master in Economics from the University of Münster in 1985 with concentration on macroeconomics, input–output economics, national accounting and econometrics. After five years as teaching and research assistant of Bernd Meyer at the Department of Economics, University of Osnabrück, he received his doctorate in 1991 for his work in empirical financial markets modelling. Afterwards he was appointed assistant professor (C1) of the department and is working in the field of integrated economic and environmental modelling together with Bernd Meyer.

Gerard Gie

Gerard Gie, born in 1945, studied economics and statistics at ENSAE, the French school for administrators in statistics. From 1978 to 1992 he worked as an economic planning and development adviser in several countries. From 1992 to 1996 he was an executive manager at Planistat Europe, a Paris-based consultancy firm specialized in statistics and economics. Since 1992 he has worked on environmental accounting, mainly for Eurostat, including work on SERIEE, environment industry and natural resource accounts. Since 1996 he has been president of Planistat France.

Kirk Hamilton

Kirk Hamilton is a senior environmental economist in the Environment Department of the World Bank. Prior to joining the Bank, Dr Hamilton was Assistant Director, National Accounts and Environment Division, Statistics Canada, and a senior research fellow at the Centre for Social and Economic Research on the Global Environment, University of London. Dr Hamilton is one of the principal authors of *Expanding the Measure of Wealth* (World Bank, 1997) and develops analysis and policy advice on the linkages between macroeconomic policies and the environment. His research interests include the treatment of pollution emissions within extended or 'green' national accounts and exploring the relationship between growth theory and sustainable development.

Stephanie Howell

Stephanie Howell is a research economist at the US Bureau of Economic Analysis (BEA). She has worked on various projects and authored and co-authored articles in the areas of national accounting and integrated economic and environmental accounting. Ms. Howell has also represented the Bureau on a number of national and international working groups. She received a Masters degree in economics from George Mason University in 1991.

Steven Keuning

Steven Keuning is currently director of the National Accounts Department at Statistics Netherlands. He holds a PhD in Economics from Erasmus University Rotterdam and has published various books and articles on economic, environmental and social accounting issues. His most recent book (1996) is entitled *Accounting for Economic Development and Social Change* (IOS Press, Amsterdam). He has drafted the chapter on Social Accounting Matrices in the 1993 System of National Accounts and the 1995 European System of Accounts. He is on the Council of the International Association for Research in Income and Wealth and on the Editorial Board of the *Review of Income and Wealth.*

Seung-Woo Kim

Seung-Woo Kim is a research fellow at the Korea Environmental Technology Research Institute in Seoul, which he joined in 1992. He gained a PhD degree from the Department of Agricultural and Resource Economics, Oregon State University in 1991. His main research areas are environmental accounting, environmental statistics and water quality management. He has implemented several domestic and international environmental research projects such as Research on Integrated Environmental and Economic Accounting (UNDP regional project), Systems of Integrated Environmental and Resource Accounting (ESCAP project) and Collaborative Research to Develop a Korean Acidic Component Emission Model (a joint project by KETRI and NIES of Japan).

J Steven Landefeld

J Steven Landefeld is Director of the Bureau of Economic Analysis (BEA). Prior to becoming Director, he served in a number of other capacities at the Bureau, including Deputy Director and Associate Director for International Economics. Before coming to BEA, Dr Landefeld held a number of academic, policy, and statistical positions, including Special Assistant to the Chairman and Senior Staff Economist for the President's Council of Economic Advisers.

While at BEA, he has led a number of pioneering efforts in national and international accounting. He has authored numerous professional articles and served on expert groups of the OECD, IMF, and other international organizations. Dr Landefeld has received the Henri Willem Methorst Medal from the International Statistical Institute, two Abramson Scroll Awards from the National Association of Business Economists, and the Gold, Silver, and Bronze Awards from the Department of Commerce for his work.

Glenn-Marie Lange

Glenn-Marie Lange is a Research Scientist at the Institute for Economic Analysis, New York University founded by Nobel-Laureate, Wassily Leontief. She has worked for the past 15 years on issues related to technological change, the environment, and sustainable development. Her publications include *The Future of the Environment: Ecological Economics and Technological Change*, co-authored with Faye Duchin. Over the last 5 years she has worked extensively on environmental and natural resource accounts (NRA) as a means to integrate environmental and economic policy analysis. She is currently directing the construction of NRA and their use for the design of sustainable development strategies in Namibia. In the next stage of this work, she and her colleagues will begin to construct NRA for neighbouring countries in Southern Africa in order to address management of resources, such as water, which transcend national boundaries. Dr Lange received a PhD degree in economics from New York University.

Christian Leipert

Christian Leipert, born in 1944, studied economics and sociology at the University of Hamburg and the Free University of Berlin. He received his PhD from the University of Augsburg in 1974. From 1971 to 1982 he worked as a researcher and associate professor at the economics faculty of the University of Augsburg where he specialized in environmental economics. Since 1982 he has worked at Berlin-based institutions, first at the Science Centre Berlin for Social Research and since 1994 at the Institute for Economic–Ecological Research. His current areas of research include 'green' national accounting and the topic of economic instruments for environmental control in developing countries.

Bernd Meyer

Bernd Meyer, born in 1946, is full professor for economics at the University of Osnabrück (Germany). He started studying economics at the University of Münster in 1965, graduated in 1970, achieved the Dr.rer.pol. in 1973 and the Habilitation in 1980 at the University of Münster. In 1981 he started teaching as a professor at the University of Augsburg and left in 1984 for Osnabrück. In 1992 he refused a call to the University of Eichstätt. Meyer's research fields are macroeconomic theory, input–output analysis, environmental economics and econometric modelling. He is a member of the Advisory Board of the German Environmental-Economic Accounting at the Federal Ministry of the Environment and of the international research group INFORUM. In 1995 he was elected chairman of the section Evolutionary Economics in the Gesellschaft für Wirtschafts- und Sozialwissenschaften.

Deborah Vaughn Nestor

Deborah Vaughn Nestor is an economist at the US Environmental Protection Agency. She received her PhD in economics from the University of Tennessee in 1991. Her current areas of research include computable general equilibrium modelling of environmental policy and incentive-based approaches to environmental regulation. She has published papers in a variety of journals, including *Land Economics, The Review of Income and Wealth, Economics Letters*, and *Economic Modelling*.

K Oda

K Oda (born in1954) is currently Director of Research and Coordination Division, Department of National Accounts at the Japan Economic Planning Agency which he joined as a government official in 1980. He studied economics at Osaka University where he received his MA in 1978.

Jyoti K. Parikh

Professor Jyoti K. Parikh obtained her MSc from the University of California, Berkeley, and her PhD in Nuclear Physics from the University of Maryland, College Park.

She worked at the International Institute for Applied Systems Analysis for nearly eight years on energy and environment problems of the developing countries. She has served as energy consultant to the World Bank, the US Department of Energy, the European Economic Community, Brussels and UN agencies such as UNIDO, FAO, UNU, and UNESCO. She has authored papers on environmental information systems, standards for environmental pollution, and sustainable development in agriculture. Her current interests range from resource utilisation policy to technology assessment and national and sectoral modelling for power systems and energy and environment.

She was the convener of the Sixth National Power Systems Conference held in 1990. She was a member of the Tariff Committee of Uttar Pradesh State Electricity Board and the Consultant for Kerala State Electricity Board. She was a senior consultant for the World Bank study on long-term issues for Power Sector in India and has written a number of papers on power systems for IEEE transactions for power systems.

Presently, she works in the area of global change, natural resource accounting, trade and environment, policies for global warming at national and international level with GEF, Ministry of Environment, UNCTAD, etc. She is the convening lead author for the second assessment report of IPCC.

Carl Pasurka

Carl Pasurka is an economist with the Economy and Environment Division, Office of Economy and Environment at the US Environmental Protection Agency. He received his PhD from the University of Illinois at Urbana-Champaign in 1981. He was an associate editor of the *Journal of Environmental Economics and Management* from 1994 through 1996. His current areas of research include environmental accounting, and measuring the impact of pollution regulations on traditional and adjusted measures of productivity. His research has appeared in a variety of journals including the *Review of Economics and Statistics, Journal of Environmental Economics and Management, Review of Income and Wealth,* and *Economic Modelling.*

Henry M Peskin

Dr Peskin has over 30 years experience as a research economist and is a specialist in cost-benefit analysis, environmental policy analysis, environmental statistics, and resource accounting. He is internationally known for his research on methods of incorporating environmental degradation in national economic accounting systems. Currently, he serves as a member of the Panel on Integrated Environmental and Economic Accounting of the National Academy of Sciences. Dr Peskin was educated at the Massachusetts Institute of Technology (chemical engineering) and Wesleyan University (government) and has his PhD (economics) from Princeton University.

Walter Radermacher

Walter Radermacher studied economics in Aschen and Münster from 1970 to 1975. After serving for two years as a member of scientific staff at Münster University in the field of business mathematics/operational research, he joined the Federal Statistical Office in 1978 as staff member. Until 1985 he was responsible for trade statistics coordination; since 1985 responsible for setting up a geographical information system based on remote sensing. Since 1991 he is also head of the Environmental–Economic Accounting Division. Since 1982 he has teaching assignments in statistics, economics and environmental economics at *Fachhochschule* Wiesbaden and Lüneburg University.

Dieter Schäfer

Dieter Schäfer, born in 1955, studied macroeconomics and sociology at the Universities of Stuttgart Hohenheim, Paris 1 and Heidelberg. He has worked since 1981 at the Federal Statistical Office of Germany. From 1981 to 1994 he worked on national accounts, input–output calculations and satellite accounts, including household production and environmental expenditure. Since 1994 he has worked on the German Economic-Environmental Accounting.

Carsten Stahmer

Dr. Carsten Stahmer studied Economics at the Universities of Tübingen, Güttingen and Nuremberg. He obtained certificate 1968 and PhD in Economics 1972 at the University of Heidelberg. He has worked since 1973 in the Federal Statistical Office, Wiesbaden. His special working areas are input–output tables, national balance sheets and environmental-related extensions of national accounts. Since 1993, he has given lectures in statistics at the University of Heidelberg.

Anton Steurer

Anton Steurer (MBA), born in 1961, studied business administration and commerce at the Vienna University of Economics and Business Administration. He has worked on environmental accounting and environmental economics since 1988 as a researcher and statistician. In 1992 and 1993 he worked on environmental expenditure accounts and environmental taxes at the Austrian Central Statistical Office. Since 1994 he has worked on SERIEE, environment industry, economic instruments for environmental protection, material flow accounts and environmental accounting at Eurostat.

Andrew Stott

Dr Andrew Stott has held the post of senior scientific adviser in the UK Department of the Environment for four years. He is responsible for strategic research and long-term monitoring in support of countryside and wildlife policy. Before this Dr Stott held a similar post in the Department of the Environment for Northern Ireland. Dr Stott was educated at University College, Oxford and was awarded a BA Hons Degree in Geography in 1980. His PhD was awarded by the University of Manchester in 1985 for studies in land use change and water resources. Before joining the civil service in 1987, Dr Stott held a lectureship in geography and environmental studies at the College of St Paul and St Mary in Cheltenham.

André Vanoli

André Vanoli, born in 1930, is a former director at INSEE, the French Central Statistical Office. After studies in law, political sciences and economics, he joined in 1957 the Ministry of Economy and Finance and in 1961 INSEE, in both cases in the national accounts department. From 1972 to 1987, he was in charge of statistical coordination in the French statistical system. He was, and is still involved intensively in international activities. This includes inter alia leading technical cooperation projects in a number of countries (Algeria, Colombia, Ecuador, Peru, Brasil, Tunisia, Greece). From 1986 to 1993, he worked actively on the preparation of the 1993 SNA (earlier, in 1964, he drafted a report that was used as a basis for the elaboration of the first European System of Accounts, the ESA). He published various articles, mostly on national accounting, turning his attention in recent years to environmental accounting issues. He served as Chairman of the Council of the International Association for Research in Income and Wealth (IARIW) and in 1977–1979, he was president of the French National Accounting Association (Association de comptabilité nationale) since its creation.

Michael Ward

Michael Ward is a Principal Economist in the World Bank and Team Leader of the Statistical Advisory Services Group. Prior to joining the Bank in March 1985, he was in charge of the International Comparisons Project at the OECD. He moved to Paris in 1982 from his position as Director of the Statistical Program at the Institute of Development Studies in Sussex where he was a Fellow from 1975 to 1982 and before that he was a senior Research Officer in the Department of Applied Economics at the University of Cambridge. During his time in Cambridge he was elected a Fellow of Selwyn College, becoming Dean 1970–1972. He has published widely in the area of national accounts and development economics and has a special interest in the role of capital in development.

Index

Economy & Environment

1. F. Archibugi and P. Nijkamp (eds.): *Economy and Ecology: Towards Sustainable Development.* 1989 ISBN 0-7923-0477-2
2. J. Bojö, K.-G. Mäler and L. Unemo: *Environment and Development: An Economic Approach.* 1990 ISBN 0-7923-0802-6
3. J. B. Opschoor and D. W. Pearce (eds.): *Persistent Pollutants: Economics and Policy.* 1991 ISBN 0-7923-1168-X
4. D.J. Kraan and R. J. in 't Veld (eds.): *Environmental Protection: Public or Private Choice.* 1991 ISBN 0-7923-1333-X
5. J.J. Krabbe and W.J.M. Heijman (eds.): *National Income and Nature: Externalities, Growth and Steady State.* 1992 ISBN 0-7923-1529-4
6. J. Bojö, K.-G. Mäler and L. Unemo: *Environment and Development: An Economic Approach* (revised edition). 1992 ISBN 0-7923-1878-1
7. T. Sterner (ed.): *Economic Policies for Sustainable Development.* 1994 ISBN 0-7923-2680-6
8. L. Bergman and D.M. Pugh (eds.): *Environmental Toxicology, Economics and Institutions.* The Atrazine Case Study. 1994 ISBN 0-7923-2986-4
9. G. Klaassen and F.R. Førsund (eds.): *Economic Instruments for Air Pollution Control.* 1994 ISBN 0-7923-3151-6
10. K. Uno: *Environmental Options: Accounting for Sustainability.* 1995 ISBN 0-7923-3513-9
11. K. Uno and P. Bartelmus (eds.): *Environmental Accounting in Theory and Practice.* 1997 ISBN 0-7923-4559-2
12. J.C.J.M. van den Bergh, K.J. Button, P. Nijkamp and G.C. Pepping: *Meta-Analysis in Environmental Economics.* 1997 ISBN 0-7923-4592-4

KLUWER ACADEMIC PUBLISHERS – DORDRECHT / BOSTON / LONDON